Geometry of Derivation

Geometry of Derivation, Volume III: Classification of Skewfield Flocks is the third book in a series of books on the topic. This book continues establishing the techniques, examples, and future directions of the specifics of flock theory over skewfields. Like its predecessors, it will primarily deal with connections to the theory of derivable nets and translation planes in both the finite and infinite cases.

Translation planes over non-commutative skewfields have not traditionally had a significant representation in incidence geometry, and derivable nets over skewfields have only been marginally understood. Both are deeply examined in this volume, while ideas of non-commutative algebra are also described in detail, with all the necessary background given a geometric treatment.

Since the work is valid for finite fields, infinite fields, non-commutative skewfields, there are left and right flocks over generalized hyperbolic quadrics, and generalized quadratic cones, there is a number of possibilities. The contribution of this volume is the main classification.

The book continues the presentation in *Geometry of Derivation with Applications, Volume I,* Johnson (2023), and *Geometry of Derivation, Volume II: Theory of Skewfield Flocks* (2026) is also available. This is the seventh work in a longstanding series of books on combinatorial geometry by the author, including *Subplane Covered Nets,* Johnson (2000); *Foundations of Translation Planes,* Biliotti, Jha, and Johnson (2001); *Handbook of Finite Translation Planes,* Johnson, Jha, and Biliotti (2007); and *Combinatorics of Spreads and Parallelisms,* Johnson (2010), all published by CRC Press.

Geometry of Derivation
Volume III: Classification of Skewfield Flocks

Norman L. Johnson

CRC Press
Taylor & Francis Group
Boca Raton London New York

CRC Press is an imprint of the
Taylor & Francis Group, an **informa** business
A CHAPMAN & HALL BOOK

Designed cover image: Norman L. Johnson and Bonnie L. Hemenover

First edition published 2026
by CRC Press
2385 NW Executive Center Drive, Suite 320, Boca Raton FL 33431

and by CRC Press
4 Park Square, Milton Park, Abingdon, Oxon, OX14 4RN

CRC Press is an imprint of Taylor & Francis Group, LLC

ISBN: 978-1-041-29088-9 (hbk)
ISBN: 978-1-041-29091-9 (pbk)
ISBN: 978-1-003-76797-8 (ebk)

DOI: 10.1201/9781003767978

Typeset in CMR12
by KnowledgeWorks Global Ltd.

Publisher's note: This book has been prepared from camera-ready copy provided by the authors.

For my mother, Catherine Elizabeth Gabriel Lamb – for sharing her talent

Contents

ABOUT THE AUTHOR

Norman L. Johnson is an Emeritus Professor (2011) at the University of Iowa where he has had ten PhD students. He received his PhD at Washington State University as a student of T.G. Ostrom. He has written 597 research items including articles, books, and chapters available on Researchgate.net. Additionally, he has worked with approximately 40 coauthors and is a previous Editor for International Journal of Pure and Applied Mathematics and Note di Matematica. Dr. Johnson plays ragtime piano and enjoys studying languages and 8-ball pool.

ACKNOWLEDGMENTS

I gratefully acknowledge the collaboration and ideas hidden in all three of these three volumes of Mauro Biliotti, Lecce, Italy; Vikram Jha, Glasgow, Scotland; and Rolando Pomareda, Santiago, Chile. Also, the author is generally indebted to the Mathematics Department of the University of Iowa and, in particular, to Brian Bacher of the Computer Support Group for great help in the preparation of these works. Finally, many thanks to my wife, Bonnie, for her advice and neverending support, especially during these last six years for this series of texts.

Part 1

The classification of flocks

In this volume on classification, the term is meant to indicate what sort of possible flocks, both hyperbolic and conical, can exist over skewfields and to construct as complete a list of examples as possible. Since this area began with the two finite general types of flocks, hyperbolic and conical, we begin here with other possible types of flocks. Things are much different in infinite field flocks and even more different in non-commutative skewfield flocks. But, here it is realized that there must be left and right flocks, and then that when considering automorphisms involved in the flocks, inner automorphisms play a much different role. Therefore, there are quite a number of different types of flocks of "generalized hyperbolic and conical flocks" over skewfields; something like 109.

In volume II, we provided a glance at the basic idea of these two volumes. Similarly, in this volume III, we will try to give an understanding of how to classify flocks over skewfields. There is a complexity in understanding what sorts of examples should exist and registering these examples. Therefore, we give another brief version of what will be covered in this text.

We shall be considering translation planes with spreads in $PG(3, D)$, where D is a skewfield, and the spreads consist of a set of derivable nets that share either one component (the conical flock spreads) or two components (the hyperbolic flock spreads) that admit what are called pseudo-regulus-inducing elation groups or homology groups, respectively. We shall be considering skewfields D, which are infinite-dimensional or finite-dimensional vector spaces over the field $K = Z(D) \cap Fix\sigma \cap \{u; A(u) = 0\}$, where σ is an automorphism of D, and A is a $\sigma - A$ derivation function. The dimension over K is finite if and only if D is finite-dimensional over the center.

The flock spreads over D and spreads in $PG(3, D)$, which contain a set of pseudo-regulus derivable nets that share two components (hyperbolic flock spreads) or one component (conical flock spreads) and admit homology or elation groups, as outlined.

DEFINITION 1. A Baer subplane of a projective plane is required to be both "point-Baer" and "line-Baer", to be "Baer". This means each point of the plane is incident on a line of the subplane, and each line of the plane is incident on a point of the subplane. That is, a derivable net can admit subplanes that are Baer with respect to the net but not Baer with respect to the projective plane. The author [**29**] has constructed a set of dual translation planes that admit derivable nets, but these nets are not derivable within the plane itself. If the dual translation plane would be a semifield plane; the ideas could impact the material in this volume.

For translation planes, this text is normally considered over finite-dimensional division ring planes. Therefore, D is a division ring of finite dimension over a field K. Under these conditions, we shall show later in the text that the Baer subspaces of the nets are Baer subspaces of the plane. The theory of deficiency one also requires aspects of when certain subspaces become Baer subplanes, which can be included in the general situation under our assumptions. However, it is an interesting question when flock spreads over division rings of infinite dimensions over their centers can produce hyperbolic and conical flocks, for in this case, the Baer subspaces of each pseudo-regulus net would not be Baer subspaces of the spread. In particular, if D is the infinite-dimensional division ring, admitting an automorphism σ and a $\sigma - A$ derivation function, we would know that we may form a 4-dimensional vector space V_4 over D and also a 4-dimensional vector space V_4^* over $D^{\sigma-A}$=$\begin{bmatrix} u^{\sigma} & A(u) \\ 0 & u \end{bmatrix}$; u in D, as isomorphic division rings. The question is when a component M not in the $D^{\sigma-A}$-generic pseudo-regulus net, which is a left 2-dimensional D-subspace and a component of the generic pseudo-regulus net N, which is a right 2-dimensional subspace over $D^{\sigma-A}$, have the property that $M \oplus N = V_4$? This question arises when considering the derivation of the generic pseudo-regulus net and when considering the theory of deficiency one over a division ring flock spread.

CHAPTER 1

The classes of flocks

It was mentioned in the preface in volume II [**35**] that for any infinite division ring D, there are infinitely many ways to define a generalized hyperbolic quadric, all of which correspond and are associated with spreads in $PG(3, D)$. This will mean that there are infinitely many different cones and flocks of cones. In order to deal with this situation in the non-commutative case, we may group the infinite classes into sets of left and right. This will provide a short version of what is to follow.

We discussed previously in the first volume [**34**] that there are four types of derivable nets, where we used the term "generic left derivable net" for types $i = 1, 2, 3$ to be when the i-shared Baer subspaces allow the representation of the derivable net to share $x = 0, y = 0, y = x$ and there is/are i common Baer subspaces with the classical right derivable net (with respect to a given skewfield D). Furthermore, generic derivable nets require a 0 in one of the two off-diagonal matrix entries, called "upper" and "lower". Specifically, the generic left derivable nets have the following "upper" form:

$$x = 0, y = 0, y = x \begin{bmatrix} u^{\sigma} & A(u) \\ 0 & u \end{bmatrix},$$

for all u in D, where σ is 1 or an inner or outer automorphism, and A is a σ-derivative function, which is a 0, inner or outer function. Therefore, there are nine different versions of left generic pseudo-regulus nets.

The Baer subspaces $B^R(a, b)$ are

$$B(a, b) = \left\{ \left((a, b) \begin{bmatrix} v^{\sigma} & A(v) \\ 0 & v \end{bmatrix}, (a, b) \begin{bmatrix} w^{\sigma} & A(w) \\ 0 & w \end{bmatrix} \right) \right\}$$

for all v, w in D.

The classical reference right pseudo-regulus net has the following form:

$$x = 0, y = 0, y = \delta x$$

for all δ in D. The Baer subspaces $B^L(a, b)$ are as follows:

$$B^L(a, b) = \{(v(a, b), w(a, b))\}.$$

DOI: 10.1201/9781003767978-1

for all v, w in D. When A is non-zero and outer, we see that

$$B^R(0,1) = B^L(0,1).$$

And this is the unique common Baer subspace. When A is zero, and σ is an outer automorphism, we also see that there are exactly two common Baer subspaces: $B^L(0,1)$ and $B^L(1,0)$. When A is zero, and σ is 1, in the most usual situation, there are three (at least) common Baer subspaces: $B^L(0,1)$, $B^L(1,0)$ and $B^L(1,1)$.

In this and the previous text, we have realized that when either σ or A is "inner" then what might look like a type 1 is really a "fake type 1"= a type 2, and what might look like a type 2 is a "fake type 2"= a type 3. A more complete version of the nine possibilities is presented shortly. In the first text, it was shown that for every derivable net R of type i, there is a mapping $(x, y) \to (xK, yK)$, where K is a non-singular 2 by 2 matrix over D that maps the associated generic net to R and leaves the classical right derivable net invariant. These K-mappings have also been discussed in volume II.

In this text, we have realized that we wish to allow the automorphism type to include the inner automorphisms and to allow the $\sigma - A$ derivation function to be inner.

Now assume that there is a translation plane/spread of hyperbolic or conical type that contains the generic derivable net, with the additions just mentioned. Then there is an isomorphic translation plane containing D of hyperbolic or conical type obtained by applying the K-mapping, where the spread is still in the same 3-dimensional projective space. Since our approach uses transforms and then sets of Baer subspaces as the points of the flocks, it would seem to follow that associated flocks corresponding to D and k_D would also be isomorphic to the flocks corresponding to the generic derivable nets. However, this cannot be ascertained, as the isomorphism property between the flocks associated with the two isomorphic spreads cannot be proved for either hyperbolic or conical flocks.

We shall discuss the general isomorphism questions later in the text.

We shall use the notation F(i) to denote the three types of flocks corresponding to the three types of derivable nets of types 1, 2 and 3. For type F(2), there is also a type $F(2)^{trans}$, called the "transpose" flock, arising from the transpose of the associated spread. For infinite fields IF, we use the analogous notation IF(i), for the four types $i = 1, 2, 3$ of derivable nets and use the notation $IF(2)^{trans}$ to denote the associated transpose flock. Similarly, for the non-commutative skewfields, we shall use the notation $N(i)$ for $i = 0, 1, 2, 3$ and $N(2)^{trans}$ for the transpose flock, assuming it exists. Furthermore, the transpose flock is only defined for conical flocks. For all of the notations Z, we use Z_{hy} and Z_{co} to denote the hyperbolic and conical flocks, respectively. For the transposed conical flock to exist, for the classes of flocks, we demand that the transposed conical flock not be isomorphic to the conical flock. However, for the connection between left and right flocks, we simply demand the transpose spreads exist. The type $Z_{hy}(1)^F$ and $Z_{co}(1)^F$ are the special type 1 spreads that admit a type 1 pseudo-derivable net that uses an inner σ-derivation function, and which turn into $Z_{hy}(2)$ and $Z_{co}(2)$ in the finite field and infinite field cases. The superscript F stands for fake.

DEFINITION 2. (**The Kaleidoscope of Derivable nets.**) Here are generic types of left derivable nets, where R is a generic left pseudo-regulus net of the form:

$$x = 0, y = x \begin{bmatrix} u^{\sigma} & A(u) \\ 0 & u \end{bmatrix},$$

where $A(uv) = u^{\sigma}A(v) + A(u)v$, for all u,v in D, where D is a division ring, either finite, an infinite field, or a non-commutative skewfield, which is of finite or infinite dimension over the center $Z(D)$. The set of Baer subspaces $B(a, b)$ is

$$B(a, b) = \{(a, b) \begin{bmatrix} v^{\sigma} & A(v) \\ 0 & v \end{bmatrix}, (a, b) \begin{bmatrix} w^{\sigma} & A(w) \\ 0 & w \end{bmatrix}\},$$

for all v,w in D, and (a, b) is written over the basis for D.

There are types 1, 2, 3. When considering spreads containing derivable nets, we consider all of these types. Occasionally, as mentioned, some of these will be called fake type i types, which mean type $i + 1$ types. This occurs when the representations appear to be type i but are really type $i + 1$. And, there is a double fake type i, which means type $i + 2$.

Here is the result from volume II regarding the typing of left generic derivable nets: The "Nine" Type i and Fake Type i Derivable Nets.

THEOREM 3. *The nine types of left generic derivable nets will be given in three groups of three: (σ, A), first entry, outer, zero, inner. We note for reference the* 1*-outer when coupled with an inner automorphism: $c^{-1}uc$ is $c^{-1}A$, for a $1 - A$ outer or inner derivative function.*

(1) σ is outer:

(a) Type 1: (outer, σ-outer), shared Baer subspace denoted by (0,1),

(b) Type 2:(outer, 0), shared Baer subspaces denoted by (0,1) and (1,0),

(c) Type 2: (outer, σ-inner : $u^{\sigma}e - eu$), shared Baer subspaces denoted by $(0,1)$ *and (1,e); fake type 1.*

(2) σ *is inner:* $c^{-1}uc$*:*

(a) Type 1: (inner, c^{-1}*(1-outer), shared Baer subspace* $(0,1)$,

(b) Type 3 *:(inner, 0), shared Baer subspaces denoted by* $(0,1),(1,0)$ *and* $(1,c)$*, fake type 2,*

(c) Type 3: (inner, c^{-1}*(1-inner : $ue - eu$), Baer subspaces denoted by* $(0,1),(1,c^{-1}e)$*. (1,*$c^{-1}(e-1)$*); double fake type 1.*

(3) $\sigma = 1$*:*

(a) Type 1*: (1,1-outer), shared Baer subspace denoted by* $(0,1)$*.*

(b) Type 3: (1,0), shared Baer subspaces denoted by $(0,1),(1,0)$*, and* $(1,1)$*.*

(c) Type 3: (1,1-inner:$ue - eu$), shared Baer subspaces denoted by $(0,1)$*, (1,e), (1,1-e); fake type 2.*

- Kaleidoscope: What appears to be just a few possible types of derivable nets is actually a kaleidoscope of different types. A K-mapping that preserves the set of (upper) generic types is defined for the left spreads as

$$(x,y) \to (xK, yK) = (x^*, y^*).$$

where $K = \begin{bmatrix} e & f \\ 0 & h \end{bmatrix}$, for eh non-zero and e, f, h in a division ring D. Note that

$$K^{-1} = \begin{bmatrix} e^{-1}, & -e^{-1}fh^{-1} \\ 0 & h^{-1} \end{bmatrix}$$

the K-mapping will map

$$x = 0, y = x \begin{bmatrix} u^{\sigma} & A(u) \\ 0 & u \end{bmatrix}$$

to the following derivable net

$$x^* = 0, y^* = x^* K^{-1} \begin{bmatrix} u^{\sigma} & A(u) \\ 0 & u \end{bmatrix} K,$$

for all u in D. This works out to have the following form:

$$x^* = 0, y^* = x^* \begin{bmatrix} w^{\sigma\tau_{e^{-1}}\rho_h}, & e^{-1}(w^{\rho_h\sigma}f - fw) + e^{-1}A(w^{\rho_h})h \\ 0 & w \end{bmatrix},$$

for all w in D. Let

$$\sigma^* = \sigma\tau_{e^{-1}}\rho_h,$$

and

$$A^*(w) = e^{-1}(w^{\rho_h \sigma} f - fw) + e^{-1}A(w^{\rho_h})h.$$

The inner automorphism $\tau_{e^{-1}}$ is defined by $u^{\tau_{e^{-1}}} = e^{-1}ue$, and the inner automorphism ρ_h is defined by $u^{\rho_h} = huh^{-1}$.

- By definition, σ^* is an outer automorphism if and only if σ is an outer automorphism and inner otherwise.
- Similarly, $\sigma - A^*$ is inner (outer) if and only if $\sigma - A$ is inner (outer).
- Then the following derivable net form is obtained as:

$$x^* = 0, y^* = x^* \begin{bmatrix} w^{\sigma^*} & A^*(w) \\ 0 & w \end{bmatrix},$$

for all w in D. By isomorphism of derivable nets, it follows that we obtain a $\sigma^* - A^*$-derivation function. For full details, see volume II, part XII, The Kaleidoscope of Derivable Nets.

- For fields, by results of [**34**], $\sigma - A$ are inner of the form: $A_c(w) = (w^\sigma - w)c$. Then, the change is as follows:

$$x^* = 0, y^* = x^* \begin{bmatrix} w^\sigma, & A_{(e^{-1}(f+ch)}(w) \\ 0 & w \end{bmatrix}.$$

We are interested in translation planes/spreads in $PG(3, D)$ that contain a given generic pseudo-regulus net. In general, the form would be expressed as follows for left hyperbolic flocks:

$$x = 0, y = x \begin{bmatrix} u^\sigma & A(u) \\ 0 & u \end{bmatrix}, y = x \begin{bmatrix} f(t) & g(t) \\ 1 & t \end{bmatrix} \begin{bmatrix} v^\sigma & A(v) \\ 0 & v \end{bmatrix}.$$

- For left conical spreads, the form would be

$$x = 0, y = x \begin{bmatrix} u^\sigma + F(t) & , G(t) + A(u) \\ t & u \end{bmatrix}.$$

for all u, t in D. The conical spreads admit a pseudo-regulus generating elation group.

- We shall assume the previous would give all varieties of flocks with the exception of the left "lower" flocks when the derivable net is of type 1 or fake type 1, which typically would refer to transpose forms in the left flock spread case. So the derivable net would have the form, when A is non-zero, there are the following types: A is outer, and σ is 1 or non-trivial outer automorphism, or A is inner with non-trivial outer automorphism σ.

$$x = 0, y = x \begin{bmatrix} u & 0 \\ A(u) & u^\sigma \end{bmatrix}.$$

- For right generic conical flock spreads, the form is
$$x^T = 0^T, y^T = \begin{bmatrix} u + R(t) & , t \\ S(t) + A^*(u), & u^\tau \end{bmatrix} x^T,$$
where v^T denotes the transpose of the 2-row vector v.
- For "right" hyperbolic spreads, the form would be
$$x^T = 0, y^T = \begin{bmatrix} u & , 0 \\ A^*(u) & u^\tau \end{bmatrix} x^T, y^T = \begin{bmatrix} v & 0 \\ A^*(v) & v^\tau \end{bmatrix} \begin{bmatrix} t & , 1 \\ s(t), & r(t) \end{bmatrix} x^T,$$
for all u, t in D and for all v in D^*, τ an automorphism. The right hyperbolic spreads admit a pseudo-regulus generating homology group acting on the left. If we consider the right generic pseudo-regulus nets, the action of the set of K-mappings, with non-singular
$$K = \begin{bmatrix} e & 0 \\ g & h \end{bmatrix},$$
where e, g, h in D. In analogous manner, there is an associated "right Kaleidoscope" for right flock spreads.
- For left spreads that admit both pseudo-regulus generating elation groups and homology groups, we assume that the axis of the elation group and the coaxis of the homology group are the same components and that the two groups fix the classical left pseudo-regulus net. These fixed components are also coordinatized by $x = 0$ for axis of the elation group and $y = 0$ for the axis of the homology group, with coaxis $x = 0$. We call these spreads "simultaneous spreads". The general form usually would then be
$$x = 0, y = x \begin{bmatrix} u^\sigma & , A(u) \\ 0 & u \end{bmatrix},$$
$$y = x \begin{bmatrix} t^\sigma + F(1) & , G(1) + A(t) \\ 1 & t \end{bmatrix} \begin{bmatrix} v^\sigma & , A(v) \\ 0 & v \end{bmatrix},$$
which would then imply certain consequences on the spreads that we consider in the chapters on simultaneous spreads. In our notation $f(t) = t^\sigma + F(1)$ and $g(t) = G(1) + A(t)$.
- We have determined a variety of classes of simultaneous spreads in the material and infinite classes of conical flock spreads, which are not hyperbolic flock spreads.
- For shorthand notation, in the following description of the possible flocks, we shall use only the forms
$$x \begin{bmatrix} u^\sigma + F(t) & , G(t) + A(u) \\ t & u \end{bmatrix}$$

to denote the left generic conical spreads or

$$x\begin{bmatrix} f(t) & ,g(t) \\ 1 & t \end{bmatrix}\begin{bmatrix} v^{\sigma} & ,A(v) \\ 0 & v \end{bmatrix}$$

to denote the left generic hyperbolic spreads.

- When $A(u) = (u^{\sigma}c - cu)$, for c of D, A is an inner *σ-derivation* function. All σ-derivations are inner in fields from [**34**]. A change in basis will return the fake type 1 derivable net to a type 2 derivable net; the K-mappings preserve type. This will mean that if a spread contains an inner pseudo-regulus derivable net, then there is a corresponding isomorphic spread that then contains the type 2 pseudo-regulus net. Because of the potentially interesting spreads and flocks that could be obtained, we have decided to keep the inner-type pseudo-regulus nets and spreads as interesting cases on their own.
- In the following, we shall provide the complete set of flocks or quasiflocks. In the finite case, all quasiflocks are flocks, but in any infinite quasifibration, the question of whether there is a covering of the affine points by the set of components of the quasifibration is always a consideration. Generally, there are left and right flocks, using left spreads and right spreads, required when considering non-commutative division rings/skewfields. While left and right flocks are complementary they cannot be considered equal in the sense that given a left flock, there need not be a right flock. On the other hand, given a left quasifibration, there is a right quasifibration (the fundamental theorem of quasifibrations), so it is the covering condition that will determine right from left spreads.
- To reiterate, we shall also be mindful that it is possible for two left flock spreads to be isomorphic without the corresponding two left flocks being isomorphic. The forms for left and right flock spreads are quite different, and the finite field and infinite fields are also sufficiently different to have separate lists. And therefore, initially, we shall provide separate lists for the field cases and the right and left non-commutative cases. Furthermore, we are also considering flock quasifibrations and quasiflocks. For the convenience of reading, we shall then provide the lists for finite fields, infinite fields, left non-commutative division rings and right non-commutative division rings. The 2 by 2 matrix sets over skewfields that represent the flock quasifibrations have four different determinants, any of which determine unique inverses, if any determinant is non-zero.

- If the components and their non-zero differences all have a non-zero determinant, then this indicates that the matrix sets are quasifibrations, which are also maximal partial spreads.
- The associated sets of left or right planes (projectively) covering the points corresponding to the maximal partial spreads are called "quasiflocks". The terms "flock" and "flock spread" are reserved when there is a covering condition satisfied for the quasifibrations.
- We then apply the K-mappings: $(x, y) \to (xK, yK)$ and their column analogues. These mappings will then provide the flock spreads arising from the set of all type i pseudo-regulus nets $i = 1, 2, 3$. When $K = \begin{bmatrix} e & f \\ 0 & h \end{bmatrix}$, the generic left flock spreads are permuted. However, although there are two corresponding constructed flocks, isomorphism of flocks is generally not implied.
- It may be somewhat counter intuitive but proper quasiflocks are much more rare than flocks. We shall review the existing situation in due course. In the field case, there are some known proper quasiflocks, but in the non-commutative skewfield cases, given what looks like a quasiflock, there is no easy proof to determine if that quasiflock, is a flock, and it would remain an open question.
- In preliminary versions, we thought to provide a unified list of the possible types of flocks, but this proved to be too complicated. We have mentioned above that there are left and right flocks of two types: hyperbolic and conical. There are type 1, 2, 3 and subtypes of derivable nets. There are right and left actions of homology groups, leading to the "inverse types" of flock spreads. There are three types of automorphisms σ, and three types of σ-derivative functions. There are transpose matrix sets often leading to non-isomorphic flocks, and there are finite fields, infinite fields, infinite-dimensional division rings and finite-dimensional division rings. However, our examples do not include infinite-dimensional division rings.
- Finally, we have decided that we would give several lists of the possible flocks, beginning with the most familiar, the finite flocks, the progression to the infinite field flocks, then to the left flocks and right flocks over non-commutative skewfields but with trivial or outer automorphisms. All of these flocks can

then be used to create the final two lists of left and right flocks with non-trivial inner automorphisms. There is an inevitable repetition, of notation, but this provides a more readable theory. Finally, there are type (0) flocks. Then, we provide some theory to construct from what are called left Baer flocks to right Baer flocks. A Baer flock is either a maximal partial flock or a flock. When left and right Baer flocks are connected, one may be a maximal left partial flock, and the corresponding right Baer flock, could be a flock. When both of these are conical flocks, and the transpose matrix sets are flock spreads, we have a direct connection between left and right conical flocks that shows they are anti-isomorphic. Our final lists concern the possible anti-isomorphic left and right flocks.

- Since transpose matrix sets of conical flocks seem to be rare over non-commutative skewfields, we make a point of listing for both of the following conditions: (i) their existence as spreads and (ii) if they are not isomorphic to the original conical flock. In referring to an example, as 1/2 indicates the example satisfies (i) but not (ii). This existence becomes important as it shows a left conical flock can define an anti-isomorphic right conical flock.
- The choice of the classes was motivated by the discovery of examples. Therefore, it is not claimed that any list is definitive. Furthermore, our lists are concerned and separated by the type of skewfield, finite field, infinite field, left and right non-commutative skewfields, when automorphisms are inner, left and right inner flocks, and when left and right conical flocks are anti-isomorphic.
- **Unwrapping** is a procedure that allows additive quasifibrations containing the generic pseudo-regulus net to become a spread. For left flock spreads where the automorphism must be either 1 or "strictly outer", in the sense that the automorphism cannot be a composition of an outer automorphism and a set of inner automorphisms. The reason for this is that the elements of the matrix must be unwrapped into left linear mappings, which can never be the case when there are inner automorphisms. Therefore, application of the kaleidoscope mappings will be applied only to known flocks. This will come up later with the consideration of "inner flocks".

1.1. Finite flocks

All derivable nets are regulus nets, and there are eleven total types of corresponding flocks: Three hyperbolic, three conical and a transpose conical. Then there is what is called an "inverse" hyperbolic, as well as an "inverse simultaneous", which considers a combination of a conical and an inverse hyperbolic. Inverse hyperbolic examples exist when the pseudo-regulus homology group is acting on the left instead of the right. There is an inversing action (change of coordinates) that then places it on the right side. Therefore, we distinguish between left- and right-acting homology groups that can produce non-isomorphic hyperbolic flocks. There is an unusual inversing hyperbolic combined with conical; this is the "inverse simultaneous".

We add the symbol σ, when a non-trivial outer automorphism is indicated; the *trans* stands for transpose and the F stands for fake. For all left flock spreads, which include finite field, infinite field and left non-commutative skewfield spreads, we use the following representation.

- **Hyperbolic Flock Spread:** The general forms for hyperbolic spreads and hyperbolic flocks are as follows:

$$x=0, y=0, y=x\begin{bmatrix} u^{\sigma} & A(u) \\ 0 & u \end{bmatrix}, y=x\begin{bmatrix} f(t) & g(t) \\ 1 & t \end{bmatrix}\begin{bmatrix} v^{\sigma} & A(v) \\ 0 & v \end{bmatrix},$$

for all non-zero u, v and for all t in $GF(q)$, where σ is 1 or an outer automorphism, A is identically zero or an inner or outer (not possible over fields)

σ-derivation function: $A(u)=u^{\sigma}c-cu$, for any non-zero c in $GF(q)$.

f and g are functions on a left division ring D, in this case, D is $GF(q)$.

- **Hyperbolic Flock:** The associated and equivalent hyperbolic flock is given by:

$$\Sigma_t^L : a_1(g(t)-A(f(t)^{\sigma^{-1}})-a_2 f(t)^{\sigma^{-1}}+a_3 t-a_4=0,$$
$$\&\varrho : a_2=a_3.$$

for all t in $GF(q)$.

- **Conical Flock Spread:** The conical flock spread is represented as

$$x=0, y=x\begin{bmatrix} u^{\sigma}+F(t), & G(t)+A(u) \\ t & u \end{bmatrix},$$

for all t, u in $GF(q)$, F, G are functions on $GF(q)$.

- **Conical Flock:** The associated and equivalent conical flock is given by:

$$\Sigma_t^{left}:$$

$$-a_1(F(t)^{\sigma^{-1}} + A(t^{\sigma^{-1}})) - a_2 t^{\sigma^{-1}} + a_3(G(t) - A(F(t)^{\sigma^{-1}}) - a_4 = 0,$$

for all t in $GF(q)$.

- In general, we will discuss "conical" flock spreads: left conical, left hyperbolic, right conical and right hyperbolic. The left flock spreads, and the images of these spreads are under the left mappings

$$(x,y) \to (x\begin{bmatrix} 0 & 1 \\ 1 & 0 \end{bmatrix}, y\begin{bmatrix} 0 & 1 \\ 1 & 0 \end{bmatrix})$$

define isomorphic left flocks, the so-called "double-transpose".

- Similarly, the right flock spreads, and the images of these spreads are under the right mappings

$$(x^T, y^T)^T \to (\begin{bmatrix} 0 & 1 \\ 1 & 0 \end{bmatrix} x^T, \begin{bmatrix} 0 & 1 \\ 1 & 0 \end{bmatrix} y^T)^T$$

define isomorphic right flocks.

- Using these mappings, we may define the transpose conical spreads, should they exist as spreads, as follows:
- Transpose left conical flock spreads:

$$x = 0, y = x\begin{bmatrix} v^{\sigma^{-1}} - F(t)^{\sigma^{-1}}, & A^*(v) - A^*(F(t)) + G(t) \\ t & v \end{bmatrix},$$

where $A(v^{\sigma^{-1}})$=$A^*(v)$.

- The reader should note that these spreads do not look like the standard transpose matrix sets, as they are isomorphic to the transpose matrix sets and produce the same/isomorphic flock as do the transpose matrix sets. It is also pointed out that it is always necessary to check that the transpose matrix sets form spreads; they must be quasifibrations (and satisfy the covering condition in the infinite cases).

The element classes for finite flocks are represented in a short version using the flock spreads.

(1) $F_{hy}(2)^{\sigma - A}$:$x\begin{bmatrix} f(t) & ,g(t) \\ 1 & t \end{bmatrix}\begin{bmatrix} v^\sigma & ,A(t) \\ 0 & v \end{bmatrix}$, σ a non-trivial automorphism, A an inner $\sigma{-}A$ derivation. A "fake type 1" derivable net is defined.

(2) $F_{hy}(2)^{F,\sigma-A,inverse}$: $x\begin{bmatrix} f(t), & g(t) \\ 1 & t \end{bmatrix}^{-1}\begin{bmatrix} v^{\sigma} & A(t) \\ 0 & v \end{bmatrix}$, σ a non-trivial automorphism, A an inner $\sigma-A$ derivation. A "fake type 1" derivable net. This flock spread originates as admitting a pseudo-regulus homology group with axis $x=0$ and coaxis $y=0$, instead of the normal homology group with axis $y=0$ and coaxis $x=0$ spread.

(3) $F(2)^{\sigma}_{hy}$: $x\begin{bmatrix} f(t) & ,g(t) \\ 1 & t \end{bmatrix}\begin{bmatrix} v^{\sigma} & ,0 \\ 0 & v \end{bmatrix}$, σ is a non-trivial automorphism.

(4) $F(2)^{\sigma,Inverse}_{hy}$: $x\begin{bmatrix} t\Delta_t^{-1} & ,1-t\Delta_t^{-1}f(t) \\ -\Delta_t^{-1} & \Delta_t^{-1}f(t) \end{bmatrix}\begin{bmatrix} v^{\sigma} & ,0 \\ 0 & v \end{bmatrix}$, σ is a non-trivial automorphism, the matrix inverse of $\begin{bmatrix} f(t) & ,g(t) \\ 1 & t \end{bmatrix}$, using the notation of the general case, Δ_t is the determinant of the matrix.

(5) $F(3)_{hy}$: $x\begin{bmatrix} f(t) & ,g(t) \\ 1 & t \end{bmatrix}\begin{bmatrix} v & ,0 \\ 0 & v \end{bmatrix}$; original hyperbolic flocks spreads. These are completely determined (see [**48**] of Thas, [**1**] Bader-Lunardon – the planes are all nearfield planes, either regular nearfield or irregular of orders $11^2, 23^2$ *or* 59^2 (see Bader, Baker-Ebert, Bonisoli, and Johnson for the irregular near fields in the references of [**34**]).

(6) $F(2)^{F,\sigma-A}_{co}$: $x\begin{bmatrix} u^{\sigma}+F(t) & ,G(t)+A(u) \\ t & u \end{bmatrix}$, σ *a non-trivial* automorphism, A an inner $\sigma-A$ derivation. When A is an inner σ-derivation set, the type is a "fake 1" type, which becomes a type 2 type.

(7) $F(2)^{trans,F,\sigma^{-1}-A^*}$ $x\begin{bmatrix} u^{\sigma^{-1}}-F(t)^{\sigma^{-1}} & ,L \\ t & u \end{bmatrix}$, where $A^*(u) = A(u^{\sigma^{-1}})$, the transpose of #6, as the spread exists, in the field case, if the transpose is not isomorphic to the original. Where

$$L = A^*(v) - A^*(F(t)) + G(t).$$

(8) $F(2)^{\sigma}_{co}$: $x\begin{bmatrix} u^{\sigma}+F(t) & ,G(t) \\ t & u \end{bmatrix}$, σ a non-trivial automorphism.

(9) $F(2)^{trans,\sigma}_{co}$: $x\begin{bmatrix} v^{\sigma^{-1}}-F(t)^{\sigma^{-1}} & ,G(t) \\ t & v \end{bmatrix}$, σ a *non-trivial*, automorphism. The transpose spread of the previous, provided these spreads are not isomorphic to the previous.

(10) $F(3)_{co}$: $x\begin{bmatrix} u+F(t) & ,G(t) \\ t & u \end{bmatrix}$; the original quadratic cone spreads. The transpose spreads here must be isomorphic to the original spread by a result of Gevaert and Johnson [**15**]. Furthermore, this is more generally true for this form over arbitrary skewfields, by use of the double-transpose mapping.

(11) $F(2)_{hy\&co}^{F,\ \sigma-A,Inverse}$: This is the case where a $\sigma - A$ conical spread admits a left-acting regulus-inducing homology group, thereby producing a hyperbolic flock spread. The derivable net is a "fake type 1" net.

There are examples of eleven out of eleven flocks.

Also, the $\sigma - A$ examples are connected to the σ-examples using "resolution mappings", which are isomorphisms of the flock spreads.

The isomorphism questions between flock spreads and flocks have not been completely answered, so it is not yet clear if some of the flock examples obtained are isomorphic in these cases. This statement also applies to all other possible skewfield examples.

1.2. Infinite field flocks

All derivable nets are regulus nets, and for $i = 1, 2, 3$, there are the following total types: There are the eleven general types analogous to those for finite flocks. Then there is $1 - A$, type 1; one hyperbolic, one conical. For inner $\sigma - A$ spreads/flocks, a change of basis can reduce to $A = 0$. Then, there is type 2; there are three types: one hyperbolic, one conical and a transpose conical. There are also the left and right hyperbolic flocks and the combination "inversing simultaneous". For type 3, there is one hyperbolic and one conical. And, finally, there is a transpose conical type 3. There are the following sixteen possible types.

The hyperbolic flocks and conical flocks have the same representation as in the finite case, where D is now an infinite field.

The sixteen cases of infinite flock spreads are as follows:

We have the eleven cases as in the finite field cases, but over infinite fields, there are five additional cases, valid only in the infinite case, with one open case. As noted in the finite cases, a considerable theory of flocks has developed. There has been essentially no theory developed for flocks of infinite structures, whether fields or skewfields. The intersecting cases when a conical flock and the transpose are not isomorphic are known very sparsely. We list separate cases when they are not isomorphic, and these classes are almost completely open.

(1) $IN_{hy}(2)^{F,\sigma-A}$: $x\begin{bmatrix} f(t) & ,g(t) \\ 1 & t \end{bmatrix}\begin{bmatrix} v^{\sigma} & ,(v-v^{\sigma})c \\ 0 & v \end{bmatrix}$. A "a fake type 1" derivation set.

(2) $IN_{hy}(2)^{F,\sigma-A,inverse}$: $x\begin{bmatrix} f(t) & ,g(t) \\ 1 & t \end{bmatrix}^{-1}\begin{bmatrix} v^{\sigma} & A(v) \\ 0 & v \end{bmatrix}$, σ a non-trivial automorphism, A an inner $\sigma-A$ derivation. A "fake type 1" derivable net. This flock spread may also originate as a double cover spread, where the regulus-inducing homology group acts on both the left and right, or as just a left action regulus-inducing homology group.

(3) $IN(2)^{\sigma}_{hy}$: $x\begin{bmatrix} f(t) & ,g(t) \\ 1 & t \end{bmatrix}\begin{bmatrix} v^{\sigma} & ,0 \\ 0 & v \end{bmatrix}$.

(4) $IN(2)^{\sigma,Inverse}_{hy}$: Δ_t is the determinant.

$$x\begin{bmatrix} t\Delta_t^{-1} & ,1-t\Delta_t^{-1}f(t) \\ -\Delta_t^{-1} & \Delta_t^{-1}f(t) \end{bmatrix}\begin{bmatrix} v^{\sigma} & 0 \\ 0 & v \end{bmatrix},$$

σ a non-trivial automorphism, the matrix of the inverse of

$$\begin{bmatrix} f(t) & ,g(t) \\ 1 & t \end{bmatrix}.$$

(5) $IN(3)_{hy}$: $x\begin{bmatrix} f(t) & ,g(t) \\ 1 & t \end{bmatrix}\begin{bmatrix} v & ,0 \\ 0 & v \end{bmatrix}$.

(6) $IN(2)^{F,\sigma-A}_{co}$: $x\begin{bmatrix} u^{\sigma}+F(t) & ,G(t)+(u^{\sigma}-u)c) \\ t & u \end{bmatrix}$,

σ non-trivial.

A "a fake type 1" derivable net.

(7) $IF(2)^{trans,F,\sigma^{-1}-A^*}$ $x\begin{bmatrix} u^{\sigma^{-1}}-F(t)^{\sigma^{-1}} & ,L \\ t & u \end{bmatrix}$, the transpose of #6, should the spread exist, where,

$A^*(u)=A(u^{\sigma^{-1}}), L=A^*(v)-A^*(F(t))+G(t)(Open)$.

(8) $IN(2)^{\sigma}_{co}$: $x\begin{bmatrix} u^{\sigma}+F(t) & ,G(t) \\ t & u \end{bmatrix}$, σ non-trivial.

(9) $IN(2)^{trans,\sigma}_{co}$:$x\begin{bmatrix} u^{\sigma^{-1}}-F(t)^{\sigma^{-1}} & ,G(t) \\ t & u \end{bmatrix}$, *$\sigma$ non-trivial.* The transpose spread of the previous, provided these spreads are not isomorphic to the previous.

(10) $IN(3)_{co}$: $x\begin{bmatrix} u+F(t) & ,G(t) \\ t & u \end{bmatrix}$.

(11) $IN(2)_{hy\&co}^{F,\sigma-A,Inverse}$: This is the case where a $\sigma - A$ (inner) conical spread admits a left-acting regulus-inducing homology group, thereby producing a hyperbolic flock spread. The derivable net is a "fake type 1" .

(12) $IN_{hy}(1)^{1-A}$: $\begin{bmatrix} f(t) & ,g(t) \\ 1 & t \end{bmatrix} \begin{bmatrix} v & ,A(v) \\ 0 & v \end{bmatrix}$; A an outer derivation function.

(13) $IN_{hy}(1)^{1-A,inverse}$: $\begin{bmatrix} f(t) & ,g(t) \\ 1 & t \end{bmatrix}^{-1} \begin{bmatrix} v & ,A(v) \\ 0 & v \end{bmatrix}$; A an outer derivation function.

(14) $IN_{hy\&co}(1)^{\,1-A,inverse}$; A an outer derivation function. The situation where there is a conical type admitting a left-acting regulus-inducing homology group, thereby obtaining also a hyperbolic type.

(15) $IN(1)_{co}$: $x\begin{bmatrix} u+F(t) & ,G(t)+A(u) \\ t & u \end{bmatrix}$, A an outer derivation function.

(16) $IN(1)_{co}^{trans}$: $x\begin{bmatrix} u-F(t) & ,G(t)+A(u) \\ t & u \end{bmatrix}$, the transpose of the #15, if (i) exists (yes) and (ii) is not isomorphic (Open).

There are examples of fifteen of the sixteen possible types.

Four of the five extra types not valid in finite fields are obtained by forming a quadratic but not a Galois extension $F(z)$ over F, where $F(z)$ is the function field over F. When the characteristic is 2, use du/dz as an A-derivative function to create the four examples #12,13,14,15. As such examples are also valid when F is $GF(q)$, it might also be reasonable to consider examples as "generated from finite fields" and included in the finite field case.

1.3. Non-commutative left skewfield flocks

In the non-commutative skewfield case, we may add the $\sigma - A$, functions, where $\sigma = 1$ or not 1. In the field case, the σ not 1 case becomes an inner $\sigma - A$ derivation function. Thus, the $\sigma - A$ -derivable net is contained in a translation plane with spread in $PG(3, K)$, for K a non-commutative skewfield. For this notation, we shall use $N(i)^{\sigma}$, where $\sigma = 1$ or not 1. Likewise, as in the finite and infinite field types, there are special types. But, also, we have the added complication that what might appear to be a spread is not a spread but a quasifibration. The associated flock types then are not flock types but what we call "quasiflocks". The actual flocks that we are able to find are always quasifibrations that satisfy either the right or left unwrapping principle or originate from this type. It is always an open question whether a quasifibration that does not satisfy an unwrapping principle is actually a proper quasifibration. However, we have determined examples of left additive quasifibrations that do not satisfy the left unwrapping principle but are, in fact, spreads.

The spreads and flocks for the left non-commutative cases are represented exactly as in the infinite field case, where D is a non-commutative skewfield, and the spreads and flocks are represented using row vectors. However, the situation uses non-commutative terms, and the arguments must be correspondingly different both for the generalized hyperbolic quadrics and the generalized quadratic cones.

The first sixteen cases are exactly as in the infinite field version, with ten additional cases.

The twenty-six cases for left non-commutative flock spreads and flocks are as follows:

(1) $N_{hy}^{L}(2)^{F,\sigma-A}$: $x\begin{bmatrix} f(t) & ,g(t) \\ 1 & t \end{bmatrix}\begin{bmatrix} v^{\sigma} & ,(v^{\sigma}c-cv) \\ 0 & v \end{bmatrix}$. A "a fake type 1" derivation set.

(2) $N_{hy}^{L}(2)^{F,\sigma-A,inverse}$: $x\begin{bmatrix} f(t) & ,g(t) \\ 1 & t \end{bmatrix}^{-1}\begin{bmatrix} v^{\sigma} & A(v) \\ 0 & v \end{bmatrix}$, σ a non-trivial an inner $\sigma - A$ derivation. A "fake type 1" derivable net. This flock spread may also originate as a double cover spread, where the regulus-inducing homology group acts on both the left and right, or as just a left action regulus-inducing homology group.

(3) $N_{hy}^{L}(2)^{\sigma}$: $x\begin{bmatrix} f(t) & ,g(t) \\ 1 & t \end{bmatrix}\begin{bmatrix} v^{\sigma} & ,0 \\ 0 & v \end{bmatrix}$.

(4) $N_{hy}^{L}(2)^{\sigma,inverse}$:

$$x\begin{bmatrix} t\Delta_t^{-1} & ,1-t\Delta_t^{-1}f(t) \\ -\Delta_t^{-1} & \Delta_t^{-1}f(t) \end{bmatrix}\begin{bmatrix} v^{\sigma} & 0 \\ 0 & v \end{bmatrix},$$

σ a non-trivial automorphism, the matrix is the inverse of,

$$\begin{bmatrix} f(t) & ,g(t) \\ 1 & t \end{bmatrix}$$

and

$$\Delta_t \textit{ the determinant.}$$

(5) $N_{hy}^{L}(3)$: $x\begin{bmatrix} f(t) & ,g(t) \\ 1 & t \end{bmatrix}\begin{bmatrix} v & ,0 \\ 0 & v \end{bmatrix}$.

(6) $N_{co}^{L}(2)^{F,\sigma-A}$: $\begin{bmatrix} u^{\sigma}+F(t) & ,G(t)+(u^{\sigma}c-cu)) \\ t & u \end{bmatrix}$, *σ non-trivial.*
A "a fake type 1" derivable net.

(7) $N_{co}^{L,trans,F,\sigma^{-1}-A^{*}}$ $x\begin{bmatrix} u^{\sigma^{-1}}-F(t)^{\sigma^{-1}} & ,A^{*}(v)-A^{*}(F(t))+G(t) \\ t & u \end{bmatrix}$,
where $A^{*}(u)=A(u^{\sigma^{-1}})$, the transpose of #6, should the spread exist. (Open).

(8) $N_{co}^{L}(2)^{\sigma}$: $\begin{bmatrix} u^{\sigma}+F(t) & ,G(t) \\ t & u \end{bmatrix}$, σ non-trivial.

(9) $N_{co}^{L}(2)^{\sigma,trans}$: $x\begin{bmatrix} u^{\sigma^{-1}}-F(t)^{\sigma^{-1}} & ,G(t) \\ t & u \end{bmatrix}$, *σ non-trivial.*

The transpose spread of the previous. These spreads are not isomorphic to the previous.

(10) $N_{co}^{L}(3)$: $x\begin{bmatrix} u+F(t) & ,G(t) \\ t & u \end{bmatrix}$.

(11) $N_{hy\&co}^{L}(2)^{F,\sigma-A,Inverse}$: This is the case where a $\sigma-A$ (inner) conical spread admits a left-acting regulus-inducing homology group, thereby producing a hyperbolic flock spread. The derivable net is a "fake type 1" .

(12) $N_{hy}^{L}(1)^{1-A}$: $x\begin{bmatrix} f(t) & ,g(t) \\ 1 & t \end{bmatrix}\begin{bmatrix} v & ,A(v) \\ 0 & v \end{bmatrix}$; A an outer derivation function.

(13) $N_{hy}^{L}(1)^{1-A,inverse}$: $x\begin{bmatrix} f(t) & ,g(t) \\ 1 & t \end{bmatrix}^{-1}\begin{bmatrix} v & ,A(v) \\ 0 & v \end{bmatrix}$;

A an outer derivation function.

(14) $N_{hy\&co}^{L}(1)^{\,1-A,inverse}$; the situation where there is a conical type admitting a left-acting regulus-inducing homology group. In this way, we also obtain a hyperbolic type. A an outer derivation function.

(15) $N_{co}^{L}(1)^{1-A}$: $x\begin{bmatrix} u+F(t) & ,G(t)+A(u) \\ t & u \end{bmatrix}$, A an outer derivation function.

(16) $N_{co}^{L}(1)^{trans.,1-A}$: $x\begin{bmatrix} w-F(t) & ,G(t)+A(w)-A(F(t)) \\ t & w \end{bmatrix}$, A an outer derivation function. The transpose of #15 is not isomorphic. (i) existence (yes), (ii) non isomorphic (Open).

(17) $N_{hy}^{L}(2)^{F,1-A}$: $x\begin{bmatrix} f(t) & ,g(t) \\ 1 & t \end{bmatrix}\begin{bmatrix} v & ,A(v) \\ 0 & v \end{bmatrix}$;

A an inner derivation function;

a"*faketype*1".

(18) $N_{hy}^{L}(2)^{F,1-A,inverse}$: $x\begin{bmatrix} f(t) & ,g(t) \\ 1 & t \end{bmatrix}^{-1}\begin{bmatrix} v & ,A(v) \\ 0 & v \end{bmatrix}$; A an inner derivation function; a " fake type 1".

(19) $N_{hy\&co}^{L}(2)^{F,\,1-A,inverse}$; the situation where there is a conical type admitting a left-acting regulus-inducing homology group, thereby obtaining also a hyperbolic type. A an inner derivation function; a "fake type 1".

(20) $N^{L}(2)_{co}^{F,1-A}$: $x\begin{bmatrix} u+F(t) & ,G(t)+A(u) \\ t & u \end{bmatrix}$, A an inner derivation function; a "fake type 1".

(21) $N^{L}(2)_{co}^{trans,F,1-A}$: $x\begin{bmatrix} w-F(t) & ,G(t)+A(w)-A(F(t)) \\ t & w \end{bmatrix}$, A an inner derivation function; a "fake type 1". Transpose of #20. (Open).

(22) $N_{hy}^{L,\sigma-A,inverse}(1)$: $x\begin{bmatrix} f(t) & ,g(t) \\ 1 & t \end{bmatrix}^{-1}\begin{bmatrix} v^{\sigma} & ,A(v) \\ 0 & v \end{bmatrix}$; A is a non-trivial σ-outer derivation function, and there is a left-acting pseudo-regulus-inducing homology group. (Open).

(23) $N_{hy}^{L}(1)^{\sigma-A}$: $x\begin{bmatrix} f(t) & ,g(t) \\ 1 & t \end{bmatrix}\begin{bmatrix} v^{\sigma} & ,A(v) \\ 0 & v \end{bmatrix}$; $\sigma-A$-outer derivation function, σ non-trivial outer automorphism. (Open case.)

(24) $N_{co}^{L}(1)^{\sigma-A}$: $x\begin{bmatrix} u^{\sigma}+F(t) & ,G(t)+A(u) \\ t & u \end{bmatrix}$; A is a non-trivial outer σ-derivation function for σ, a non-trivial outer automorphism. (Open case.)

(25) $N_{co}^{L}(1)^{trans,\sigma-A}$: $x\begin{bmatrix} w^{\sigma^{-1}}-F(t)^{\sigma^{-1}} & ,L \\ t & w \end{bmatrix}$;

$$A^{*}(u)=A(u^{\sigma^{-1}}), L=G(t)+A^{*}(w)-A^{*}(F(t)),$$

is a non-trivial outer σ-derivation function for σ, a non-trivial outer automorphism. The transpose of #24 (Open case.)

(26) $N^{L}(3)_{hy}^{inverse}$: $\begin{bmatrix} f(t) & ,g(t) \\ 1 & t \end{bmatrix}^{-1}\begin{bmatrix} v & ,0 \\ 0 & v \end{bmatrix}$. (Open case).

There are examples of twenty/twenty-one of the twenty-six possible left cases.

1.4. Non-commutative right skewfield flocks

Here are the twenty-six non-commutative right skewfield flocks, as represented by the associated right spreads. The function notation of the left is also used for the right, but there is not necessarily any general connection other than perhaps sometimes flock to quasiflock. However, we will show that using the unwrapping theory with central constants will allow the use of exactly the same functions in the transposed setting. However, in all cases, the left generalized hyperbolic quadric and the right generalized cone are not identical to the right generalized hyperbolic quadric and the right generalized cone, respectively. If, instead of a non-commutative skewfield D, one would take a field, these varieties would then be identical.

For the right flock spread and right flocks, we use the column vector representation. The forms for the right correspond to what would be expected if the left forms are transposed. However, it is not the case that a right form may be obtained by a transposition of a left form.

We use T to denote transpose to denote column vectors and trans to indicate transpose matrix sets.

- **Right hyperbolic spread:**

$$x^T = 0^T, \; y^T = \begin{bmatrix} u & 0 \\ A^*(u), & u^\tau \end{bmatrix} x^T,$$

$$y^T = \begin{bmatrix} v & 0 \\ A^*(v) & v^\tau \end{bmatrix} \begin{bmatrix} r(t) & 1 \\ s(t) & t \end{bmatrix} x^T,$$

where T denotes transpose, $\tau - A^*$ is a τ-derivation function, for τ the identity or outer automorphism, A^* is identically zero, inner or outer τ-derivation function, and r, s are functions on D, a non-commutative skewfield.

- **Right hyperbolic flock:** $\Pi_t^R : (s(t) - A(r(t))a_1 - r(t)^\tau a_2 + ta_3 - a_4 = 0, and\, \varrho^R : x_2 = x_3.$
- **Right conical flock spread:**

$$x^T = 0^T, y^T = \begin{bmatrix} u + R(t) & t \\ S(t) + A^*(u), & u^\tau \end{bmatrix} x^T,$$

for all t, u in D, R, S functions on D.

- **Right conical flock:**

$$\Sigma_t^R :$$

$$-(R(t)^\tau + A^*(t))x_2 - t^\tau x_4 + (S(t) - A^*(R(t))x_3 - x_4 = 0.$$

- **Transpose of the right conical flock spreads.** We use the mappings

$$(x^T, y^T)^T \to (\begin{bmatrix} 0 & 1 \\ 1 & 0 \end{bmatrix} x^T, \begin{bmatrix} 0 & 1 \\ 1 & 0 \end{bmatrix} y^T)^T$$

to construct an isomorphic transpose as follows:

$$x^T = 0^T, y^T = \begin{bmatrix} w - R(t)^\tau & t \\ S(t) + A^{**}(w) - A^*(R(t)), & w^{\tau^{-1}} \end{bmatrix} x^T,$$

where $A^{**}(w) = A^*(w^{\tau^{-1}})$.

The twenty-six right flock cases:

(1) $N_{hy}^R(2)^{F,\tau - A^*}$: $\begin{bmatrix} v & ,0 \\ (vc - cv^\tau), & v^\tau \end{bmatrix} \begin{bmatrix} r(t) & ,1 \\ s(t) & t^\tau \end{bmatrix} x^T$, τ is a non-trivial outer automorphism, "fake type 1" derivable net.

(2) $N_{hy}^R(2)^{\tau - A^*, inverse}$:

$$\begin{bmatrix} v & ,0 \\ (vc - cv^\tau), & v^\tau \end{bmatrix} \begin{bmatrix} t\Delta_t^{*^{-1}} & , -\Delta_t^{*^{-1}} \\ 1 - \Delta_t^{*^{-1}} f(t)t, & \Delta_t^{*^{-1}} f(t) \end{bmatrix} x^T,$$

τ a non-trivial outer automorphism. Here we use the non-commutative "upper determinant" $\Delta_t^* = (t^\tau r(t) - s(t))$, and the associated left inverse of $\begin{bmatrix} r(t) & ,1 \\ s(t) & t^\tau \end{bmatrix}$. (There are four ways to use determinants over 2 by 2 matrices over skewfields, check the notes in the text).

(3) $N_{hy}^R(2)^\tau$: $\begin{bmatrix} v & ,0 \\ 0 & v^\tau \end{bmatrix} \begin{bmatrix} r(t) & ,1 \\ s(t) & t^\tau \end{bmatrix} x^T$, τ a non-trivial outer automorphism.

(4) $N_{hy}^R(2)^{\tau, inverse}$:

$$\begin{bmatrix} v & ,0 \\ 0 & v^\tau \end{bmatrix} \begin{bmatrix} t\Delta_t^{*^{-1}} & , -\Delta_t^{*^{-1}} \\ 1 - \Delta_t^{*^{-1}} r(t)t, & \Delta_t^{*^{-1}} r(t) \end{bmatrix} x^T,$$

τ a non-trivial outer automorphism. Here we use the non-commutative "upper determinant" $\Delta_t^* = (t^\tau r(t) - s(t))$, and the associated left inverse of $\begin{bmatrix} r(t) & ,1 \\ s(t) & t^\tau \end{bmatrix}$.

(5) $N_{hy}^R(3)$: $\begin{bmatrix} v & ,0 \\ 0 & v \end{bmatrix} \begin{bmatrix} r(t) & ,1 \\ s(t) & t \end{bmatrix} x^T$.

(6) $N_{co}^R(1)^{F,\tau - A^*}$: $\begin{bmatrix} u + R(t) & ,t \\ S(t) + A^*(t), & u^\tau \end{bmatrix} x^T$, τ a *non-trivial* outer automorphism.

(7) $N_{co}^R(1)^{trans.,F,\tau^{-1} - A^{**}}$: $\begin{bmatrix} w - R(t)^\tau & t \\ S(t) + A^{**}(w) - A^*(R(t)), & w^{\tau^{-1}} \end{bmatrix} x^T$. The transpose of # 6, assuming the spread exists and is not isomorphic to previous. (Open).

(8) $N_{co}^{R}(2)^{\tau}$: $\begin{bmatrix} u+R(t) & ,t \\ S(t) & u^{\tau} \end{bmatrix} x^T, \tau$ *a non-trivial* outer automorphism.

(9) $N_{co}^{R}(2)^{\tau,trans}$: $\begin{bmatrix} w-R(t)^{\tau} & t \\ S(t), & w^{\tau^{-1}} \end{bmatrix} x^T, \tau$ *a non-trivial* outer automorphism. The transpose spread of the previous. These spreads are not isomorphic to the previous. (Open).

(10) $N_{co}^{R}(2)$: $\begin{bmatrix} u+R(t) & ,t \\ S(t) & u \end{bmatrix} x^T$.

(11) $N_{hy\&co}^{R}(2)^{\tau,inverse}$, a union of $N^{right}(1)_{hy}^{\tau,Inverse}$ and $N^{right}(2)_{co}^{\tau}$, may be determined.

(12) $N_{hy}^{R}(1)^{1-A^*}$: $\begin{bmatrix} v & ,0 \\ A^*(v) & v \end{bmatrix} \begin{bmatrix} r(t) & ,1 \\ s(t) & t \end{bmatrix} x^T$; A^* is an outer derivation function.

(13) $N_{hy}^{R}(1)^{1-A^*,inverse}$: $\begin{bmatrix} v & ,0 \\ A^*(v) & v \end{bmatrix} \begin{bmatrix} f(t) & ,1 \\ g(t) & t \end{bmatrix}^{-1} x^T$, the right-inverse type. A^* is an outer derivation function.

(14) $N_{hy\&co}^{R}(1)^{1-A^*,inverse}$: The situation where there is also a conical type admitting a right-acting regulus-inducing homology group, thereby obtaining also a right hyperbolic type. A^* is an outer derivation function.

(15) $N_{co}^{R}(1)^{1-A^*}$: $\begin{bmatrix} u+R(t) & ,t \\ S(t)+A^*(t), & u \end{bmatrix} x^T$; A^* is an outer derivation function.

(16) $N_{co}^{R}(1)^{1-A^*}$: $\begin{bmatrix} w-R(t) & t \\ S(t)+A^{**}(w)-A^*(R(t)), & w \end{bmatrix} x^T$;

A^* is an outer derivation function.

The transpose of #15. (i) existence (yes), (ii) not isomorphic (Open).

(17) $N_{hy}^{R}(2)^{F,1-A^*}$: $\begin{bmatrix} v & ,0 \\ A^*(v) & v \end{bmatrix} \begin{bmatrix} f(t) & ,1 \\ g(t) & t \end{bmatrix} x^T$; A^* is an inner derivation function; a "fake type 1".

(18) $N_{hy}^{R}(2)^{F,1-A^*,inverse}$: $\begin{bmatrix} v & ,0 \\ A^*(v) & v \end{bmatrix} \begin{bmatrix} f(t) & ,1 \\ g(t) & t \end{bmatrix}^{-1} x^T$, the right-inverse type. A^* is an inner derivation function; a "fake type 1".

(19) $N_{hy\&co}^{R}(2)^{F,1-A^*,inverse}$: The situation where there is also a conical type admitting a right-acting regulus-inducing homology group, thereby obtaining also a right hyperbolic type. A^* is an inner derivation function; a "fake type 1".

(20) $N_{co}^R(1)^{F,1-A^*}$: $\begin{bmatrix} u+F(t) & ,t \\ G(t)+A^*(t), & u \end{bmatrix} x^T$; A^* is an inner derivation function; a "fake type 1".

(21) $N_{co}^R(1)^{trans.,F,1-A^*}$: $\begin{bmatrix} w-R(t) & t \\ S(t)+A^{**}(w)-A^*(R(t)), & w \end{bmatrix} x^T$,

A^* is an inner derivation function;

a "fake type 1". (Open).

(22) $N_{hy}^R(1)^{\tau-A^*,\ inverse}$: $\begin{bmatrix} v & ,0 \\ A^*(v) & v^\tau \end{bmatrix} \begin{bmatrix} f(t) & ,1 \\ g(t) & t^\tau \end{bmatrix}^{-1} x^T$;

A^* is a non-trival outer τ-derivation function,

τ non-trivial outer automorphism. (Open case.)

(23) $N_{hy}^R(1)^{\tau-A^*}$: $\begin{bmatrix} v & 0 \\ A^*(v), & v^\tau \end{bmatrix} \begin{bmatrix} r(t) & ,1 \\ s(t) & t \end{bmatrix} x^T$, $\tau-A^*$ outer derivation function. (Open case).

(24) $N_{co}^R(1)^{\tau-A^*}$: $\begin{bmatrix} u+F(t) & ,t \\ G(t)+A^*(u), & u^\tau \end{bmatrix} x^T$; A^* is a non-trivial outer τ-derivation function for τ, a non-trivial outer automorphism. (Open case.)

(25) $N_{co}^R(1)^{\tau^{-1}-A^{**}}$: $\begin{bmatrix} w-R(t)^\tau & t \\ S(t)+A^{**}(w)-A^*(R(t)), & w^{\tau^{-1}} \end{bmatrix} x^T$; A^* is a non-trivial outer τ-derivation function for τ, a non-trivial outer automorphism. The transpose of #24 (Open case.)

(26) $N_{hy}^R(3)^{inverse}$: $\begin{bmatrix} v & 0 \\ 0 & v \end{bmatrix} \begin{bmatrix} f(t) & ,g(t) \\ 1 & t \end{bmatrix}^{-1} x^T$. (Open case).

There are examples of twenty/twenty-one of the twenty-six possible cases.

For the hyperbolic and conical flock types with automorphism, a non-trivial inner automorphism, there are most likely interesting examples of quasifibrations and/or flock spreads of this type. Here are the types of inner automorphism types, both left and right, each with fifteen total possible types.

1.5. The left inner and right inner types

In this section, we consider the corresponding "left inner automorphism types" of spreads or quasifibrations over a non-commutative skewfield D. We illustrate certain reductions, but we may also regard the types as non-reductive, as reductions can change the isomorphism type of the associated flocks. The reader might be familiar with certain division rings that may be reduced if using inner derivation functions. All outer automorphisms are assumed to be "strict".

The real problem with determining examples of inner flock spreads, where there is always a non-trivial inner automorphism, is that outside of skewfield spreads themselves, such matrix sets never satisfy either the left or right unwrapping principle. In other words, there is, at present, no way to cover the affine points unless there is some outside isomorphism mapping from a spread that takes you into this set, thereby providing a natural cover. That is, there would be isomorphic flock spreads, but we would wish to find non-isomorphic flocks. In the material on "the kaleidoscope of derivable nets" in part IX, the use of the K-maps, where $K = \begin{bmatrix} e & f \\ 0 & h \end{bmatrix}$, eh non-zero and e, f, h in the associated non-commutative skewfield, it may be shown that it is possible to change the left isomorphic spread to the point where the two associated flocks are not isomorphic. To obtain, if e is not in the center of D and f is non-zero, this idea works for all three types of flocks; conical, hyperbolic and inverse hyperbolic. If there is an example where the outer automorphism $\sigma = 1$, we may then create a set. There are fifteen possible types of left non-commutative flocks with $\sigma = 1$. All of these then become 15 types of left inner flocks. There are four open cases. For the examples arising from kaleidoscopic mappings, there are infinitely many examples, as the K-mapping may use infinitely many possible entries e, f, h, for all e, h for at least one element in $D - Z(D)$, and for f non-zero.

We shall then indicate the number for the left inner with the same number as the left non-commutative flock spreads. We note that the general derivable net type cannot change under the kaleidoscopic mapping.

It is also possible that an inner flock example may arise from an "elation double spread". When this occurs, the inner arises by a change of coordinates; there are two derivable nets that act as the derivable net that defines the flock. There are often two such derivable nets that form the spread using a sort of "direct product". We shall display examples of this type of construction of inner flocks in the study of "flock and quasiflock strings" in part XI. These types will be called the "direct sum" types of inner flocks".

Here is how to locate the general types of inner flocks. In all of the following cases, assume that D is a non-commutative skewfield.

Kaleidoscopic:

1.5.1. For left inner flocks: Go to the list for left non-commutative flock spreads. There are fifteen possible left non-commutative flock spreads with σ equal to 1. The numbers are

as follows:

$$5, 10, 12, 13, 14, 1, 15, 16, 17, 18, 19, 20,$$
$$21(open), (24)(open), (25)(open), (26)(open).$$

Take any left kaleidoscopic mapping with e not in the center. This will construct a left inner flock. If f is non-zero the associated left flocks are not isomorphic. It is possible to find $\sigma - A$ outer derivation functions here, where σ is inner but A is not!

1.5.2. For right inner flocks: Go to the list for right non-commutative flock spreads. There are fifteen possible left non-commutative flock spreads with σ equal to 1. The numbers are as follows:

$$5, 10, 12, 13, 14, 1, 15, 16, 17, 18, 19, 20,$$
$$21(open), (24)(open), (25)(open), (26)(open).$$

Take any right kaleidoscopic mapping with e not in the center. This will construct a right inner flock. If f is non-zero the associated right flocks are not isomorphic. It is possible to find $\sigma - A$ outer derivation functions here, where σ is inner but A is not!

Direct Product.

1.5.3. Take any left or right conical flock spread that is an elation double spread with σ equal to 1. Change bases to change the fundamental derivable net. This may be done either pre- or post-mapping. If one of these produces a left or right spread or an inner left or right conical flock, this then creates another example. Finally, all of the different types of inner flocks (Inn,0), (Inn,Inn)(two types) and (Inn,Out) will be represented.

1.6. Type (0) flocks

The type (0) derivable nets arise in the following situations: there is a left L-derivable net in the affine space created by the embedding-contraction method by a right classical K-pseudo-regulus net, where L and K are mutually non-isomorphic skewfields. There are the following situations:

(i) L and K are both fields. (ii) L is a non-commutative skewfield and $K$is a field, (iii) L is a field, and K is a non-commutative skewfield or (iv) L and K are both non-commutative skewfields. In all of situations, the tension between left and right spreads produces irreducible groups to act faithfully on the common subspaces of the two derivable nets, which then creates a spread in $PG(3,K)$ obtained by the tensor product of L and K, in a technical manner explained in the skewfield extension theorem, in [**34**], Theorem 24 p. 65. This spread extends both L and K and thus may be embedded in both 3-dimensional projective spaces $PG(3,L)$ and $PG(3,K)$. What this means is that, especially when L or K is a field and one is a non-commutative skewfield, that there are four flocks constructed, two hyperbolic and two conical. The spreads are of the type considered here and, as extensions of skewfields, are linear flocks. There are then twelve possible left flocks, constructed in three ways of four each.

Of course, there is an equivalent method for studying the right type (0) flocks, for which there are also twelve possible right flocks.

Although such type (0) flocks are not necessarily of the Thas-Walker extension types, the quaternion division rings $(a,b)_F^{F(\theta)}$-spreads do have this structure and will be used in this text. There are two interesting ways to consider quaternion division rings. As a division ring spread, the matrix representation is given as

$$x=0, y=x\begin{bmatrix} u^{\sigma} & bt^{\sigma} \\ t & u \end{bmatrix},$$

for all t,u in $F(\theta)$, a quadratic Galois extension of the field F, where θ is connected to a in the irreducible quadratic; for example, if the characteristic is not 2 case, $F(\theta)=F(\sqrt{a})$. This is a Desarguesian spread, and is a left conical spread, and a left hyperbolic spread. Moreover, the transpose is of the form

$$x=0, y=x\begin{bmatrix} u^{\sigma} & b^{-1}t^{\sigma} \\ t & u \end{bmatrix}.$$

The determinant is non-zero because the determinant of

$$\begin{bmatrix} u^{\sigma} & bt^{\sigma} \\ t & u \end{bmatrix}$$

is non-zero. So, we have another quaternion division ring. This means that the left and right flocks are anti-isomorphic, in the section on anti-isomorphic flocks. This is a type 2 form when realizing the quaternion division ring $(a,b)_F$ in $PG(3,F(a,b)_F)$, but it is a type (0) when considered in $PG(3,F(\theta,\delta)$, where $F(\theta)$ and $F(\delta)$ are quadratic extensions over the center F,$F(\theta)$ is Galois, and $F(\delta)$ is not required to be Galois, but $F(\theta,\delta)$ is a 4-dimensional F vector space over F that admits σ as an involutory automorphism. This creates a quaternion division ring extension by $F(\delta)$. Extending t,u to $F(\theta,\delta)$, creates a Desarguesian spread containing a type 0 regulus-net (pseudo-regulus over a field are regulus nets).

Going back to the type 2, regulus net and Desarguesian spread, we may apply the set of K-mappings, for $K = \begin{bmatrix} e & f \\ 0 & g \end{bmatrix}$, ef non-zero and e, f, g in $F(\theta, \delta)$, produces an isomorphic Desarguesian spread; the original flock is linear, and the flocks obtained from the K-mappings of the above form are probably linear but not necessarily isomorphic.

REMARK 4. Therefore, there are eleven possible finite flocks, sixteen possible infinite field flocks, twenty-six possible non-commutative left skewfield flocks and twenty-six right skewfield flocks, and fifteen left and fifteen right inner flocks. Then there are the type 0 flocks. The twelve possible type (0) flocks will fall into the other categories but provide an important type of flock construction. Finally, if we were to include infinite-dimensional division rings, this would require a completely new type of theory, as we would essentially have no examples. That is, we know that there are classical pseudo-reguli over any skewfield K. If σ is an automorphism of K, we could use the multiple replacement theorem to construct Ostrom/André's type translation planes. We have not been able to determine an example of a $\sigma - A$ pseudo-regulus net where σ is a non-identity outer automorphism and A is an outer σ-derivation function. The cases where σ is an inner automorphism may be obtained using the K-mappings where $K = \begin{bmatrix} e & f \\ 0 & h \end{bmatrix}$, where eh is non-zero. Futhermore, the multiple replacement theorem still constructs spreads that look like André's replacements, using inner automorphisms. To understand these replacements would be an interesting theory that connects to the theory of infinite dimensional division rings.

For the 109 types, we shall use the previously given notation for the finite, infinite field and non-commutative skewfield examples when these are constructed. It should be noted that many of our constructions use simultaneous spreads corresponding to both conical and hyperbolic flocks. So, while we provide examples of 90/109 flocks, there is a wealth of open problems concerning the possible non-simultaneous spreads and the associated flocks. We note that the theory given here is valid over any skewfield K; our examples are over finite-dimensional skewfields. While the multiple replacement theorem works for inner automorphisms, or any automorphism σ, if there is an automorphism of order 2, obtained from a central Galois quadratic extension, we would then lift the skewfield to a semifield spread, which provides a simultaneous spread and hence a hyperbolic and a conical flock.

PROBLEM 5. This list is for all division rings, whereas all of our examples are for finite-dimensional division rings. In [**34**] , p. 316 (see problem # 77 and discussion on bullet item above), a promising option to find examples for open cases above would be to find an example of infinite-dimensional skew polynomial rings over division rings that have non-trivial $\sigma - A$ derivative functions, where σ is an outer automorphism and A is an outer σ-derivative function. *Let* $R = D[t;\sigma, A]$. Then there would be a division ring of fractions $D(t;\sigma, A)$.

Consider the following putative spread in $PG(3,D(t;\sigma, A))$:

$$x = 0, y = x \begin{bmatrix} u^\sigma + A(t), & A(u) + \gamma t^\sigma \\ t & u \end{bmatrix};$$

for all u, t in $D(t;\sigma, A)$, where γ is in the center $Z(D(t;\sigma, A))$, σ is an automorphism, and the σ-derivative function extends to $D(t;\sigma, A)$, and σ fixes the center pointwise. Then this would be a spread provided the determinant is non-zero for (t, u) not $(0, 0)$. And the determinant is non-zero if and only if for t, we would have:

$$(u^\sigma + A(t)t^{-1}u - (A(u) + \gamma t^\sigma))t.$$

Using the fact that $A(t(t^{-1}u) = A(u) = t^\sigma A(t^{-1}u) + A(t)t^{-1}u$, and assuming that there is a (t, u) such that the determinant is zero, we would have:

$$(t^{-1}u)^\sigma(t^{-1}u) - A(t^{-1}u) = \gamma.$$

Therefore, if there exists an element γ that is the center, where σ fixes γ and for all k non-zero,

$$k^\sigma k - A(k)$$

is not γ, then we would have a semifield spread that produces a conical flock. Show that this can occur. This will solve problem #22 $N_{co}^L(1)^{\sigma-A}$.

We shall show that spreads of this particular type also admit a left pseudo-regulus-inducing homology group, so the inversing mapping will create another non-semifield hyperbolic flock, which would solve problem #21-left. It might look like we would also obtain right and left flocks of this type.

When it is indicated that there are 109 possible classes of hyperbolic and conical flocks, this is really to try to provide something of an understanding of how the finite field, infinite field, left and right non-commutative skewfields fit into the generalization of Thas-Walker. These examples arise from the left and right generic examples. But there are interesting spreads containing pseudo-regulus nets that are not of these forms. In particular, there are the open-form derivable nets covered in the original volume [**34**] (p. 72, chapter 8), whose look is completely different.

The general result describing the total picture is relisted in the following chapter.

CHAPTER 2

General theorems of flocks

We provide two very similar results, the first of which also appears in volume II, regarding the general ideas of how flock and spread fit together.

2.1. Flocks from generic and non-generic spreads

In this section, we consider how to create flocks from generic or non-generic derivable nets contained in spreads and look at the more general situation.

THEOREM 6. ***(The 1st General Flock Theorem)*** *Let D be a skewfield and let a derivable net be contained in a spread S in $PG(3, D)$. Assume that the derivable net contains the components $x = 0, y = 0, y = x$. Then there exists an index set λ, which is also a skewfield such that the representation is as follows. There are two possible situations, called the hyperbolic and conical. In both cases, the derivable net is represented*

$$x = 0, y = xM$$

such that M is in λ, the Baer subspaces $B(a, b)$ of the derivable net are:

$$B(a, b) = \{(a, b)M, (a, b)J; \text{for all} M, J \text{ in } \lambda\}$$

(1) Then

(i) In the conical situation, there is an elation group E with axis $x = 0$, with mapping elements $(x, y) \to (x, xM + y)$ for all M in λ that acts on S. By the axiom of choice, choose representatives N_i for i in an index set ρ so that the orbits of E in S are given by $y = x(N_i+M)$ for all M in λ. The spread would then have the following form:

$$x = 0, y = x(M + N_i),$$

for all M in λ and for all i in ρ, where ρ contains the element indicating

$$y = 0.$$

DOI: 10.1201/9781003767978-2

Each orbit defines a derivable net with Baer subspaces

$$B_i(a,b) = \{((a,b)M, (a,b)(J - N_i)); \textit{for all}\, M, J \textit{ in } \lambda\}.$$

Hence, the spread is a set of λ-isomorphic derivable nets nets sharing one component $x = 0$.

(ii) In the hyperbolic situation, there is a homology group with axis $y = 0$ and coaxis $x = 0$, M with mapping elements $(x, y) \to (x, xM)$ for all M in λ. S then has the following form:

$$x = 0, y = 0, y = xM,$$

for all M in λ, and denoting the representatives of the homology group orbits by L_i, where i is an index set μ,

$$y = xL_iN$$

for all N in λ and for all i in μ. The Baer subplanes $B_i(a,b)$ are

$$\left\{(a,b)ML_i^{-1}, (a,b)N; \textit{ for all } M, N \textit{ in } \lambda\right\}.$$

Therefore, the spread is a union of λ-isomorphic derivable nets sharing two components.

(2) There are four types, $0, 1, 2, 3$(at least), which indicates the number of common Baer subspaces shared by the derivable net admitting the central collineation group, elation or homology, with the classical right pseudo-regulus net R obtained using the embedding/contracting method. That is, with

$$x = 0, y = \delta x; \delta \textit{ in } D,$$

with Baer subspaces R(a,b):

$$\{\alpha(a,b), \beta(a,b); \alpha, \beta \textit{ in } D\}.$$

It should be noted that the 2-(vector) is represented using the standard D-basis.

(3) If λ and D are not isomorphic, there is a type 0 obtained. If we have a commutative and a non-commutative pair, and λ is non-commutative, then λ is a quaternion division ring. Since then, D is a field; the theory of derivable nets in [**34**] *creates a quaternion division ring extension defining a spread. This spread may not be the spread S. For example, there is a lifted quaternion division ring semifield spread also containing a quaternion division ring derivable net. The first representation is called an open-form, and the second representation is a closed-form, or also a left upper generic form. There is a third representation as a closed-form, as a left lower generic form.*

(4) Considering the open-form derivable nets contained in spreads in $PG(3, D)$, let $D = E(\theta)$, which is a central Galois quadratic extension of E, for E a division ring of finite dimension over the center F. Let $x^2 + x\delta - \gamma$ be the irreducible polynomial, implying that δ, γ are both central. The definition is

$$x = 0, y = x \begin{bmatrix} u + t\delta, & t\gamma \\ t & u \end{bmatrix}$$

for all t, u in E. So, λ is isomorphic to $E(\theta)$. There are non-generic pseudo-regulus nets of type 2. The common Baer subspaces with the classical right pseudo-regulus net are the Baer subspaces $B(1, b)$ where

$$b^2 + b\delta - \gamma = 0.$$

(5) In general, when λ is isomorphic to D, the K-mappings for

$$K = \begin{bmatrix} e & f \\ g & h \end{bmatrix},$$

K non-singular and e, f, g, h in D are sharply, order-preserving, triple transitive on the Baer subspaces of λ. This implies that there are K-mappings that map λ to a left upper generic pseudo-regulus net and then map the spread S to a corresponding hyperbolic or conical flock spread.

(1) At this point, the transform theory applies to create an associated flock. The converse from the flock to the left generic flock spread, and then applying the K^{-1}-mapping recreates the spread in the given original form.

2.2. The 2nd general flock theorem

Here we give the most general theorem relative to flock-type spreads and flocks over skewfields.

THEOREM 7. ***(The 2nd General Flock Theorem)*** *Let D be a division ring. Consider a matrix spread as follow: Let M for M in λ be a set of 2 by 2 matrices over D, such that the set*

$$x = 0, y = xM$$

for all M in λ, and $y = xM$ is a set of left 2-dimensional D-subspaces over the left kernel mappings of D, which is also isomorphic to D. Then this matrix set is a left pseudo-regulus net.

Assume that

$$x = 0, y = x(M + R)$$

for R in an index set ϕ such that $M+R$ is a D-matrix set forming a spread in the left 4-dimensionl D-subspace, so each such element is a left 2-dimensional D-vector space. By use of $y=xM$, using only non-zero affine points, it is possible to define a generalized hyperbolic quadric. In general terms, this defines the non-commutative generalized hyperbolic quadric. We consider two possible flock types: hyperbolic and conical.

(1) If the spread admits the pseudo-regulus-inducing homology group defined by the following mappings

$$(x,y)\to(x,yM)$$

for all M non-zero elements in λ, with axis $=0$ and coaxis $x=0$, then the orbits acting on the spread create a set of pseudo-regulus nets, each sharing two components $x=0, y=0$ and covering the spread. The spread may be reconfigured into the following form:

$$x=0, y=0, y=xM$$

for all non-zero M in λ and

$$y=x\left[\begin{array}{cc} f(t) & g(t) \\ 1 & t \end{array}\right]N$$

for all non-zero N in λ, where f and g are functions on D. This part requires the Axiom of Choice for the construction.

There is a transform function that maps the affine points into corresponding points that eliminate λ. Using the transform on Baer subspaces points, then the transformed points satisfy the right homogeneous group defined by the following right mappings:

$$(x,y)\to(xM,yM).$$

The transform points also satisfy the generalized hyperbolic quadric. This is obtained using certain generators of the right homogeneous points. These transforms are called the non-commutative transforms, and all but two of the transforms of Baer subspaces of each of the pseudo-regulus nets are non-commutative.

The remaining two transforms per pseudo-regulus net sharing two components, the commutative transforms of the Baer subspaces, are found to satisfy the commutative form of the generalized hyperbolic quadric. The commutative is the natural complete form if D were a field. The union of these non-commutative and commutative transforms is a covering set of the projective elements of the defined generalized hyperbolic quadric.

By eliminating λ, a projective set of left planes is determined, and the non-commutative transform points become right 1-dimensional D'-subspaces that represent right points of the left planes that form a subset of a corresponding hyperbolic flock, where D' is the division ring of the right homogeneous group and is then isomorphic to D. The union of the commutative transforms and non-commutative transforms as projective 1-dimension D'-subspaces are the right points incident with the left planes, form an M-hyperbolic flock.

(2) If the spread admits a pseudo-regulus-inducing elation group with the following mappings:

$$(x, y) \to (x, xM + y),$$

for all M in λ, a modification of the generalized hyperbolic quadric becomes a generalized quadratic cone embedded within. Then $M + R$, with $x = 0$, shows that the same pseudo-regulus net also lives within the spread, where, in this realizing, the additive group as an elation group whose orbits define a set of pseudo-regulus nets sharing one component, $x = 0$. For each R in ϕ, there is an orbit containing, R, and a set of orbits may be chosen so that the set of pseudo-regulus forms a set of derivable nets that pairwise intersect in $x = 0$, and are otherwise disjoint. This set requires the Axiom of Choice. Just as in the hyperbolic part, there are transforms of the affine points that both satisfy the generalized quadric cone form and are right homogeneous points defined in part (1).

Eliminating λ again provides a set of left planes in the associated projective space whose intersections with the 1-dimensional homogeneous points are covered by the union of the non-commutative and commutative generalized quadratic cone points and are disjoint from the generalized cone vertex v_0, formed in an analogous manner as in the hyperbolic case. In this way, an M-conical flock is constructed.

(3) Conversely, retaining the algebraic structure an M-hyperbbolic flock or an M-conical flock reconstructs the same M-hyperbolic flock spread or M-conical flock spread that constructed it.

(4) If an elimination process cannot be determined for a non-generic pseudo-regulus net, then the $*$-elimination process of writing (xK, yK) as (x^*, y^*) and computing the transforms of the generic pre-image. Then substituting

$$(x_1^*, x_2^*, x_3^*, x_4^*)$$

to the original affine point set

$$(x_1, x_2, x_3, x_4)$$

will create a flock isomorphic to the generic flock.

COROLLARY 8. *(1) Using the above theorem, all left pseudo-regulus nets when embedded within a left spread of 2-dimensional D-vector subspaces and admitting either an* $M = K^{-1}\begin{bmatrix} u^{\sigma} & A(u) \\ 0 & u \end{bmatrix}K$ *pseudo-regulus where* $K = \begin{bmatrix} e & f \\ g & h \end{bmatrix}$, *where K is non-singular, and e,f,g,h are elements in D may be used in this theorem. This means all generic and non-generic pseudo-regulus nets can theoretically construct M-hyperbolic and/or M-conical flocks over skewfields.*

(2) When $g = 0$ *in the above K-mapping, left generic flocks are permuted. The two spreads are isomorphic, but the two flocks may not be isomorphic. The set of such left generic pseudo-regulus nets is called the "kaleidoscope of derivable nets." This set has been previously considered. In this situation, the "elimination process" is known.*

(3) When g is not 0 in the above K-mapping, the set of non-generic left pseudo-regulus nets also produce M-hyperbolic and conical flocks. The pairs of flocks obtained from an original generic flock spread and an isomorphic generic flock spread (which may not then be generic) may not be isomorphic flocks.

(4) Given the associated classical right pseudo-regulus net

$$x = 0, y = \delta x$$

for all δ *in D, is invariant under any K-mapping as*

$$(x, \delta x) \to (xK, (\delta x)K); xK = x^*,$$

then

$$(\delta x)K = (\delta x^* K^{-1})K = \delta x^*,$$

for all x^*, δ *represents a left, acting scalar over the right vector space. So, all of the associated left spreads contain left imposter derivable nets.*

This result, combined with the kaleidoscope of derivable nets, both left and right, gives something of an understanding of the possible forms of pseudo-regulus nets and how they are extended to flock spreads and connected to flocks over skewfields.

CHAPTER 3

The isomorphism questions

Here, we state the four isomorphism questions that are important to the study of flock spreads and flocks. To read these questions requires understanding of how flock spreads and how flocks are connected. We place these questions here for two important reasons. If isomorphic spreads can give rise to non-isomorphic flocks, then the use of isomorphic mappings and coordinate changes would be an important method for the construction of examples of the 109 possible types of flock spreads and thereby of flocks. The second reason is that infinite flock spreads, the "covering criterion" is difficult to establish even for infinite fields and has had no theory for spreads over non-commutative division rings. In fact, other than Desarguesian spreads, there does not seem to be any theory regarding spreads over non-commutative division rings. If D is a division ring, all of the flock spreads have components in $PG(3, D)$. We have found a rather simple procedure, the "unwrapping criterion", that provides a way to determine covering for both the left and right flock spreads when the quasifibration is additive. Furthermore, we have four possible determinants, the (i, j)-determinants for i, j either 1 or 2, that may be used combined with the unwrapping criterion as a construction theory of semifield flock spreads. Then we may use a variety of mapping construction methods to then construct a variety of related flocks, many of which are not isomorphic. By these techniques some of our examples are not semifield flocks. So, here are the questions that we initially asked regarding isomorphism.

These are somewhat overlapping but shall be stated in four parts:

Given two isomorphic flock spreads:

1. If the flock spreads are hyperbolic, the hyperbolic isomorphism question: Are the associated hyperbolic flocks always isomorphic?

2. If the flock spreads are conical, the conical isomorphism question: Are the associated conical flocks always isomorphic?

3. Given a hyperbolic flock spread and coordinates are changed, but a hyperbolic form is retained, the hyperbolic coordinate question is: Are the two corresponding hyperbolic flocks isomorphic?

DOI: 10.1201/9781003767978-3

4. Given a conical spread and coordinates are changed, but a conical form is retained, the conical coordinate question is: Are the two corresponding conical flocks isomorphic?

The answer to each of the hyperbolic questions is no! Here are the examples that verify this. There are probably many other examples; the ones given here follow the pattern of showing that two isomorphic spreads can correspond to linear and non linear flocks, which will show that the two flocks cannot be isomorphic.

In the text, there are other questions, that in situations where the two flocks are both linear, the arguments suggest that the flocks also cannot be isomorphic. We shall give the proofs of these results in later parts when the connections between flock spreads and flocks are well established. Here we also deal with two linear flocks. There are situations where the two flock spreads are isomorphic, but the flocks are not isomorphic. This is true, for example, in the simultaneous flock situations. This last part is shown in part XI.

Although it is possible to consider isomorphisms between flocks that are different types of cones or hyperbolic quadrics, generally we restrict isomorphisms to be "strict", that is, between the two hyperbolic flocks using the same generalized hyperbolic quadric or two conical flocks using the same generalized quadratic cone. It is also interesting to consider other types of isomorphic flocks, and we shall illustrate the complexity of such situations. What can normally be proved is that if two flock spreads are isomorphic by a mapping m, but are not of the same type, then the associated flocks cannot be isomorphic by the mapping induced by m. The only exception to this is that the double-transpose mapping will map a left "upper" generic flock spread into a left "lower" generic flock spread where the two flocks are isomorphic, as noted in the theory developed in volume II.

3.1. The fundamental theorem of quasifibrations

In volume II, we have seen how we may use the "upper form theory" for left flock spreads and the "lower form theory" for right flock spreads. The alternate theory of right/left would provide an equivalent study of flock theory. Furthermore, it has been shown that a study of flocks and quasiflocks could be considered simultaneously using the concept of Baer flocks given in this volume; Baer flocks are either flocks or quasiflocks. It is shown that every left generic Baer flock constructs a right generic Baer flock. But a left flock may be considered a Baer flock and does construct a right Baer flock, which is not necessarily a flock.

Also in this volume, we classify the situations when a left flock does induce a right flock. When this occurs, the two flocks are not isomorphic but are, in fact, anti-isomorphic, if the transpose matrix sets are spreads. The full class of examples is constructed.

3.2. The quasifibration question

We have just noted that left Baer flocks are equivalent to right Baer flocks. And, in this volume III and the original first volume, there were infinitely many quasifibrations over infinite fields. Quasifibrations over non-commutative skewfields appear often in this volume, and it seems clear that they are or should be proper. But this is the open question: Is there an example of a proper quasifibration over a non-commutative skewfield?

3.3. The anti-isomorphic flocks

In this class, we consider the possible left conical flocks, which are anti-isomorphic to the right conical flocks by the fundamental theorem of quasifibrations, part 2. In parts VIII and IX, we discuss what we call the conical flock square, or more generally, the conical Baer flock square, when the Baer flocks are not flocks. When the pairs of left and right flocks on the two diagonals of the conical flock square are anti-isomorphic, we consider this an example in this class. In the two parts, it is shown that when the conical flocks both have central unwrapping procedures, and the transposes of the conical flocks are also isomorphic to generic conical flocks, the diagonals are, in fact, anti-isomorphic. In our previous classes, we required that transposes of conical flock spreads be non-isomorphic from the original flock spread. But for anti-isomorphic left and right flocks, this is not a requirement. The problem is that, as mentioned above, such situations are extremely rare. It is therefore of interest when such anti-isomorphic flocks exist.

Here is a brief description of the situation using Baer flocks, which are either maximal partial flocks or flocks. We then intersect the main theory when Baer flocks are flocks. This is taken from parts VIII and IX.

There is a "square of four conical Baer flocks" constructed as follows: On the Northwest point, there is a left conical Baer flock. On the Northeast point, there is a right conical Baer flock, obtained using the Fundamental Theorem, part 1. On the Southwest point, is the transposed left Baer flock, and on the Southeast point is the right conical

Baer flock obtained from the transposed left conical Baer flock by the transposition of the Northeast right conical Baer flock.

To travel east or west on the square, use the Fundamental Theorem applied to the corresponding generic quasiflock quasifibrations. To travel north or south on the square, use the transpose on the corresponding generic quasiflock quasifibrations.

When Baer flocks are all flocks, then the NW left conical flock is anti-isomorphic to the SE right conical flock, and the SW left conical flock is anti-isomorphic to the NE right conical flock, and travel on the diagonals is by using anti-isomorphism.

The most important examples are valid for every finite-dimensional division ring D, obtained from the Multiple Replacement Theorem – Redux, part VI, which shows that there is lifted skewfield conical flock spread obtained from every such division ring D. Thus, for these division rings the left and right semifield flocks are anti-isomorphic. Since there are infinitely many possible ways to lift such skewfields, we have infinitely many ways to find such anti-isomorphisms.

There are other examples, as there are examples obtained in the case that a central cyclic extension of degree n exists for a division ring of finite degree n over a field.

Here are the possibilities: We note that when a left-hand conical flock spread exists, whose transpose matrix set is also a spread, then there is a conical flock square. Therefore, we may list the pairs of conical flock spreads, and transposed flock spreads, relaxing the requirement that the two flock spreads be non-isomorphic. Again, noting that a left-hand conical flock spread and existing transpose is sufficient, we list the six pairs for the left conical side where the associated automorphism σ is either 1 or outer and the three pairs for the inner left conical side. Since the lifted skewfields obtained by central Galois quadratic extensions and the central Galois cyclic extensions of degree n of cyclic skewfields of degree n over the centers are in the same class of conical flock spreads, we obtain exactly one class of examples of the nine possibilities.

The classes of anti-isomorphic left and right conical flocks are given by their pairs of classes of left conical flock spreads and their transpose spreads, as right flock spreads. These come in two varieties: when the isomorphism is σ or 1 and then when σ is an inner automorphism. This is also the case where one may also find the Desarguesian quaternion left and right anti-isomorphic flocks. Therefore, there are eight cases.

When σ is outer or 1: There are three examples out of the four possibilities.

1. $N(2)^{F,\sigma-A} : x\begin{bmatrix} u^{\sigma}+F(t) & ,G(t)+(u^{\sigma}c-cu)) \\ t & u \end{bmatrix}$, *$\sigma$ non-trivial*. A "a fake type 1" derivable net.

$\& \; N_{co}^{L,trans,F,\sigma^{-1}-A^{*}}(2)$: $x\begin{bmatrix} u^{\sigma^{-1}}-F(t)^{\sigma^{-1}} & ,L \\ t & u \end{bmatrix}$,

where, $A^{*}(u)=A(u^{\sigma^{-1}})$, $L=A^{*}(v)-A^{*}(F(t))+G(t)$, indicating the transpose. When σ has order 2 and $A(w)=(w^{\sigma}-w)c$, where $c^{\sigma}=-c, F(t)=A(t), G(t)=t^{\sigma}\theta$, there is an example where the transpose is isomorphic to the original. There is a central Galois quadratic extension $D_1(\theta)$, D_1 a finite-dimensional division ring. Furthermore, in this case, the transpose admits the left-acting pseudo-regulus-inducing homology group.

2. $N_{co}^{L}(2)^{\sigma}$: $x\begin{bmatrix} u^{\sigma}+F(t) & ,G(t) \\ t & u \end{bmatrix}$, σ non-trivial,

$\& \; N_{co}^{L}(2)^{\sigma,trans}$: $x\begin{bmatrix} u^{\sigma^{-1}}-F(t)^{\sigma^{-1}} & ,G(t) \\ t & u \end{bmatrix}$, *$\sigma$ non-trivial.*

The transpose spread of the previous. In this case, if $F(t)=0$ and $G(t)=\theta t^{\sigma^{-1}}$, there is a solution. In this case, the transpose exists and admits the left-acting pseudo-regulus-inducing homology group. Here, there is a central Galois cyclic extension of degree n $D_1(\theta)$, for $n=3$ or 4.

3. $N_{co}^{L}(1)^{1-A} : x\begin{bmatrix} u+F(t) & ,G(t)+A(u) \\ t & u \end{bmatrix}$, A an outer derivation function

$\& \; N_{co}^{L}(1)^{trans.,1-A}$: $x\begin{bmatrix} w-F(t) & ,G(t)+A(w)-A(F(t)) \\ t & w \end{bmatrix}$,

A an outer derivation function. For characteristic 2, $F(t)=A(t)$,$G(t)=t\sqrt{z}$,

4. $N^{L}(2)_{co}^{F,1-A}$: $x\begin{bmatrix} u+F(t) & ,G(t)+A(u) \\ t & u \end{bmatrix}$,

A an inner derivation function;

a "fake type 1"

$\& \; N^{L}(2)_{co}^{trans,F,1-A}$: $x\begin{bmatrix} w-F(t) & ,G(t)+A(w)-A(F(t)) \\ t & w \end{bmatrix}$,

A an inner derivation function; a "fake type 1". There is a central non-Galois extension of a quaternion division ring $(a,b)_F$ of characteristic 2 by $F(z)(\sqrt{z}))$, where $F(z)$ is rational function field of the field F, turning the quaternion division ring into $(a,b)_{F(z)(\sqrt{z})}$. $A(w) = dw/d\sqrt{z}$, where $F(t) = A(t)$, and $G(t) = t\sqrt{z}$. The transpose is identical to the original flock spread and admits the left-acting pseudo-regulus-inducing homology group.

5. $N_{co}^{L}(1)^{\sigma-A} : x\begin{bmatrix} u^{\sigma}+F(t) & ,G(t)+A(u) \\ t & u \end{bmatrix}$; A is a non-trivial outer σ-derivation function for σ a non-trivial outer automorphism

& $N_{co}^{L}(1)^{trans.,\sigma-A}$: $x\begin{bmatrix} w^{\sigma^{-1}}-F(t)^{\sigma^{-1}} & ,G(t)+A^{*}(w)-A^{*}(F(t)) \\ t & w \end{bmatrix}$;

$A^{*}(u) = A(u^{\sigma^{-1}})$ is a non-trivial outer σ-derivation function for σ a non-trivial outer automorphism.

When σ is inner. There are no examples.

1.$^{Inner}N_{co}^{L}(2)^{F,\sigma-A}$: $\begin{bmatrix} u^{\sigma}+F(t) & ,G(t)+A(u) \\ t & u \end{bmatrix}$, *$\sigma$ non-trivial.* A "a double fake type 1" derivable net for $\sigma - A$, where both σ and A are inner &$^{Inner}N_{co}^{L}(2)^{F,\sigma-A}$:$x\begin{bmatrix} u^{\sigma^{-1}}-F(t)^{\sigma^{-1}} & ,L \\ t & u \end{bmatrix}$;

$$L = A^{*}(v) - A^{*}(F(t)) + G(t),$$

the transpose.

2. $^{Inner}N_{co}^{L}(2)^{\sigma}$: $\begin{bmatrix} u^{\sigma}+F(t) & ,G(t) \\ t & u \end{bmatrix}$, σ non-trivial. A fake type 2

&$^{Inner}N_{co}^{L}(2)^{trans.,\sigma}x\begin{bmatrix} w^{\sigma^{-1}}-F(t)^{\sigma^{-1}} & ,G(t) \\ t & w \end{bmatrix}$, *$\sigma$ non-trivial.*

The transpose spread of the previous.

3. $^{Inner}N_{co}^{L}(1)^{\sigma-A}$: $x\begin{bmatrix} u^{\sigma}+F(t) & ,G(t)+A(u) \\ t & u \end{bmatrix}$; A is a non-trivial outer σ-derivation function for σ a non-trivial inner automorphism. A fake type 1

&$^{Inner}N_{co}^{L}(2)^{F,\sigma^{-1}-A^{*}}$:$x\begin{bmatrix} u^{\sigma^{-1}}-F(t)^{\sigma^{-1}} & ,A^{*}(v)-A^{*}(F(t))+G(t) \\ t & u \end{bmatrix}$,

the transpose, $A^{*}(u) = A(u^{\sigma^{-1}})$, A is an outer $\sigma - A$ derivation function, for σ a non-trivial inner automorphism.

3.4. The hyperbolic flock square

Now the question is, there a "hyperbolic flock square", a square where at the two left corners stand left hyperbolic flocks and on the right corners stand right hyperbolic flocks, and when do these two pairs of diagonal points become anti-isomorphic? It was mentioned that the creation of an appropriate form for the right flocks comes from the transpose matrix over fields. Since the transpose conical matrix is possible to describe, but the transpose hyperbolic matrix is not – over skewfields – what is necessary for a hyperbolic flock square to have the NW corner be a simultaneous flock? The "same" matrix set can be configured in two ways, one conical and one hyperbolic. Then, if the conical transform exists and this matrix set admits a pseudo-regulus-inducing homology group acting on the left, then rewrite this exact matrix set on the right side at the SE point. This form will be generic. Now we know that this form also arises from a conical left form, so we are getting the same structure. But now we see that the SE point may be represented in the left-acting homology group form that defines a right hyperbolic flock. It then follows that the original left hyperbolic flock is anti-isomorphic to the right hyperbolic form. Also, since the original NW point is simultaneous, so is the SW point; it just has the homology group acting on the "wrong"/right side for a right hyperbolic flock. It can be made into a generic form using the right inversing method. Therefore, we now have a hyperbolic flock square, and the hyperbolic flocks on the diagonals are anti-isomorphic.

In order to have a hyperbolic flock square, we need to have a left simultaneous flock to take up the NW point, and we need to have the conical transpose of that point to be a conical flock that admits the left-acting pseudo-regulus-inducing homology group. Looking back at the possible choices for such flocks. These would occur exactly when the conical flock square exists, provided the conical flock spread is simultaneous and provided the transpose admits the left action. When working through this material, read back, where a left simultaneous flock exists as it appears in all of the examples listed above. Since there are three types of examples that are simultaneous and three possibilities, the hyperbolic anti-isometric left and right flocks exist for all of the examples that do in the conical anti-isometric conical flocks. The problems on the inner flock are difficult, and they appear to arise from mappings from other flock types. The transform problems within an inner flock spread are particularly difficult, and this is the reason that there are open classes.

3.5. Generic flock spreads and K-mappings

In the first text [**34**], we provide a classification of all possible forms of pseudo-regulus nets over all division rings D. Given what we are calling a "generic" left and "generic" right flock spread and given any other non-generic pseudo-regulus net R, there is a mapping $(x, y) \to (xK, yK)$ for left nets and $(x^T, y^T)^T \to (Kx^T, Kx^T)^T$ for right nets so that these mappings construct isomorphic spreads of the generic flock spreads but now containing R, where R is left or, respectively, right. However, the mappings called K-mappings can distort the form to the point that underlying flocks can not be determined. It is not the case that just because the generic flock spread and the non-generic flock spread are isomorphic, that there are two associated isomorphic flocks. Therefore, it remains an open question how to determine flocks from non-generic pseudo-regulus derivable nets. What is the case for the left generic flock spreads is that when $K = \begin{bmatrix} e & f \\ 0 & g \end{bmatrix}$, the various generic flock types are permuted but not necessarily creating isomorphic flocks. For right generic flock spreads we would use $K^T = \begin{bmatrix} e & 0 \\ f & g \end{bmatrix}$. Therefore, given any generic flock spread, there are infinitely (finitely, in the finite case) many other flock spreads constructing possibly infinitely many different non-isomorphic flocks. We have proved that this possibility works for generic hyperbolic flock spreads, in all cases of skewfields.

We note that if $K = \begin{bmatrix} 0 & 1 \\ 1 & 0 \end{bmatrix}$, that this will change the left generic form, which is also termed "upper" to "lower" form, and change the right generic form, which is also termed "lower" to "upper" but does, in fact, produce isomorphic flocks.

This mapping, called the "double-transpose", allows the transpose of a matrix, then applies double-transpose to construct an associated generic form from transpose spreads.

3.6. Summary of classes of flocks and spreads

In conclusion, counting the number of cases of flocks, separately by type of skewfield and not counting the twenty-four type (0) examples, there are:

- eleven possible finite flocks, all of which have examples,
- sixteen possible infinite field flocks, with fifteen and 1/2-examples,

- twenty-six possible left and twenty-six possible right flocks, where the defining automorphism is 1 or is an outer automorphism. And, for which it would seem that every left flock gives rise to a right flock, but this is not the case. Therefore, the right and left flock systems are independent. There are twenty+1/2 left and twenty+1/2 right examples due to the central structure of our examples.
- There are fifteen left and fifteen right cases where the defining automorphism is an inner automorphism, and there are eleven examples, each. Using the kaleidoscopic mappings that turn the fundamental derivable net into a/the kaleidoscope of derivable nets, the actual enumeration of types is only to give a rough idea of the possible types involved in the noncommutative skewfield cases. There are other types of inner flocks called the "direct sum" type.
- There are seven possible examples of anti-isomorphic pairs of left and right conical flocks, with three examples, with infinitely many different non-Desarguesian semifield examples, and the class of Desarguesian examples containing all quaternion division ring planes.
- And there are seven possible examples of anti-isomorphic pairs of left and right hyperbolic flocks, with three examples, as these are paired with the conical flocks.
- In addition to flocks and spreads, André's theorem over any finite 2-dimensional division rings is proved. In particular, if D is any finite-dimensional division ring of degree k over the center K, there is a double transcendental extension $D_2 \otimes_K K(z_1, z_2)$, when $D_1 = D \otimes_K Fix\sigma$, for $\sigma = (z_1 z_2)$ as a permutation and σ fixes D pointwise, and D_1 is of index 2 in D_2 induces a central Galois quadratic extension $D_2 = D_1(\theta)$. Therefore, each such division ring $D_2(\theta)$ admits a set S of mutually disjoint pseudo-regulus nets such that any proper subset M of S mutually derives the set of pseudo-reguli in M creating infinitely many spreads whose kernel is $PG(2k^2 - 1, K(z_1, z_2))$.
- Concerning spreads of larger dimension, Ostrom's theorem on hyper-reguli is generalized to all finite division rings. This provides spreads of large dimension obtained by net replacement in the spirit of Ostrom but over arbitrary finite skewfields and hints at more general work using subplane-covered nets in place of derivable nets.

Part 2

Multiple replacement – redux

Although we are interested in the connection between spreads over skewfields that may be written as a union of pseudo-regulus nets sharing either one or two components and admitting central collineation groups, elation or homology, that are pseudo-regulus inducing, we are also mindful that there are few classes of non-Desarguesian spreads over non-commutative skewfields. These spreads are connected to generalized conical and hyperbolic flocks. So, this text intends on changing that scenario. There is one such spread found twenty-five years ago by Jha and Johnson [**34**]. In that study, non-Desarguesian spreads called "Skew-Desarguesian" spreads mentioned "simultaneous spreads", a rare spread, indeed. So, many of the ideas of these two volumes on "skewfield flocks and derivable nets" have had a certain history, mostly in the papers of V. Jha and N.L. Johnson.

In the first volume, the geometry of derivation and applications, there was a general replacement theorem given. T.G. Ostrom in the 60s was probably the first mathematician to use such machinery to create new affine planes/translation planes.

DEFINITION 9. A "replacement net" is simply a subnet of an affine plane together with a net that when switched with the original net, creates another affine plane.

The derivable net concept, also invented by Ostrom in the finite case that lies in an affine plane of order q^2, is a net of $q+1$ parallel classes of the q^2+1 parallel classes for which there is a set of subspaces, each of cardinality q^2, each of which intersects each line of the net in unique set of q points (think of a 1-dimensional $GF(q)$-subspace). The subspaces are called "Baer subplanes" and the process of interchanging one set for the other is called "derivation". The famous André's theorem, shows that if one considers a Desarguesian plane (coordinated by $GF(q^2)$, odd order, with automorphism σ of order 2), then for the Desarguesian plane coordinatized by $GF(q^2)$, the following spread (the set of subspaces incident with the zero vector) has the following form:

$$x = 0, y = 0 \cup \{y = xm; m^{q-1} = \delta\}$$

where δ varies over $GF(q)^*$, is a set of $q-1$ set of nets of degree $q+1$ each, that has the following set of replacement nets; derivable nets:

$$\{y = x^{\sigma}m; m^{q-1} = \delta\},$$

where, in this setting, $x^{\sigma} = x^q$. Note also that $m^{q-1} = m^{\sigma}m^{-1}$. A regulus in $PG(3, GF(q))$ is a set of $q+1$ lines that are covered by an opposite set of $q+1$ lines. It is proved in Johnson [**30**] that every finite derivable net has the affine form of a regulus. Furthermore, in general, all derivable nets are classified as either pseudo-regulus over either field (regulus nets) or non-commutative skewfields (pseudo-regulus nets). What this means is that we could choose any subset of the $q-1$ nets, 2^{q-1} such sets, and derive each derivable net of that subset, and leave the other derivable nets as they are. Each of the replacements, called "multiple derivation", produces one of the so-called André's planes. Of course, there will be isomorphic planes among the multiply derived sets, but the potential is vast. One very interesting set consists of all of the square nets (where δ is a square); the resulting André's plane is a Dickson nearfield plane, coordinatized by a nearfield. This net plays a very big role, which is what we are to discuss in this volume III, and which we mentioned in the original volume.

Ostrom also considered more general replacement settings when the order of the affine plane could be q^{kn}, and a replacement set would be a net of cardinality q^k-1 and there are $(\frac{q^{kn}-1)}{q^k-1})$ such nets in a finite Desarguesian affine plane and here there are also multiple replacements, whereas in derivable net, each replacement net has a unique replacement.

So, the natural question is whether any of these constructions, whether there are division rings that admit quadratic Galois extensions, or whether there are division rings that admit cyclic Galois extensions.

Here is the original multiple replacement theorem, which says that any division ring that admits a non-trivial automorphism σ may be written as a union of mutually disjoint replaceable nets union two components. The automorphism can be an inner automorphism or an outer automorphism of infinite or finite order. And the division ring may be finite-dimensional or infinite-dimensional, there is no restriction.

THEOREM 10. *(**The Multiple Replacement Theorem**) (Theorem 126 of* [**34**]*) Let Σ denote a Desarguesian translation plane coordinatized by a skewfield/division ring S that admits a non-trivial automorphism σ.*

(1) Then writing the spread as for $x,y\epsilon S$,

$$x=0,y=xm;m\epsilon S$$

denotes a spread where all components are left subspaces over S and where elements of the kernel mappings S_L of Σ have the following form:

$$S_L\colon (x,y)\to(\beta x,\beta y);\beta\epsilon S^*.$$

There is a corresponding spread Σ_σ with spread of the following form:

$$x=0,y=mx^\sigma;m\epsilon S.$$

The kernel mappings of Σ_σ have the following form:

$$S_R^\sigma : (x, y) \to (xe, ye^\sigma); e\epsilon S^*.$$

(2) Then $S_L S_R^\sigma = S_R^\sigma S_L$ is a collineation group of both Σ and Σ_σ.

(3) Σ_σ is a Desarguesian plane, and both Σ and Σ_σ cover each other.

(4) The intersections of $y = xm$ and $y = mx^\sigma$ are isomorphic subspaces over Fix$\sigma\cap$Z(S), where $Z(S)$ is the center of S. As a result, we have a vector space that is a left/right bimodule, where the components are isomorphic vector spaces over the field Fix$\sigma\cap$Z(S).

(5) Let the subskewfield $Fix\sigma$ be denoted by F, and note that Σ is a set of left F-subspaces, and Σ_σ is a set of right F-subspaces.

We emphasize that the skewfield Fappears in $S_L S_R^\sigma$ as the set of left mappings of the form $(x, y) \to (fx, fy)$ for $f\epsilon F^*$ that fixes all components of Σ and as the set of right mappings of the form $(x, y) \to (xf, yf)$ that fix all components of Σ_σ.

(a) The set of components of Σ that are images of $y = x$ under S_R^σ is

$$\{y = xbb^{-\sigma}; b\epsilon S^*\}$$

and the set of components of Σ_σ that are images of $y = x^\sigma$ under S_L is

$$\{y = bb^{-\sigma}x^\sigma; b\epsilon S^*\}.$$

Furthermore,

$$\{y = xbb^{-\sigma}; b\epsilon S^*\} = \{y = bb^{-\sigma}x^\sigma; b\epsilon S^*\},$$

in the sense that the two sets are replacements of each other.

(b) Similarly, $y = xm$ and $y = mx^\sigma$ intersect non-trivially and have the following sets of images under $S_L S_R^\sigma$:

$$(*) : \{y = xbmb^{-\sigma}; b\epsilon S^*\} = \{y = bmb^{-\sigma}x^\sigma; b\epsilon S^*\},$$

as replacement sets.

(6) There is an index set λ such that

$$\Sigma = \{x = 0, y = 0\} \cup_{i\epsilon\lambda} \{\{y = xbm_i b^{-\sigma}; b\epsilon S^*\} = \Gamma_i\}$$

an, where the unions are disjoint unions. Then the multiply replaced translation plane has blended kernel (S_L, S_R^σ), and formal kernel

$$Z(S) \cap Fix\sigma = S_L \cap S_R^\sigma \cup \{0\}.$$

So, this repeats the statement of the theorem in the first volume. The idea in the following theorem will be repeated for arbitrary cyclic Galois extensions of division rings. We are first considering automorphisms of order 2.

We shall make use of the fact that for finite-dimensional division rings D over a field F, $D \otimes_F F(z_1, ..., z_k)$, where the z_i, for $i = 1, 2, ..., k$

are commuting indeterminants, is a division ring. This will become an important tool in using the multiple replacement theorem.

But, there are infinite-dimensional division rings that admit automorphisms. What are the spreads that are multiply-replaced? Does the collineation group of certain multiple replaced spreads permute the replaced spreads?

PROBLEM 11. Do anything with the previous questions!

CHAPTER 4

Extensions of division rings

THEOREM 12. *Let D be a finite-dimensional division ring of degree e over a field F. Let $F(z_1, z_2, ..., z_n)$, where the z_i for $i = 1, 2, ..., n$ are commuting indeterminants.*

Form

$$D_n = D \otimes_F F(z_1, ..., z_n)$$

so, is also a division ring. Let $\sigma : z_1 \to z_2 \to .... \to z_n \to z_1$. Then $F(z_1, ..., z_n)$ is a cyclic Galois extension of $Fix\sigma$. Now form

$$D_1 = D \otimes_F Fix\sigma.$$

D_1 is a division ring of index n with center $Fix\sigma$, as it is a subdivision ring of D_n. Therefore, it follows that D_1 is a division subring of D_n. Define σ restricted to D to be the identity. Therefore, $Fix\sigma_{D_n}$ (the fixed point subspace of σ over D_n), and it follows that σ is a cyclic automorphism of order n of D_n. And since we prove below that $F(z_1, ..., z_n)$ is a cyclic Galois extension of $Fix\sigma$, D_n is a central cyclic Galois extension of D_1 of degree n.

PROOF. We need to prove that $F(z_1, ..., z_n)$ is a cyclic Galois extension of $Fix\sigma$. This proof will use the theory of the elementary symmetric functions.

We claim that

$$x^n - (\Sigma_{i=n}^{n} z_i)x^{n-1} + (\Sigma_{distinct\ pairs}^{n} z_i z_j)x^{n-2} -$$

$$- ... + (\Sigma_{distinct\ k-tuples}^{n}\ z_i z_j ... z_w)x^{n-k} + ... - (z_1 z_2 ... z_n),$$

where $+, -$ alternate,

is the corresponding irreducible polynomial. Note that all coefficients are invariant under σ. That is, the coefficients are in $Fix\sigma$. Also, note that z_i for $i = 1, ..., n$ are roots of the equation, which are not fixed by σ. This is easily proved by mathematical induction. This will also prove that the irreducible polynomial may be written in an equivalent way as

$$(x - z_1)(x - z_2)....(x - z_n).$$

DOI: 10.1201/9781003767978-4

The rest of the proof follows from Jacobson [**18**], as the theory allows that finite dimension division rings may be extended to an arbitrary number of function fields over their centers, using tensor products.

A special case of this result is given when $n = 2$. Then there are some important ways to choose an irreducible polynomial and a formal generating element for the Galois quadratic polynomial.

So, we begin with

$$x^2 - (z_1 + z_2)x + z_1 z_2,$$

and divide by $(z_1 + z_2)^2$,to obtain:

$$(\frac{x}{z_1 + z_2})^2 - (\frac{x}{z_1 + z_2}) + \frac{z_1 z_2}{(z_1 + z_2)^2}.$$

Therefore, we may always assume that an irreducible polynomial is

$$x^2 - x + \alpha,$$

where

$$\alpha = \frac{z_1 z_2}{(z_1 + z_2)^2}.$$

When the characteristic is 2 and θ is a formal root, then

$$\theta^\sigma = \theta^2 = \theta + \alpha.$$

When the characteristic is not 2, we may use the radical of the quadratic formula to represent an irreducible polynomial,

$$\sqrt{(z_1 + z_2)^2 - 4z_1 z_2 = (z_1 - z_2)^2},$$

the irreducible polynomial

$$x^2 + (z_1 - z_2)$$

over K, has the form root $\theta = (z_1 - z_2)$,so that $\theta^\sigma = -\theta$. □

CHAPTER 5

Automorphism groups of division rings

THEOREM 13. *Given any positive integer* $n>1$. *Take any finite division ring* D *of degree* k *over a field* F. *Given any finite group* G, *there is a symmetric group* S_n *such that* G *is a subgroup of* S_n. *Form* $D_n = D \otimes_F F(z_1, z_2, ..., z_n)$. *The symmetric group* S_n *induces an automorphism group on* D_3. *Then form* $D_1^G = D \otimes_{FixG} FixG$ *and define* G *acting on* D *as the identity mapping. Then* G *is a Galois group of* F *over* $FixG$, *and becomes an autotopism group of the left Desarguesian spread coordinatized by* D_n. *Then* D_n *is a central Galois extension by* G *of* D_1^G, *which is* $FixG_{D_n}$.

Let Σ *denote the left translation plane obtained by forming the 2-dimensional right or left vector space over*

$$D_n \oplus D_n.$$

Consider the left Desarguesian spread

$$x = 0, y = 0, y = xm; m \text{ in } D_n.$$

Let the group G_Σ *be denoted with elements* $(x, y) \to (x^g, y^g)$, *for all* g *in* G. *Since* G *is an automorphism group of* D_n, *then* $(x, xm) \to (x^g, (xm)^g = x^g m^g)$, *which shows that* G_Σ *is a collineation group of* Σ.

Therefore, given any finite group G *and any finite-dimensional division ring* D, *there exists a finite central Galois* G *extension* D_n *of* D_1^G *of degree* $ordG$ *so that* G *is a collineation group of the left Desarguesian translation plane coordinatized by* D_2.

DOI: 10.1201/9781003767978-5

CHAPTER 6

The theorem of André

Here we employ the tensor product construction mentioned just above to prove the theorems of André and Ostrom over finite-dimensional division rings. This then explodes the sets of non-Desarguesian translation planes over division rings. There is an astonishing variety of new translation planes about which essentially nothing is known.

THEOREM 14. *Let D be any finite-dimensional division ring of dimension k^2 over the center, a field F. Then consider $D_2 = D \otimes_F F(z_1, z_2)$, where z_1 and z_2 are commuting indeterminants. Then the mapping $\tau : z_1 \leftrightarrow z_2$ induces a Galois quadratic extension of the subfield $K = Fix\tau$ in $F(z_1, z_2)$. $D \otimes_K K(z_1) = D_2 = E$ is a division ring and admits τ as an automorphism by letting τ act trivially on D, which admits $D \otimes_F Fix\tau = D_1$ is the fixed point space of τ in D_2, noting that D_1 is a division ring as it is a subring of D_2.*

Therefore, form the vector space $E \oplus E$, which is a left 4-dimensional space over D_2. That is, the components of Σ are left 2-dimensional vector spaces over D_2, and the components of Σ_τ are right 2-dimensional vector spaces over D_2, in the notation adopted in the statement of the multiple replacement theorem. Therefore, the spreads are left and right spreads in PG(3,D_2), respectively. Then the replaceable nets in the dimension 2 case are left and right 2-dimensional D_2-spaces. This is exactly the definition of a pseudo-regulus net in the text [**34**]. *For all characteristics, the following quadratic is irreducible over K:*

$$x^2 - (z_1 + z_2)x + z_1 z_2.$$

$$\alpha = \frac{z_1 z_2}{(z_1 + z_2)^2}.$$

In addition, the vector space $E \oplus E$ is also a right space over D_2.

Given any finite-dimensional division algebra D of dimension k^2 over the center $Z(D) = K$, there is an extension D_2 of D of dimension k^2 over $K(z_1, z_2) = Z(D_2)$, which is a central 2-dimensional Galois extension of $D_1^{(z_1 z_2)}$ with automorphism τ such that the corresponding left Desarguesian spread is a set of mutually disjoint pseudo-regulus nets

DOI: 10.1201/9781003767978-6

together with two components, whose sets of Baer subspace nets belong to an associated right Desarguesian spread such that both Desarguesian spreads cover each other. Also, all of this may be accomplished over the left space or over the right space.

(1) Hence, we have extended André's theorem to all finite-dimensional division rings in this manner.

(2) If the two components are written as $x = 0, y = 0$*, write components of the Galois extension in the form* $u + tz_1$*, where* u*,*t *are in* $D \otimes_F Fix\tau = D_1$.

(a) Then the components may be considered in $PG(3,D_2)$*, and there is a set of left pseudo-regulus nets that form a linear hyperbolic flock, and then the spread may be reconfigured so there is a set of left nets that form a left linear conical flock.*

(b) Furthermore, there is an analogous set of right pseudo-regulus that also form right linear hyperbolic and right linear conical flocks.

(c) The proper multiply derived spreads are in $PG(4k^2 - 1, K(z_1))$.

PROOF. Note that $Fix_1\tau \cap Z(D_2)$ is $K(z_{1,}z_2)$. With this setup, the theorem follows from the multiply replacement theorem. □

The multiple replacement theorem is valid for any division algebra and any non-trivial automorphism τ. We use a similar argument for spreads of higher dimensions.

As this previous theorem applies to any field, for any finite group G, there is an extension field that admits a Galois quadratic extension with associated automorphism group G. This is closer to André's theorem in practice, as it allows the extension of certain familiar so-called "Dickson nearfield" spreads. These were discussed in the previous volume, but they bear repeating here.

CHAPTER 7

Dickson nearfield planes

DEFINITION 15. For any field F not of characteristic 2 that admits a Galois quadratic extension of $F(\theta)$, with associated irreducible polynomial $x^2 - b$, and automorphism σ of order 2 that fixes F elementwise. Then let π denote the associated spread with components in $PG(3, F)$. Applying the multiple replacement theorem, the spread may be written as a union of mutually disjoint regulus nets R_i, for i in λ, an appropriate index set, that admits a regulus-inducing homology group, but this time the homology group is isomorphic to $F(\theta)^{*(1-\sigma)}$. Define the F-Dickson nearfield plane as the plane obtained from the associated Pappian plane coordinatized by $F(\theta)$ by replacement of the derivable nets $\{y = xm; m^{\sigma+1}$ is a square in $F\}$.

In the notation of the multiple replacement theorem, the derivable nets are

$$(*) : \{y = xbmb^{-\sigma}; b\epsilon S^*\} = \{y = bmb^{-\sigma}x^{\sigma}; b\epsilon S^*\},$$

where $S = F$, and then we obtain: $S_L S_R^{\sigma}$ is now

$$\{(x, y) \to (\beta xe, \beta ye^{\sigma}),$$

for all β, e in F^*. But we now have commutativity, and since we have the kernel mappings, we may let $\beta e = 1$, so that $\beta e^{\sigma} = e^{\sigma-1}$, for all e in F^*. Now $bmb^{-\sigma} = mb^{1-\sigma}$. We assert that the index set λ as above may be taken as the set of elements in $F(\theta)^{*(1-\sigma)}$. Since we have characteristic not 2, every element of $F(\theta)^{*(1-\sigma)}$ is an element of the form $n^{\sigma+1}$, which may be either square or non-square. Therefore, we now distinguish the replacement sets as the square nets. This is the classical method of Dickson in the finite case. We now show the F-Dickson theorem in a slightly different form. First, we remind the reader of the André's planes for any field F of characteristic not 2, which admits a quadratic extension $F(\theta)$, with automorphism σ. We recall that a "full field" of characteristic not 2 is a field with non-squares, such that the product of any two non-squares is a square. There are examples of full fields of characteristic not 2 that admit a quadratic extension $F(\theta)$. There are examples where the quadratic extension is also a full field (see, e.g., the examples of Jha and Johnson [**24**]) at the end of the article). In

DOI: 10.1201/9781003767978-7

particular, every finite field of odd order is a full field, and there are many infinite examples as well.

THEOREM 16. *(**The** F – Dickson **Theorem for full fields**) Let F be any full field of characteristic not 2 that admits a full field quadratic extension $F(\theta)$ with automorphism σ. There is a set of mutually disjoint regulus nets together with two components $x = 0, y = 0$. The net replacement of $\{y = xme^{\sigma-1}; m^{\sigma+1}$is a non-square in F^* and for all ein $F(\theta)^*\}$ produces a nearfield spread from the Desarguesian/Pappian spread coordinatized by $F(\theta)^*$if and only if the product of any two non-squares in F is a square, which is the definition of a full field of characteristic not 2.*

PROOF. The first part follows from the multiple replacement theorem. The corresponding putative nearfield $(F(\theta), +, *)$ then has the multiplication given by

$$x * m = x^\sigma m;\ if\ m^{\sigma+1} \text{ is a non-square in } F^*$$

and

$$x * m = xm;\ if\ m^{\sigma+1} \text{ is a square in } F^*.$$

Then we need to show that multiplication is associative. Consider (i): $(x * m) * n$ and $(ii) x * (m * n)$.

There are three cases for $\{ m^{\sigma+1}, n^{\sigma+1}\}$: The two elements are both squares, one is a square and the other is a non-square, and both are non-squares. Since we have a field, the product of two squares is a square. So, in this case, (i) is $(xm)n$ and (ii) is $x(mn)$, so that is since $m^{\sigma+1}n^{\sigma+1}$ and is a product of squares andis $(mn)^{\sigma+1}$ must be a square $a^2b^2 = (ab)^2$, this proves that $(i) = (ii)$, when both of the elements are squares. When one is a square, and one is a non-square, then suppose that product is a square. Then that is if $a^2b = c^2$, where b is a non-square, then $b = a^{-2}c^2 = (a^{-1}c)^2$, a contradiction. In the case where the first element $m^{\sigma+1}$ is a square and $n^{\sigma+1}$ is a non-square, then (i) is $(xm) * n = (x^\sigma\ m^\sigma)n$ and, noting that $m{*}n = m^\sigma n$, and $(m^\sigma n)^{\sigma+1} = (mn)^{\sigma+1}$, as $m^{\sigma(\sigma+1)} = m^{\sigma+1}$, then (ii) is $x^\sigma(m^\sigma n)$. The case for $m^{\sigma+1}$ non-square and $n^{\sigma+1}$ square is analogous. This leaves the case when $m^{\sigma+1}$ and $n^{\sigma+1}$ are both non-squares. Then (i) is $xm^\sigma n$ and (ii) is $x * (m^\sigma n)$. Therefore, we require that $(m^\sigma \mathrm{n})^{\sigma+1}$ is a square, which is equivalent to the product of two non-squares $in\, K^{\sigma+1}$ is a square. This completes the proof. □

REMARK 17. We have taken the hypothesis that we have a field F of characteristic not 2 that has a Galois quadratic extension, or equivalently any quadratic extension. But, when we are speaking of division rings, we usually consider division rings that admit central Galois quadratic extensions, or when these are not available, we look at finite-dimensional division rings D over a field F and for $D \otimes_F F(z_1, z_2)$. If we consider the same sort of construction when D is F then we would actually have $F(z_1 + z_2, z_1 z_2) = K$ as the initial field with the Galois quadratic extension as $K(z_1)$ with K-irreducible quadratic $x^2 - (z_1 + z_2)x + z_1 z_2$. Consider the non-squares $z_1 + z_2$ and $z_1 z_2$ then the product of these non-squares is a non-square (see problem below) and would never produce a K-*Dickson* nearfield spread. Therefore, if we are looking for Dickson nearfield spreads, this would not be a good way to construct them. However, could there be nearfield spreads here at all?

PROBLEM 18. In the above example, why is $(z_1 + z_2)(z_1 z_2)$ a non-square in K? Prove this!

Before we delve into the problem of other nearfield spreads, or rather if there are non-nearfield hyperbolic spreads, we mention again the remarkable result that for finite hyperbolic spreads, these are always nearfield spreads: the Dickson nearfield spreads and three other irregular nearfield spreads ([**34**], Theorem 59, p. 139). Furthermore, for fields K, admitting a Galois quadratic extension $K(\theta)$, and $K^- = K(\theta)^{*(\sigma+1)}$, and letting S_2 denote the cyclic group of order 2, assume that K^- contains S_2, then S_2 is a normal subgroup, and K^-/S_2 is an elementary Abelian 2-group. It is shown in ([**34**], Theorem 76, p. 170), that is, if $2^{Card(K^-/S_2)-1} > Card(K^-/S_2)$, then there is always an André's spread that is not a nearfield spread.

THEOREM 19. (***The infinite opposite of the finite hyperbolic flock theorem.***) *With the notation and assumptions,*

If $Card(K^-/S_2) > 2$, *there is always an André's plane that is not a nearfield plane. There are usually non-nearfield hyperbolic spreads over infinite fields.*

CHAPTER 8

Ostrom's theorem

Ostrom generalized André idea of replacing a derivable net by a replaceable net of larger dimension. This idea led to the extension field $GF(q^n)$, for $n > 2$, noting that this is a cyclic Galois extension, $F(\theta)$, where θ has order n. Since this may easily be extended to infinite fields. There are some similarities in the following material with those appearing in [**34**].

THEOREM 20. *We recall Hilbert's Theorem 90 ([**11**], problem 23, p. 583). Let F be a field and $F(\theta) = K$ be a Galois cyclic field extension of dimension n over F, where n is a positive integer. Let σ denote a generator of the Galois group. Using exponent notation, let a in K^*, such that $a^{1+\sigma+\sigma^2+\ldots+\sigma^{n-1}} = 1$ (the norm of K/F is 1). Then there exists an element e of K^* such that $a = e^{\sigma-1}$.*

Now we consider Ostrom's theorem over finite and infinite cyclic field extensions.

THEOREM 21. ***(Ostrom's Theorem for fields)***

Let F be a field and $F(\theta)$ be a Galois cyclic extension of dimension n over F, where θ is a generator and σ a generator of the Galois group. Considering the spread for the Desarguesian spread Σ coordinatized by $K = F(\theta)$, this may be partitioned into a set of mutually disjoint replaceable sets together with the components $x = 0, y = 0$:

$$\Sigma = \{x = 0, y = 0\} \cup_{i \epsilon \lambda} \{\{y = xm_i b^{1-\sigma}; b \epsilon S^*\} = \Gamma_i\}.$$

These replacement sets have the following replacements:

$$\{y = xm_i b^{1-\sigma}; b \epsilon S^*\} = \{y = x^\sigma m_i b^{1-\sigma}; b \epsilon S^*\}.$$

These replacement sets are called "hyper-Regulus nets". The multiple replacements of these sets lead to spreads, again called "André spreads".

PROOF. The proof of this theorem is now a special case of the multiple replacement theorem, but originally used Hilbert's Theorem 90. It is something of interest that the more general result does not seem to use that result. This completes the proof of the theorem. □

DOI: 10.1201/9781003767978-8

Now, using ideas of using extensions of finite-dimensional division algebras, Ostrom's theorem may be extended to all such division algebras.

THEOREM 22. ***(Ostrom's Theorem for finite-dimensional division algebras.)*** *Let D be any finite-dimensional division algebra of degree k^2 over a field $K = Z(D)$, the center of D. Consider $D \otimes_F K(z_1, z_2, ..., z_n) = D_n$, where the z_i are commuting in determinants, for $i = 1, 2, .., n$, so D_n is a division ring of degree k over $K(z_1, z_2, ..., z_n)$ and $Z(D_1) = K(z_1, z_2, ..., z_n)$. Consider the cyclic permutation τ of $z_1 \to z_2 \to ... \to z_n \to z_1$. Then $K(z_1, z_2, ..., z_n)$ is a cyclic Galois extension degree n over $Fix\tau$. Assume that τ fixes D element-wise. Form $D \otimes_K Fix\tau = D_1^{(z_1 z_2 ... z_n)} = Fix_{D_1}(\tau)$ (the fixed-point subspace of τ as an automorphism of D_n). Then $D \otimes_{Fix\tau} Fix\tau(z_1) = D_1$, so that D_n has index n over $D_1^{(z_1 z_2 ... z_n)}$. So D_n is a central Galois cyclic extension of $D_1^{(z_1 z_2 ... z_n)}$ of dimension n (central here means it arises using an extension of the center of $D_1^{(z_1 z_2 ... z_n)} = Fix\tau$). Therefore, $D_n = D_1^{(z_1 z_2 ... z_n)}(z_1)$, as $K(z_1, ..., z_n) = Fix\tau(z_1)$.*

Then, using the notation of the multiple replacement theorem, the corresponding left Desarguesian spread Σ and right Desarguesian spread Σ_τ. There is a set of mutually disjoint left hyper pseudo-regulus nets in Σ which have corresponding replacement nets in Σ_τ. Since $D_n \oplus D_n$ has dimension $2n$ over D. which has dimension k^2 over K, the translation planes obtained by proper replacements lie in

$$PG(2nk^2 - 1, K(z_1, z_2, ..., z_n)).$$

PROOF. Note that there is no statement about the division algebra containing primitive kth roots of 1, which one would assume since there is a cyclic Galois extension of degree k. This result uses the Jacobson result. The proof uses the fact that $D_1^{(z_1 z_2 ... z_n)}$ is a division ring and also essentially follows from the skewfield extension theorem for subplane covered nets, which appears in [**34**]. This extension is a central cyclic Galois extension, wherein the tensor product follows directly. This basically completes the proof, which is completely analogous to the pseudo-regulus, dimension 2 situation above. □

PROBLEM 23. In the finite case, Ostrom showed that there are various different replacements for a "hyper regulus" net. Is this true in the more general setting? If so, determine the replacements. This is an open problem. Also, if there are different replacements, are the various spreads obtained from using different replacements for the same hyper pseudo-regulus net isomorphic? This is also open in the infinite case, in particular, the infinite field case. Any replacements that mirror the finite "André replacements" shall then be called of André type. In the finite case, there are replacements that are not of André (to read on this subject, see the Handbook [**37**]).

Here is a restatement of the original idea for skewfield extension theorem, using geometric ideas of "subplane covered nets". Note that in the following theorem, there is a central cyclic Galois extension of a skewfield S, which then has the correct type of automorphism to apply the multiple replacement theorem in the original form. The multiply replacement theorem-redux constructs the required automorphism using extensions by commuting indeterminants. The geometric and non-commutative algebraic ideas connect strongly in these two results.

THEOREM 24. *(See Theorem 137, p. 286) – Skewfield extension theorem for division rings of degree n. Let F be a field that contains a primitive n-th root of unity and such that the characteristic of F does not divide n. Let S be a division ring with center F that contains a field $F(\theta)$ which is a cyclic Galois extension of degree n of F. Assume that S induces a subplane covered net on $PG(2n-1, F(\sigma,\tau))$, where $F(\theta,\tau)$ is the composition field of the two non-isomorphic cyclic Galois extensions of degree n of F, $F(\theta)$ and $F(\tau)$, so that $F(\theta,\tau)$ is an Abelian central field extension of degree n^2.*

(1) Then S coordinatizes a matrix skewfield of n by n matrices over $F(\theta)$, defining a left subplane covered net as a suspect subplane covered net relative to the classical (left/right) regulus net over $F(\theta,\tau)$.

(2) Also, there is a central division ring extension of S over F, which is a division ring $S(\tau)$ such that the center $Z(S(\tau))$, that coordinates a spread in $PG(2n-1, F(\theta,\tau))$, and also a spread in $PG(2n-1, S)$.

(3) This $S(\tau)$ central division ring extension defines a Desarguesian translation plane with left spread in $PG(2n - 1, F(\theta,\tau))$ and right spread in $PG(2n-1, S)$.

We have been most interested in the study of left spreads over division rings in this and the previous work. This is because the spreads look like finite or field spreads. However, there are also right spreads, which do not look exactly like the left spreads. Why not? Here is a long problem/project to consider.

PROBLEM 25. Let Π be a Desarguesian right spread with the following representation:

$$x = 0, y = 0, y = mx,$$

where m is in a division ring D. Let τ be an automorphism of D. Let Π_τ denote the following spread:

$$x = 0, y = 0, y = x^\tau m,$$

for all m in D.

(1) Reconstruct the multiple replacement theorem, but now on the right side.

Now let Σ be the Desarguesian spread with the following representation

$$x = 0, y = 0, y = xm,$$

and let Σ_σ denote the Desarguesian spread with the representation

$$x = 0, y = 0, y = mx^\sigma,$$

where σ is the inverse automorphism τ^{-1}. This is the original assumptions of the multiple replacement theorem.

(2) Assume, for example, that τ has order 2. From there, construct both the left spread and the right spread. Find the matrices describing the left spread. For the right spread, use column vectors to describe the right spread.

(3) Each right and left spread may be "lifted" to semifield spreads. The left spreads are described in the first volume. Determine that the right spreads look like the original left spreads, modulo the matrices can now be represented by use of column vectors. Are they the same, other than there are using column vectors, as the left lifted spreads. Prove that the left lifted spreads define linear flocks. Prove the right lifted spreads also define linear flocks.

(4) Now, assume the skewfields/division rings are fields. Are the common lines the same? Now recall that the Desarguesian spreads are Pappian spreads that admit a set of mutually disjoint André's regulus nets. Determine that the mutually disjoint replacements define lifted spreads that define bilinear σ-conical flocks (the planes of the flocks share one of two distinct lines, but not all planes share the same line).

(5) What goes wrong if you try this construction over non-commutative skewfields that admit automorphisms of degree 2?

Part 3

Simultaneous flock spreads

A characterization of the simultaneous spreads is given over arbitrary skewfields. These spreads admit both the pseudo-regulus-inducing groups for α-cone flocks and for α-twisted hyperbolic spreads over skewfields. A variety of new examples of semifield spreads are constructed in $PG(3, F)$ for certain skewfields F.

CHAPTER 9

Simultaneous spreads of type 2

In this chapter, we consider simultaneous spreads of various types.

THEOREM 26. *(**Main Theorem on Simultaneous Flock Spreads of Type** 2) Let π be a translation plane with spread in $PG(3,F)$, where F is a skewfield. Assume that F admits an automorphism α, which is obtained from a central Galois extension of a skewfield S and $F=S(f)$. If π admits an α-pseudo-regulus-inducing elation group with axis L as well as an α-twisted pseudo-regulus-inducing homology group with axis L and coaxis M, then π has the following general form:*

$$x=0,y=\begin{bmatrix} u^{\alpha}+gt, & ft^{\alpha^{-1}} \\ t & u \end{bmatrix};\ \textit{for all } t,u\in F,\ g,f \textit{ constants in } F.$$

There is an additive quasifibration if and only if $k^{\alpha}k+gk-f\neq 0$ for all $k\in F$. By the fundamental surjective result, as constants are on the left, α is an outer automorphism obtained from a central extension of the skewfield, we have a semifield plane when there is a quasifibration.

PROOF. Consider the $(2,1)$-row determinant and assume some matrix has zero determinant. Therefore, we have for some pair (t,u),where t is non-zero and the following equation.

$$(u^{\alpha}+gt)t^{-1}u=ft^{\alpha^{-1}}.$$

This becomes

$$u^{\alpha}t^{-1}ut^{-\alpha^{-1}}+gut^{-\alpha^{-1}}=f.$$

Let $ut^{-\alpha^{-1}}=k$, and notice that $u^{\alpha}t^{-1}=(ut^{-\alpha^{-1}})^{\alpha}$, thereby giving the condition that

$$k^{\alpha}k+gk=f.$$

Therefore, we have a quasifibration if $k^{\alpha}k+gk-f\neq 0$ for all $k\in F$.

We assume that L is coordinatized as $x=0$ and the α-pseudo-regulus-inducing homology has axis L and coaxis $y=0$. We therefore realize that we have a semifield plane with matrix

$$\begin{bmatrix} u^{\alpha}+g(t) & f(t) \\ t & u \end{bmatrix};t,u\in F,$$

DOI: 10.1201/9781003767978-9

where g and f are functions on F. Since we have an α-twisted regulus-inducing homology group, it follows that

$$\begin{bmatrix} u^\alpha + g(t), & f(t) \\ t & u \end{bmatrix} \begin{bmatrix} s^\alpha & 0 \\ 0 & s \end{bmatrix}, \text{ for all } s \neq 0 \text{ of } F,$$

preserves the spread and is

$$\begin{bmatrix} (us)^\alpha + g(t)s^\alpha & f(t)s \\ ts^\alpha & us \end{bmatrix}.$$

Then it follows that

$$g(t)s^\alpha = g(ts^\alpha) \text{ and } f(t)s = f(ts^\alpha).$$

Therefore, $g(1)s^\alpha = g(s^\alpha)$ and $f(1)s = f(s^\alpha)$. Therefore, letting $g(1) = g$ and $f(1) = f$, we see that $g(t) = gt$ and $f(t) = fs^{\alpha^{-1}}$. This completes the proof. □

DEFINITION 27. A left simultaneous flock spread is said to be "super-simultaneous" for type 2, provided the plane also admits an α^{-1}-elation group with the same axis, which intersects the α-elation group in the identity. In practice, in a translation plane, there is at most one derivable net that shares $x = 0, y = 0$ and $y = x$. However, there could be many derivable nets sharing two of the fundamental components. When a derivable net shares $x = 0$ and $y = 0$ it is called a carrier net of type 2. A matrix-derivable net is a carrier net of type $(2 - j)$ if and only if post or pre-multiplication by $\begin{bmatrix} I_2 & 0 \\ 0 & M \end{bmatrix}$ forces the net to be a derivable net of type j. In the second volume, we proved that all inner $\sigma - A$ derivations functions actually are of type 2, called "fake type 1". And, inner automorphism, are "fake type 2" and thus type 3. A super simultaneous translation of type 2 has three pseudo-regulus-inducing groups: the simultaneous α-pair or an α-cone pseudo-regulus-inducing group, and an α-twisted hyperbolic pseudo-regulus-inducing group, and, in addition, an α^{-1}–$(2 - j)$-pseudo-regulus-inducing elation group, where j is 1 or 2. This would mean that there is a pseudo-regulus-inducing α^{-1}-elation group that is standard/generic in the transformed net obtained by post- or pre-multiplication. Since such derivable nets are not necessarily of type 2 $\begin{bmatrix} u^\sigma & 0 \\ 0 & u \end{bmatrix}$; $u \in F$, but could be of type $\begin{bmatrix} u^\sigma & A(u) \\ 0 & u \end{bmatrix}$; $u \in F$ where A is a σ-derivative function. That is a more general definition of super simultaneous that could provide for a type 2 pseudo-regulus net where the shared Baer subspaces are of a more general type. Therefore, there are several possibilities. To determine the possible types of derivable nets that are possible, we may change the function of the two elation-type derivable nets, replacing the net in question with the original net. This is done using post- or pre-multiplication and will create two additional spreads, as we shall see.

The concept mentioned above will more generally be developed in the material on "Elation Double Spreads", which means a conical spread that is the union of two elation-derivable nets. We also note that a derivable net of the form

$$\begin{bmatrix} u^{\sigma} & A(u) \\ 0 & u \end{bmatrix}; u \in F,$$

where $A(u)$ is an inner σ-derivation function, is a "fake type 1" net, becoming a type 2 net.

In the following theorem, we will change the form of the spread by determining the form of the second elation group net, pseudo-regulus-inducing elation group and interchanging the original elation group net with the second elation group. This result provides the idea for the work on elation double spreads, independent of simultaneous properties.

THEOREM 28. ***(The Elation Net Interchange Theorem.)***
A super simultaneous spread of type 2 and not "fake type 2" has the following isomorphic quasifibration representations using the carrier nets associated with the second elation group net using post- and pre-matrix multiplication.

If the original spread is:

$$x = 0, y = \begin{bmatrix} u^{\alpha} & +gt, ft^{\alpha^{-1}} \\ t & u \end{bmatrix}; \text{ for all } t, u \in F, \ f, \ g \text{ constants in } F.$$

where F is a skewfield. When $k^{\alpha}k + gk \neq f$, there is a semifield plane constructed, provided α is an outer automorphism obtained from a central extension of a skewfield.

(1) The quasifibration obtained by post multiplication of $\begin{bmatrix} g & f \\ 1 & 0 \end{bmatrix}^{-1}$;
using the mapping

$$(x, y) \to (x, y \begin{bmatrix} g & f \\ 1 & 0 \end{bmatrix}^{-1})$$

is a spread isomorphism mapping the original quasifibration to the following quasifibration:

$$x = 0, y = \begin{bmatrix} w^{\alpha^{-1}}\tau + s^{\alpha^{-1}}\tau\, g^{\alpha^{-1}}\tau, & D_g(w) + s^{\alpha}f^{\alpha} + D_g(sg) \\ s & w \end{bmatrix},$$

for all s, w *in* F, *where* $D_g(w) = gw - w^{\alpha^{-1}}\tau\, g$,

$$\textit{where } D_g \textit{ is an } \alpha^{-1}\tau - \textit{derivation function.}$$

(2) The quasifibration obtained by pre-multiplication by the isomorphism

$$(x, y) \to (x\begin{bmatrix} g & f \\ 1 & 0 \end{bmatrix}, y)$$

will represent the original quasifibration in the form:

$$x = 0, y = x\begin{bmatrix} w^{\alpha} + f^{-\alpha}g^{\alpha}fs, & f^{\alpha^{-1}}s^{\alpha^{-1}} \\ s & w \end{bmatrix}$$

for all s, w *in* F.

PROOF. In this proof, we will be using the fact that the spread is a direct sum of derivable nets. We have considered this more generally in the material called "Elation Double Nets". By the previous result, we have the spread form

$$x = 0, y = \begin{bmatrix} u^{\alpha} + gt & ft^{\alpha^{-1}} \\ t & u \end{bmatrix}; \text{ for all } t, u \in F, \ g, f \text{ constants in } F.$$

It now follows that we have the carrier-derivable net of type $(2-2)$ $\alpha^{-1}\tau$-net. Therefore, the net in question is $x = 0, y = \begin{bmatrix} gt & ft^{\alpha^{-1}} \\ t & 0 \end{bmatrix}$; for all $t \in F$, g, f constants in F. Post-multiply the matrix set by $\begin{bmatrix} g & f \\ 1 & 0 \end{bmatrix}^{-1} = \begin{bmatrix} 0 & 1 \\ f^{-1}, & -f^{-1}g \end{bmatrix}$, to obtain:

$$\begin{bmatrix} gt, & ft^{\alpha^{-1}} \\ t & 0 \end{bmatrix}\begin{bmatrix} 0 & 1 \\ f^{-1}, & -f^{-1}g \end{bmatrix} = \begin{bmatrix} t^{\alpha^{-1}}\tau, & gt - t^{\alpha^{-1}}\tau\, g \\ 0 & t \end{bmatrix}.$$

Let, & $D_g(w) = gw - w^{\alpha^{-1}}\tau\, g$. In this piece of the spread is represented by the matrices:

$$\begin{bmatrix} t^{\alpha^{-1}}, & D_g(t) \\ 0 & t \end{bmatrix}.$$

Then

$$y = \begin{bmatrix} u^{\alpha} & 0 \\ 0 & u \end{bmatrix}\begin{bmatrix} 0 & 1 \\ f^{-1} & , -f^{-1}g \end{bmatrix} = \begin{bmatrix} 0 & u^{\alpha} \\ uf^{-1} & , -uf^{-1}g \end{bmatrix}.$$

Putting this together, letting $w = t - uf^{-1}g, s = uf^{-1}$, note that D_g is additive, we obtain the following representation of the spread:

$$x = 0, y = \begin{bmatrix} w^{\alpha^{-1}}\tau + s^{\alpha^{-1}}\tau\, g^{\alpha^{-1}}\tau, & D_g(w) + s^{\alpha}f^{\alpha} + D_g(sg) \\ s & w \end{bmatrix}.$$

This completes the proof. □

9.1. Quadratic super-simultaneous flock spreads

Let α have order 2, and consider

$$x=0, y=\begin{bmatrix} u^{\alpha}+gt, & ft^{\alpha^{-1}} \\ t & u \end{bmatrix}; \text{ for all } t,u\in F,\ g,f \text{ constants in } F.$$

If F is a central Galois quadratic extension $S(\theta)$, it has been proved in Johnson and Jha [**38**] that if $g=0$ and $f=\theta$, there is a semifield spread constructed in the case where $\alpha=\alpha^{-1}$. Assume then that $g=1$. First, assume that the characteristic is not 2. Assume that $f=\theta$. Assume that $k^{\alpha}k+k=\theta$. Let $k=\theta\alpha_1+\alpha_2$, for $\alpha_i\in S$, $i=1,2$.

Let $x^2-\delta$ for $\delta\in Z(S)$ be irreducible, where $\delta+4^{-1}$ is non-square in S. This would also be equivalent to $x^2-(\delta+4^{-1}\ )$ is irreducible in S, and assume that $\theta^2=\delta$ and that the automorphism α maps

$(\theta\alpha_1+\alpha_2)$ to $(-\theta\alpha_1+\alpha_2)$.

Then

$$\begin{aligned} k^{\alpha}k+k &= (-\theta\alpha_1+\alpha_2)((\theta\alpha_1+\alpha_2)+(\theta\alpha_1+\alpha_2) \\ &= -\delta\alpha_1^2+\alpha_2^2+\alpha_2+\theta(\alpha_1+\alpha_2\alpha_1-\alpha_1\alpha_2)=\theta. \end{aligned}$$

Therefore, we have:

$$\begin{aligned} (*)(1) &: \ \delta\alpha_1^2=\alpha_2^2+\alpha_2 \text{ an} \\ (*)(2) &: \ \alpha_2\alpha_1-\alpha_1\alpha_2=1-\alpha_1\ . \end{aligned}$$

Since $\delta\in Z(S)$, we see from $(*)(1)$ that α_2 commutes with α_1^2. Now form $\alpha_1((*)(2))\alpha_1$ to obtain:

$$\begin{aligned} \alpha_1\alpha_2\alpha_1^2-\alpha_1^2\alpha_2\alpha_1 &= \alpha_1^3\alpha_2-\alpha_2\alpha_1^3=\alpha_1^2(1-\alpha_1)=(1-\alpha_1)\alpha_1^2 \\ &= (\alpha_2\alpha_1-\alpha_1\alpha_2)\alpha_1^2=\alpha_2\alpha_1^3-\alpha_1^3\alpha_2. \end{aligned}$$

Therefore,

$$2\alpha_2\alpha_1^3=2\alpha_1^2(\alpha_2\alpha_1)=2\alpha_1^3\alpha_2=2\alpha_1^2(\alpha_1\alpha_2).$$

Assume that $\alpha_1\neq 0$, then $\alpha_2\alpha_1=\alpha_1\alpha_2$, so that $\alpha_1=1$. Then $\delta=\alpha_2^2+\alpha_2=(\alpha_2+2^{-1})^2-4^{-1}$ as 2^{-1} and 4^{-1} are in the center. Hence, if $\delta+4^{-1}$ is a non-square, we have a contradiction as δ is non-zero.

Now consider the other possibilities: Characteristic 2 with minimal polynomial $x^2+x+\beta$, where we and that there is a central Galois quadratic extension $S(\theta)$ where the unique involutory automorphism σ, maps θ to $\theta+1$, as $(\theta+1)^2+(\theta+1)+\beta=0$, may also be used as an extension $S(\tau)$ of S, if the characteristic is 2. If the characteristic is not 2 with the minimal polynomial $x^2-x-\beta$ used for a central Galois quadratic extension $S(\theta)$, then we may again assume that the unique involutory automorphism maps θ to $-\theta+1$. We may take $\theta^2=\theta+\beta$. We take $g=1$ and must show that $k^\sigma k+k\neq\theta$. Letting $k=\theta\alpha_1+\alpha_2$ and assuming that $k^\sigma k+k=\theta$. Note that $k^\sigma kk^{-1}=\theta k^{-1}-1=k^{-1}k^\sigma k=k^{-1}-1$. Hence, it then follows that $k^\sigma=k^{-1}k^\sigma k$, so that $kk^\sigma=k^\sigma k$. In this case, we see that

$$\begin{aligned} k^\sigma k &= ((-\theta+1)\alpha_1+\alpha_2)(\theta\alpha_1+\alpha_2) = \\ & (-\theta\alpha_1+(\alpha_1+\alpha_2)(\theta\alpha_1+\alpha_2) \\ &= -\beta\alpha_1^2+(\alpha_1+\alpha_2)\alpha_2+\theta(\alpha_2\alpha_1-\alpha_1\alpha_2). \end{aligned}$$

Since $k^\sigma k=kk^\sigma$, we see from the 1-term that

$$\alpha_2(\alpha_1+\alpha_2)=(\alpha_1+\alpha_2)\alpha_2,$$

so that

$$\alpha_2\alpha_1-\alpha_1\alpha_2=0$$

and hence

$$k^\sigma k=-\beta\alpha_1^2+(\alpha_1+\alpha_2)\alpha_2.$$
$$k^\sigma k=\theta-k=\theta(1-\alpha_1)-\alpha_2$$

so we see that

$$\alpha_1=1 \text{ and } -\beta+2\alpha_2+\alpha_2^2=0.$$

In characteristic 2, this would say that

$$\theta+(\theta+\alpha_2)^2=\beta,$$

since θ is central. This implies that $(\theta+\alpha_2)^2=\theta^2$ so $\alpha_2=0$, a contradiction. So, assume that the characteristic is not 2. We allow that $x^2+2x-\beta$ is S-irreducible. Therefore, we have the following theorem.

THEOREM 29. *Let $S(\theta)$ be a central Galois quadratic extension with irreducible polynomial $x^2+\alpha x-\beta$, and assume further that when the characteristic is not 2 that $\delta=1+\frac{\beta}{\alpha^2}$ and $\delta+\frac{1}{4}$ are both non-square. Then*

$$x=0, y=\begin{bmatrix} u^\sigma+t, & \theta t^\sigma \\ t & u \end{bmatrix};\ \textit{for all } t,u\in F.$$

is a semifield spread in $PG(3,S(\theta))$.

The skewfields exhibit the requirements that include certain central Galois quadratic extensions of quaternion division rings and various central Galois cyclic extensions of cyclic division rings of dimension $n=2^r$.

9.2. Simultaneous spreads – unified form

In this chapter, we determine the unified form for a general simultaneous spread of the form:

$$x = 0, y = \left[\begin{array}{cc} u^\sigma + F(t) & , A(u) + G(t) \\ t & u \end{array} \right],$$

for all t,u in D, where D is a division ring, and F,G are functions on D. The spread produces a conical flock spread. In order for the spread to be simultaneous, it must be able to be reconfigured as a hyperbolic spread. There, the spread must also have the following form:

$$x = 0, y = x \left[\begin{array}{cc} v^\sigma & A(v) \\ 0 & v \end{array} \right]$$

and

$$y = x \left[\begin{array}{cc} t^\sigma + f & , A(t) + g \\ 1 & t \end{array} \right] \left[\begin{array}{cc} u^\sigma & A(u) \\ 0 & u \end{array} \right],$$

for all v, t in D and for all non-zero u in D, where $f = F(1)$ and $g = G(1)$. The uniform theorem is as follows:

THEOREM 30. *The above conical flock spread is simultaneous if and only if the functions F, G have the following conditions:*
(1) $F(t) = ft{-}A(t^{\sigma^{-1}})^\sigma$,
(2) $G(t) = gt^{\sigma^{-1}} + fA(t^{\sigma^{-1}}) - A^2(t^{\sigma^{-1}})$.

PROOF. We see that

$$\left[\begin{array}{cc} t^\sigma + f & , A(t) + g \\ 1 & t \end{array} \right] \left[\begin{array}{cc} u^\sigma & A(u) \\ 0 & u \end{array} \right] =$$

$$= \left[\begin{array}{cc} (tu)^\sigma + fu^\sigma, & t^\sigma A(u) + A(t)u + fA(u) + gu \\ u^\sigma & A(u) + tu \end{array} \right]$$

Since, $t^\sigma A(u) + A(t)u = A(tu)$, now write $u^\sigma = s$, $w = A(u) + tu$. Then this implies the following:

$$F(s) = fs{-}A(s^{\sigma^{-1}})^\sigma$$

and

$$G(s) = gs^{\sigma^{-1}} + fA(s^{\sigma^{-1}}) - A^2(s^{\sigma^{-1}}),$$

for all s in D. By $A^2(r)$, we mean $A(A(r))$. This completes the proof. □

The simultaneous spread examples that have been found are basically of three types. When A is identically zero, this devolves to σ-types. When A is identically zero, and g is also zero, there are examples when

σ has orders 3 and 4. When f and g are both non-zero, there are examples only when σ has order 2. Thus, there is room for any further results here.

When f is zero and A is not zero, we now consider the examples when σ has order 2 or 1.

When σ has order 2, we know that there are $\sigma - A$ examples, the inner $\sigma - A$ derivations. The question is, what are the examples when the automorphism arises from the central Galois quadratic extension of a division ring D. Later in the text, we shall determine when

$$-A(t^{\sigma^{-1}})^{\sigma} = A(t), and\ A^2(v) = 0.$$

Part 4

Semifields over division rings

Previously, Johnson and Jha [**38**] showed that it is possible to "lift" translation planes coordinatized by central Galois quadratic extensions of arbitrary skewfields. In this part, an analogous construction is given of infinitely many semifields of dimension 2 that are constructed over central Galois cyclic skewfields of degrees 3 and 4. These planes are generalizations of translation planes corresponding to σ-flocks in the field case. Furthermore, let $S = ((\mathbb{Q}\varsigma_3)(\tau), \rho, g)$ be any cyclic division ring of degree 3; then there is a central cyclic Galois extension of degree 3, $S(\theta)$, such that

$$x = 0, y = x \begin{bmatrix} u^{\sigma}, & \theta t^{\sigma^{-1}} \\ t & u \end{bmatrix} ; t, u \in S(\theta)$$

is a derivable semifield plane with spread in $PG(3, S(\theta))$. There is an infinite variety of semifield planes of this type. These unusual semifield planes have components, which are contained in 3-dimensional projective spaces over non-commutative skewfields. There are a number of generalizations of these spreads, which we shall provide.

In Johnson and Jha [**38**], it was shown that it is possible to extend the "lifting" theory of Hiramine, Matsumoto and Oyama [**17**], which was initially considered only in finite translation planes of dimension 2. In this part, we first review the lifting over skewfields theory, which is possible for central Galois quadratic extensions of skewfields S. The term "central" extension here means the extension also extends $Z(S)$, the center of S, in the same manner. We then generalize the lifted semifields to consider semifields obtained from central Galois cyclic extensions of skewfields.

CHAPTER 10

Twisted T-copies

Referring back to lifted skewfield planes, it follows that $S_{(\delta,\gamma)}(s)(\theta)$ is a skewfield and the extension of the skewfield $S_{(\delta,\gamma)}(s)$ is a central quadratic extension of the skewfield $S(s)$; the lifted skewfield plane is

$$x = 0, y = 0, y = x \begin{bmatrix} u^{\sigma} & \theta^{\sigma} t^{\sigma} \\ t & u \end{bmatrix}; t, u \in S(s); (t, u) \neq (0, 0),$$

and there exists a family of $S_{(\delta,\gamma)}$ planes,

$\delta \neq 0$, $\theta\delta + \gamma \in Z(S(s)) = Z(S)(s)$, the center of $S(s)$,

where $\theta\delta + \gamma$ replaces θ^{σ}.

σ is always the induced automorphism of $S(s)(\theta)$,

the quadratic extension.

Therefore, we see that we have another skewfield $S(s)_{(\delta,\gamma)}(\theta)$, which is for $\theta\delta + \gamma \in Z(S)$ and is called the "twisted T-copy of $S_{(\delta,\gamma)}$". We observe the following result.

THEOREM 31. *Let S be any skewfield of finite dimension over its center $Z(S)$. And, let $S(\theta)$ be a central Galois quadratic extension with corresponding automorphism σ. Let $S(s)$ denote the skewfield extension, where s is an independent variable. Then $S(s)(\theta) = S(\theta)(s)$.*

(1) So, we construct the set of semifield planes lifted from $S(s)(\theta)$, $S(s)_{(\rho,\varepsilon)}$, where $\rho \neq 0$, and $\theta\rho + \varepsilon \in Z(S(s))$.

(2) Since $Z(S(s)) \supset Z(S)$, we see that the semifield planes $S_{(\delta,\gamma)}$ themselves have twisted T-extensions that are semifield planes.

(3) Let $\{s_1, s_2, ..., s_k\}$ be a set of mutually independent, commuting indeterminants, and let S be a skewfield of finite dimension over the center $Z(S)$. Then $S(s_1, ..., s_k)(\theta)$ is a central Galois quadratic extension of $S(s_1, ..., s_k)$ and there is a set of semifields with spreads in $PG(3, S(s_1, ..., s_k)(\theta))$, which also have twisted T-copies that are semifield spreads.

In the following, we shall see that there are interesting generalizations of lifted skewfield planes.

DOI: 10.1201/9781003767978-10

In general, again as we have noted previously, to show that a putative spread with form as above is, in fact, a spread, two requirements must be proved:

(1) that the components are non-singular; that the set forms a "quasifibration"

(2) that for any non-zero vector (x, y) for all x, y 2-vectors over S, there is a component containing (x, y); "the covering property".

In the finite case, (2) follows from (1). Furthermore, the theory of lifting of translation planes coordinatized by central Galois quadratic extensions of skewfields S has the amazing property that (1) and (2) are basically equivalent. To generalize this study to include central Galois cyclic extensions of skewfields appears to make sense if the matrix form for a more general idea of lifting also has this property. This idea will be developed in the next section. As mentioned, in general, if one has a quasifibration, it is normally a difficult problem to determine when this partial spread is also a spread. This will not be an issue in this article, but we shall mention the most general result known over quasifibrations of arbitrary finite dimension in this direction.

CHAPTER 11

General skewfield lifts to semifields

In the previous section, we mentioned lifting skewfields $S(\theta)$ that were central Galois quadratic extensions of a skewfield S to new semifields in $PG(3, S(\theta))$. The same sort of idea might construct semifields from central Galois cyclic extensions of dimension n of a skewfield S. However, there is a slight variation in the form of the semifield.

The hypothesis requires a central cyclic Galois extension of S of dimension n, where S is an arbitrary skewfield, by which the extension arises initially as a cyclic Galois extension of $Z(S)$ of dimension n. If there is a corresponding Galois extension of S of dimension n, we use the term "central" extension. In [**34**], there is a construction method for central Galois quadratic extensions of quaternion division rings and an analogous method for cyclic division rings. These methods are geometric as the ideas of derivable nets and subplane-covered nets are used for the construction. To illustrate the ideas, let $(a, b)_K$ be a quaternion division ring with center K, of characteristic $\neq 2$. There is a matrix representation and a corresponding geometric/net structure:

$$(\$) : x = 0, y = x \begin{bmatrix} u^{\sigma} & bt^{\sigma} \\ t & u \end{bmatrix}; u, t \in K(\sqrt{a}),$$

which actually is a derivable net with components in $PG(3, K(\sqrt{a}, \sqrt{c}))$, provided there are mutually non-isomorphic quadratic extensions

$$K(\sqrt{a}) \& K(\sqrt{c})$$

the latter is not isomorphic to $K(\sqrt{b})$. In order to have a derivable net, the dimension of the skewfield over K must be the dimension of the ambient vector space. Since the dimension of $K(\sqrt{a}, \sqrt{c})$ is 4 over K, there is a related derivable net

$$x = 0, y = x \begin{bmatrix} v & 0 \\ 0 & v \end{bmatrix}; v \in K(\sqrt{a}, \sqrt{c}).$$

This means that the associated 4-dimension $K(\sqrt{a}, \sqrt{c})$-vector space V_4 is also a 4-dimensional vector space over $(a, b)_K = L$ so that V_4 is a $((a, b)_K, K(\sqrt{a}, \sqrt{c}))$-bimodule. This is necessary, as otherwise, what

DOI: 10.1201/9781003767978-11

is represented in ($) is a subplane covered net but not necessarily a derivable net.

The Baer subplanes incident with the zero vector of the quaternion division ring net are as follows: Consider $V_4/K(\sqrt{a},\sqrt{c})$, with standard basis, then note that over $(a,b)_K$, we have the following basis

$$\{(0,1,0,0),(0,\sqrt{c},0,0),(0,0,0,1),(0,0,0\sqrt{c})\}.$$

The Baer subspaces are as follows for all vectors 2-vectors (d,f), written over $(a,b)_K$.

$$B(d,f)=\{((d,f)\begin{bmatrix} u^\sigma & bt^\sigma \\ t & u \end{bmatrix},(d,f)\begin{bmatrix} v^\sigma & bs^\sigma \\ s & v \end{bmatrix});t,u,s,v\in K(\sqrt{a}).$$

The theory shows that there are no common Baer subspaces shared between derivable nets written over non-isomorphic skewfields. It then follows that the two scalar groups, right and left, commute as groups and then induce irreducible groups on any common component. This then creates an extension skewfield of $(a,b)_K$ by $(a,b)_K\otimes_K K(\sqrt{c})$ which is then $(a,b)_{K(\sqrt{c})}$. For $L=(a,b)_K$, we have a central Galois quadratic extension $L(\sqrt{c})$.

The same theory is valid for cyclic division rings of dimension n over K. Assume that K contains a primitive n-th. root of unity ς_n, assuming that the characteristic does not divide n and n is a unit. Write the cyclic division ring as an n by n matrix skewfield where the n row (or column) has the form $[t_1,...,t_n]$ for all $t_i\in K(\sqrt[n]{a})$, a cyclic Galois field extension of dimension n over K. Embed geometrically as a subplane covered net in a vector space $V_{2n}/K(\sqrt[n]{a},\sqrt[n]{c})$, where this field is Galois of degree n^2, analogously to the derivable net case. Use the notation $(a,b)_{\varsigma_n K}$, just to see the similarity, which is a symbol division ring. We write the subplane-covered net as

$$x=0,y=xM;M\in(a,b)_{\varsigma_n K}$$

as a matrix division ring with spread in $PG(2n,K(\sqrt[n]{a},\sqrt[n]{c}))$. The set of little subspaces have matrix representations

$$B(\overrightarrow{C})=\{(\overrightarrow{C}M,\overrightarrow{C}N);\forall M,N\in(a,b)_{\varsigma_n K},$$

where $\overrightarrow{C}$ is an n-vector written over $(a,b)_{\varsigma_n K}$. Using the classical regulus net

$$x=0,y=xvI_n;v\in K(\sqrt[n]{a},\sqrt[n]{c}),$$

there is a corresponding set of left/right and commuting scalar mappings, which ultimately generates an irreducible group that creates a left and a right version of a central cyclic Galois extension of

$(a,b)_{\varsigma nK} \otimes K(\sqrt[n]{c})$, which is $(a,b)_{\varsigma_n K(\sqrt[n]{c})}$ and if $L = (a,b)_{\varsigma nK}$, we have the central cyclic Galois extension $L(\sqrt[n]{c})$ of degree n.

The main point of this review is that if a cyclic division ring of degree n exists relative to $K(\theta)$ and there is a cyclic Galois extension $K(\theta,\tau)$ of $K(\theta)$ exists, where $K(\theta)$ and $K(\tau)$ are mutually non-isomorphic cyclic Galois extensions of degree n of K, then there is a central Galois extension of a cyclic division ring of degree n.

When $n = 2$, this theory is called 'lifting skewfields', since is it similar to the field version of the concept of lifting a 2-dimensional spread over a field. The lifted structure is a semifield plane of the form

$$(\$) : x = 0, y = x \begin{bmatrix} u^\sigma, & \theta t^\sigma \\ t & u \end{bmatrix}; u, t \in S(\theta),$$

where S is any skewfield that admits a central Galois quadratic extension $S(\theta)$, where the components lie in $PG(3, S(\theta))$, a non-commutative skewfield 3-dimensional projective space. It is known that given any quaternion division ring $(a,b)_K$ there is an extension skewfield $(a,b)_{K(z)}$, where $K(z)$ is the function field of K, where z is an indeterminant. And then, for characteristic $\neq 2$, there is a central Galois quadratic extension $(a,b)_{K(z)(\sqrt{z})}$ (there is another extension for characteristic 2 that has a central Galois quadratic extension).

Then $n = 2^r > 2$; there are cyclic Galois extensions of degree n for which the same ideas may be formulated. There are cyclic Galois quadratic extensions of division subrings that then construct sets of semifield planes but within a much different 3-dimensional projective geometry over a non-commutative skewfield.

In this part, we find that there are probably many more sets of semifield planes that may be constructed from central cyclic Galois extensions of skewfields S, and it is possible to prove this when $n = 3$ and 4. Furthermore, we show the transposed matrix set is a spread and we provide another simultaneous spread. We point out that we are trying to construct spreads that are either unions of a set of pseudo-regulus nets sharing two components or unions of a set of pseudo-regulus nets sharing one component. We call such spreads "hyperbolic spreads" or "conical spreads", as such spreads will be shown to construct flocks of (generalized) hyperbolic quadrics or of flocks of (generalized) quadratic cones. We also show that the transpose of the matrices forms a spread, called the "transposed spread". We shall obtain that there are four distinct spreads: two hyperbolic and two conical.

In the next chapter, we consider another type of lifting again of degree $n > 2$, but having a connection with dimension 2 flock spreads.

CHAPTER 12

Central extensions of degree 3,4

In this section, we show that it is possible to construct classes of semifield planes from division rings of degree n^2 over the center. When $n = 3$, any division algebra of dimension 3^2 over the center is known to be cyclic (e.g., Jacobson [**18**] (2.9.17)).

Assume that $F(\varsigma_3)$ is a field that contains a 3rd. root of unity ς_3 and has characteristic $\neq 2$ or 3. Then, $\varsigma_3 = \frac{-1\pm\sqrt{-3}}{2}$. Then a cyclic Galois field of degree 3 over F is obtained, provided $x^3 - \delta$ is irreducible if and only if δ is not a cube. Then $\{1, \delta^{\frac{1}{3}}, \delta^{\frac{2}{3}}\}$ is an $F(\varsigma_3)$-basis. For an easy proof of this, let $\varsigma_3 = \frac{-1+\sqrt{3}i}{2}$, a 3rd. root of unity, consider the polynomial $x^3 - \delta$, over $F(\varsigma_3)$, where δ is an element in F, and assume that $\sqrt{3}i$ is not in F. Then consider $\alpha + \beta(\frac{-1+\sqrt{3}i}{2})$, for all α, β in F. Since the characteristic is not 2, replace β by 2γ and $\alpha - \gamma$ by λ to obtain the set of elements $\{\lambda + \gamma\sqrt{3}i; \lambda, \gamma \in F\}$ is clearly equal to $F(\varsigma_3)$ so that $\{1, \sqrt{3}i\}$ is a basis for $F(\varsigma_3)$. We now compute the cubic elements of $F(\varsigma_3)$ in F. If $\delta = \rho^3$, then

$$(x^3 - \rho^3) = (x - \rho)(x^2 + \rho x + \rho^2) =$$

$$(x - \rho)(x - \rho(\frac{-1 + \sqrt{3}i}{2})(x - \rho\frac{-1 - \sqrt{3}i}{2},$$

$$= (x - \rho)(x - \rho\varsigma_3)(x - \rho\varsigma_3^2).$$

And, for $\rho^3 \in F(\varsigma_3)$, and, of course

$$(\lambda + \gamma\sqrt{3}i)^3 = \lambda^3 + 3\lambda^2\gamma\sqrt{3}i) - 9\lambda\gamma^2 - 3\gamma^3\sqrt{3}i) = \rho^3 = \delta,$$

shows that either $\gamma = 0$, or $\lambda = \pm\gamma$. If $\gamma = 0$, then δ is a cube in F. If $\lambda = \pm\gamma$ then $\delta = (\mp 2\gamma)^3$. Therefore, if δ is not a cube in F, it is not a cube in $F(\varsigma_3)$. Therefore, if δ is not a cube in F, we have $x^3 - \delta$ is irreducible over $F(\varsigma_3)$ and, letting $\theta^3 = \delta$, we have a central cyclic extension $F(\varsigma_3)(\theta)$ of degree 3 over $F(\varsigma_3)$ with basis $\{1, \delta^{\frac{1}{3}}, \delta^{\frac{2}{3}}\}$. The above argument shows that the roots are $\{\theta, \varsigma_3\theta, \varsigma_3^2\theta\}$. For the $k^\sigma k$-theorem to work, the proof will show that we require a generator that is a non-square in $F(\varsigma_3)(\theta)$. If δ is a non-square, we are finished (with the assumption). It has been shown in the text that if θ is central and the

(2,1)-determinant is always non-zero, we have a spread. The reference to the $k^{\sigma}k$-refers to the results that state is $k^{\sigma}k$ is never θ; spread is obtained.

Let $F(\varsigma_3) = (\mathbb{Q}\varsigma_3)$, the cyclotomic extension of the field of rational numbers $\mathbb{Q}$.

LEMMA 32. *(1) Let $F(\varsigma_3) = (\mathbb{Q}\varsigma_3)$, the cyclotomic extension of the field of rational numbers $\mathbb{Q}$. If d is a non-cube in $\mathbb{Q}$ then $x^3 - d$ is irreducible and $d^{1/3}$ is a generator of a cyclic Galois extension of $(\mathbb{Q}\varsigma_3)$.*

(2) There is a generator of $(\mathbb{Q}\varsigma_3)(d^{1/3})$, such that d is a non-square.

(3) If d is an odd prime, then $F(d^{1/3})$ is a cyclic Galois extension of degree 3. In general, there are infinitely many mutually non-isomorphic extensions $(\mathbb{Q}\varsigma_3)\sqrt[3]{e}$ of degree 3, where e is a non-square.

PROOF. It remains to show that δ may be chosen as a non-square. First choose δ to be any positive integer not equal to $(2\gamma)^3$, then choose ρ any non-square positive integer such that $(\mathbb{Q}\varsigma_3)(\sqrt[3]{\delta})$ and $(\mathbb{Q}\varsigma_3)(\sqrt[3]{\rho})$ are mutually non-isomorphic cyclic Galois extensions of $(\mathbb{Q}\varsigma_3)$ of degree 3. Then there are infinitely many non-mutually central cyclic Galois extensions of degree 3. Given any integer δ not equal to $(2\gamma)^3$, then $x^3 - \delta$ is irreducible, and there is a generator $\sqrt[3]{\delta^*}$ of $(\mathbb{Q}\varsigma_3)(\sqrt[3]{\delta})$, such that δ^* is a non-square. Let $\delta = \rho^2$, then since $\delta^{2/3}$ is also a generator, we see that $\rho^{1/3}$ is a generator. So, it follows that there is always a generator θ, which is a non-square, from this reduction is finite. More generally, let $d = \alpha + \beta\sqrt{3}i$, $\alpha, \beta \in \mathbb{Q}$, which is a non-cube. assume d is a square. First case: If $\beta = 0$, then α is a rational square and by prime decomposition theorem, there is a maximum element n such that $\alpha = c^{2^e}$. Then $\alpha^{\frac{1}{3}}$ is a generator of the cyclic extension of degree 3. If $e = 1$, then $\alpha^{\frac{2}{3}}$ is a generator $c^{\frac{4}{3}} = cc^{\frac{1}{3}}$, which shows that $c^{\frac{1}{3}}$ is a generator, and c is not a square. If $e > 1$, then $c^{\frac{4+2^{d-2}}{3}}$, shows that $c^{\frac{4+2^{d-2}}{3}}$ produces $c^{2^{d-2}}c^{\frac{2^{d-2}}{3}}$, which shows that $c^{\frac{2^{d-2}}{3}}$ is a generator. The previous proof shows that by a finite number of applications, assume that $c^{\frac{1}{3}}$ is a generator and c is non-square. Second case: $\beta \neq 0$. If $d = \alpha+\beta\sqrt{3}i$, is a square and β is not zero and d is a square $(\delta+\gamma\sqrt{3}i)^2$, for δ and γ rational numbers, then the square is $\delta^2 - 3\gamma^2 + 2\delta\gamma\sqrt{3}i$. This implies that β is even, as $\delta\gamma \neq 0$. Hence, there is a maximum integer e such that $\beta = 2^e\beta^*$, where β^* is a rational number and 2^{e+1} does not divide β. So, by a finite number of at most e consecutive squares and corresponding generator reductions, it follows that there is a reduction of d, d^* such that $d^{*\frac{1}{3}}$ is a generator and d^* is non-square. This completes the proof of the lemma. □

In the first volume [**34**], there are theorems called "skewfield extension theorems", one for derivable nets and one for subplane-covered nets. The skewfield extension theorem for derivable nets discusses type 0 derivable nets and the extension theorem for subplane covered nets, discussing type 0 subplane-covered nets. The conclusions of these two theorems, characterizes the extensions in terms of tensor product constructions. In the following result, we assume that there is a central Galois extension of a cyclic division ring of degree n. There are always such extensions, by using extensions by rational function fields of many variables. This was noticed in the first volume, in particular, when analyzing quaternion division rings and their central Galois extensions. So, while it may appear that we are using algebraic extensions initially, we could also be using a combination of transcendental and algebraic extensions. Here is the skewfield extensions theorem for division rings.

In [**34**], we recall

THEOREM 33. ***(Skewfield Extension Theorem for Division Rings of Degreen***). *Let F be a field that contains a primitive n-th root of unity and so that the characteristic of F does not divide n; n is a unit. Let S be a division ring of degree n with center F that contains a field $F(\theta)$, which is a Galois cyclic field extension of dimension n over F. Assume S induces a subplane covered net on $PG(2n-1, F(\theta,\tau))$, where $F(\theta,\tau)$ is an Abelian central field Galois extension of dimension n^2, the composition of two non-isomorphic cyclic extensions of dimension n, containing $F(\theta)$ and $F(\tau)$:*

(1) Then S coordinatizes a matrix skewfield of n by n matrices over $F(\theta)$, defining a left subplane covered net as a suspect subplane covered net relative to the classical (left/right) regulus net over $F(\theta,\tau)$.

(2) Also, there is a central division ring extension of S over F, which is a division ring $S(\tau)$ such that $Z(S(\tau)) = F(\tau)$, that coordinatizes a spread in $PG(2n-1, F(\theta,\tau))$ and a spread in $PG(2n-1, S)$.

(3) This $S(\tau)$ central division ring extension defines a Desarguesian translation plane with left spread in $PG(2n-1, F(\theta,\tau))$. Additionally, the translation plane may be written in $PG(2n-1, S)$.

In the remarks preceding this theorem in [**34**], on page 286, we notice that the central division ring extension is isomorphic to $S_{right} \otimes_F F(\tau)$, and there are two centralizer division rings that are not isomorphic as groups when n>2. In the following, the notation has been reversed to align with that of [**34**].

THEOREM 34. *Let S be a cyclic division ring of degree 3 with center $(\mathbb{Q}\varsigma_3)$, $S = ((\mathbb{Q}\varsigma_3)(\tau)/(\mathbb{Q}\varsigma_3), \rho, g)$, and let $S(\theta)$ be a central Galois* *extension of dimension* 3. $(\mathbb{Q}\varsigma_3)(\theta)$ *be a cyclic Galois field extension of* $(\mathbb{Q}\varsigma_3)$*, such that* $\mathbb{Q}\varsigma_3(\tau)$ *is not isomorphic to* $(\mathbb{Q}\varsigma_3)(\theta)$*, with associated automorphism group* $\langle\sigma\rangle$*. Then* θ *may be taken as* $\sqrt[3]{\delta}$ *for* $\delta \in \mathbb{Q}\varsigma_3$ *and* δ *is non-square, and* τ *may be taken as* $\sqrt[3]{\gamma}$*, where* γ *is a non-square. Then the center of* $S(\theta)$ *is* $(\mathbb{Q}\varsigma_3)(\theta)$*. Note that the* $S(\theta)$ *is* $((\mathbb{Q}\varsigma_3)(\tau,\theta)/(\mathbb{Q}\varsigma_3)(\theta),\rho,g)$*, and the automorphism of order* 3 *is defined to fix* S *elementwise.*

PROOF. Using the skewfield extension theorem, there is a central cyclic Galois extension $S(\theta)$, where the center of $S(\theta)$ is now $(\mathbb{Q}\varsigma_3)(\theta)$, and σ is defined to fix S pointwise. Since $(\mathbb{Q}\varsigma_3)(\theta)$ is cyclic. Galois extension field of degree 3, we see, using the Lagrange resolvent ([**29**]), that θ may be replaced by $\sqrt[3]{\delta}$, where $\delta \in \mathbb{Q}\varsigma_3$. Assume that there is an element $k \in S$, such that $k^\sigma k = \theta^\sigma$, then $\theta^\sigma = \varsigma_3\theta$, and we may assume that $x^3 - \delta$ is irreducible over $Z(S)$ (See e.g., [**34**]) and $\theta^3 = \delta$ and ς_3 is a primitive 3rd. root of unity and δ is non-square in $(\mathbb{Q}\varsigma_3)$ by the lemma. Since θ commutes with S, we see that $k^\sigma k = \theta^\sigma$, implying that $k^\sigma k$ commutes with k^{-1}, so that $k^\sigma k k^{-1} = k^\sigma = k^{-1}k^\sigma k = k^\sigma$, so that $k^\sigma k = kk^\sigma$. □

PROOF. As a guide to the proof (the long version), we first consider k within $S(\theta)$, and write the terms relative to the basis $\{1,\theta,\theta^2\}$, with the note that $(\mathbb{Q}\varsigma_3)(\theta)$ is the center of the division ring $S(\theta)$. Step (i). We obtain that δ is a square in S. Step (ii) writes S in either matrix of what we will call the w-form. This will show that δ is a square in $(\mathbb{Q}\varsigma_3)(\tau)$. This is a contradiction by degrees, as $(\mathbb{Q}\varsigma_3)(\tau)$ cannot admit a subfield of dimension 2 over $(\mathbb{Q}\varsigma_3)$. But another option is: Step (iii), we write $(\mathbb{Q}\varsigma_3)(\tau)$, over $(\mathbb{Q}\varsigma_3)$ relative to the basis $\{1,\tau,\tau^2\}$, and show that δ is a square in $(\mathbb{Q}\varsigma_3)$, which is a contradiction.

Step (i): Let

$$\begin{aligned}
&k = (\alpha_0 + \alpha_1\theta + \alpha_2\theta^2),\\
&\alpha_i \in S = ((\mathbb{Q}\varsigma_3)(\sqrt[3]{\gamma}),\rho,g)\ i = 0,1,2,\\
&\rho : \sqrt[3]{\gamma} \to \varsigma_3(\sqrt[3]{\gamma}).
\end{aligned}$$

so that

$$k^\sigma = (\alpha_0 + \alpha_1\varsigma_3\theta + \alpha_2\varsigma_3^2\theta^2).$$

Since $k^\sigma k = kk^\sigma$, we claim that $\alpha_i\alpha_j = \alpha_j\alpha_i$. To see this, compute first kk^σ, to obtain:

$$\begin{aligned}
&(\alpha_0^2 + \alpha_1\alpha_2\varsigma_3^2\delta + \alpha_2\alpha_1\varsigma_3\delta) +\\
&+ (\alpha_0\alpha_1\varsigma_3 + \alpha_1\alpha_0 + \alpha_2^2\varsigma_3^2\delta)\theta +\\
&+ (\alpha_0\alpha_2\varsigma_3^2 + \alpha_2\alpha_0 + \alpha_1^2)\theta^2.
\end{aligned}$$

Computing $k^\sigma k$, we obtain then

$$\begin{aligned}&(\alpha_0^2 + \alpha_2\alpha_1\varsigma_3^2\delta + \alpha_1\alpha_2\varsigma_3\delta)+\\&+(\alpha_1\alpha_0\varsigma_3 + \alpha_0\alpha_1 + \alpha_2^2\varsigma_3^2\delta)\theta+\\&+(\alpha_2\alpha_0\varsigma_3^2 + \alpha_0\alpha_2 + \alpha_1^2\varsigma_3)\theta^2.\end{aligned}$$

Equating the coefficients on the $1, \theta$, and θ^2 terms of the previous two expressions, we first see, using the 1-terms, that

$$\alpha_1\alpha_2\varsigma_3^2\alpha_2\alpha_1\varsigma_3^2\delta + \alpha_1\alpha_2\varsigma_3\delta + \alpha_2\alpha_1\varsigma_3\delta = \alpha_2\alpha_1\varsigma_3^2\delta + \alpha_1\alpha_2\varsigma_3\delta$$

if and only if

$$\alpha_1\alpha_2(\varsigma_3^2 - \varsigma_3) = \alpha_2\alpha_1(\varsigma_3^2 - \varsigma_3),$$

and since $(\varsigma_3^2 - \varsigma_3) \neq 0$, we have $\alpha_1\alpha_2 = \alpha_2\alpha_1$. Similarly, we see from the θ-coefficients that

$$\alpha_0\alpha_1(\varsigma_3 - 1) = \alpha_1\alpha_0(\varsigma_3 - 1),$$

so that $\alpha_0\alpha_1 = \alpha_1\alpha_0$, from the θ^2-coefficient that

$$\alpha_0\alpha_2(\varsigma_3^2 - 1) = \alpha_2\alpha_0(\varsigma_3^2 - 1),$$

so $\alpha_0\alpha_2 = \alpha_2\alpha_0$. This shows that $\alpha_i\alpha_j = \alpha_j a_i$ for all $i, j = 0, 1, 2$. So, if $k^\sigma k = \theta^\sigma$, we see that

$$\begin{aligned}(i) &: \alpha_0^2 + \alpha_1\alpha_2(\varsigma_3 + 1)\varsigma_3\delta = 0,\\(ii) &: (\alpha_0\alpha_1(\varsigma_3 + 1) + \alpha_2^2\varsigma_3^2\delta)\theta = \varsigma_3\theta\\(iii) &: (\alpha_0\alpha_2(\varsigma_3^2 + 1) + \alpha_1^2\varsigma_3)\theta^2 = 0.\end{aligned}$$

Using the first and the third equations, we see that

$$\begin{aligned}\alpha_0^2 &= -\alpha_1\alpha_2(\varsigma_3 + 1)\varsigma_3\delta,\\\alpha_1^2 &= -\alpha_0\alpha_2(\varsigma_3^2 + 1)\varsigma_3^2 = -\alpha_0\alpha_2((\varsigma_3 + 1)\varsigma_3.\end{aligned}$$

Multiply the first equation by α_0 and the second equation by α_1, to obtain:

$$\begin{aligned}\alpha_0^3 &= -\alpha_0\alpha_1\alpha_2(\varsigma_3 + 1)\varsigma_3\delta,\\\alpha_1^3 &= -\alpha_0\alpha_1\alpha_2((\varsigma_3 + 1)\varsigma_3.\end{aligned}$$

Assuming that $\alpha_0\alpha_1\alpha_2 \neq 0$. Then, since $\alpha_i\alpha_j = \alpha_j\alpha_i$, we have

$$(\alpha_0\alpha_1^{-1})^3 = \delta.$$

We note that $(\alpha_0\alpha_1^{-1}) \in S$ so that $\sqrt[3]{\delta} \in S$, a contradiction to our assumptions, which is a contradiction to $x^3 - \delta$ irreducible over S. Therefore, $\alpha_0\alpha_1\alpha_2 = 0$. If $\alpha_0 = 0$, then we see by (i) that $\alpha_1\alpha_2 = 0$, and $\alpha_0 = \alpha_1 = 0$, so that $\alpha_2^2\varsigma_3\delta = 1$. Therefore, $\delta = (\varsigma_3\alpha_2^{-1})^2$. Since δ may be chosen as a non-square in $(\mathbb{Q}\varsigma_3)$, there is a Galois quadratic extension subfield $(\mathbb{Q}\varsigma_3)(\sqrt{\delta})$, but since $\sqrt{\delta} = \pm(\varsigma_3\alpha_2^{-1})$, and α_2 is an element of S, there is a corresponding subdivision ring B of S corresponding to $(\mathbb{Q}\varsigma_3)(\sqrt{\delta})$, and using the matrix representation of S, this subring is obtained by taking the first column as $\begin{bmatrix} \beta_0 \\ \beta_1 \\ \beta_2 \end{bmatrix}$ for all $\beta_i \in (\mathbb{Q}\varsigma_3)(\sqrt{\delta})$, $i = 0, 1, 2$ within $(\mathbb{Q}\varsigma_3)(\sqrt[3]{c})$. However, this cannot occur since the dimension of the superfield is 3, and the subfield has dimension 2. In particular, therefore, we have a contradiction, which completes the proof. But, we shall continue with our long proof, as outlined above.

Here is a second proof that is more traditional and uses other methods.

Step (ii): Let $l = \varsigma_3\alpha_2^{-1}$ be an element of S. Since S is a cyclic division ring, let $\{1, w, w^2\}$ denote the special symbols, where $w^3 = g$, and elements of S have the following form:

$$l = (\beta_0 + \beta_1 w + \beta_2 w^2),$$

where $w\alpha = \alpha^{\varrho} w$, and β_i in $(\mathbb{Q}\varsigma_3)(\sqrt[3]{\gamma})$ (see, e.g., [**34**]). Work out $l^2 = \delta$, to obtain the following:

$$l = (\beta_0^2 + (\beta_1\beta_2^{\varrho^2} + \beta_2\beta_1^{\varrho})g +$$

$$+(\beta_1(\beta_0^{\varrho} + \beta_0) + \beta_2^{1+\varrho^2} g)w +$$

$$+(\beta_2(\beta_0 + \beta_0^{\varrho^2}) + \beta_1^{1+\varrho})w^2.$$

This is an alternative form for the matrix version:

$$\begin{bmatrix} \beta_0 & \beta_2^{\varrho} g & \beta_1^{\varrho^2} g \\ \beta_1 & \beta_0^{\varrho} & \beta_2^{\varrho^2} g \\ \beta_2 & \beta_1^{\varrho} & \beta_0^{\varrho^2} \end{bmatrix}.$$

Using the matrix notation, we notice that $\{1, w, w^2\}$ forms a basis, and since a matrix form equal to zero means the zero vector, we see that the alternative w-form $l^2 = \delta$ shows that the 1-coefficient is δ, and the w and w^2-coefficients are both 0.

Then we have:

$$(1) : \beta_0^2 + (\beta_1\beta_2^{\varrho^2} + \beta_2\beta_1^{\varrho})g = \delta,$$

$$(2) : \beta_1(\beta_0^{\varrho} + \beta_0) = -\beta_2^{1+\varrho^2} g,$$

$$(3) : \beta_2(\beta_0 + \beta_0^{\varrho^2}) = -\beta_1^{1+\varrho}.$$

Apply the automorphism ϱ to the third equation to obtain

$$(4) : \beta_2^{\varrho}(\beta_0^{\varrho} + \beta_0) = -\beta_1^{\varrho+\varrho^2} = 0.$$

Multiply the second equation (2) by β_2^{ϱ} and the fourth equation (4) by β_1 to obtain the following two equations:

$$\beta_2^{\varrho}\beta_1(\beta_0^{\varrho} + \beta_0) = -\beta_2^{1+\varrho+\varrho^2} g = \beta_1\beta_2^{\varrho}(\beta_0^{\varrho} + \beta_0) = -\beta_1^{1+\varrho+\varrho^2}.$$

All coefficients are in a field, so there are two cases: $\beta_1 = 0$ if and only if $\beta_2 = 0$, or both terms are non-zero, and we have

$$g = (\beta_1\beta_2^{-1})^{1+\varrho+\varrho^2}.$$

But this means that g is in the norm of S, contrary to the definition of a cyclic division ring. Therefore, $\beta_1 = \beta_2 = 0$. Therefore, we have

$$\beta_0^2 = \delta.$$

To be certain that this violates the non-square assumption, we then check the last situation, as β_0 is in $(\mathbb{Q}\varsigma_3)(\tau)$. □

Step (iii): Let

$$\beta_0 = \gamma_0 + \gamma_1\tau + \gamma_2\tau^2,$$

where γ_i are in $(\mathbb{Q}\varsigma_3)$. Then consider β_0^2, to obtain the following:

$$\beta_0^2 = (\gamma_0^2 + 2\gamma_1\gamma_2 g) + (2\gamma_0\gamma_1 + \gamma_2^2 g)\tau + (2\gamma_0\gamma_2 + \gamma_1^2)\tau^2.$$

Therefore,

$$(*)_1 : (\gamma_0^2 + 2\gamma_1\gamma_2 g) = \delta,$$

$$(*)_2 : 2\gamma_0\gamma_1 = -\gamma_2^2 g,$$

and

$$(*)_3 : 2\gamma_0\gamma_2 = -\gamma_1^2.$$

If the characteristic is 2, then γ_1 and γ_2 are both zero, so that δ is a square in $(\mathbb{Q}\varsigma_3)$, a contradiction. When the characteristic is not 2, multiply the equation $(*)_2$ by γ_2 and equation $(*)_3$ by γ_1 and equate the resulting two equations to obtain $\gamma_2^3 g = \gamma_1^3$. There are two conclusions: Both terms are zero, which cannot occur by $(*)_1$, and the condition that δ is non-square, or $g = (\gamma_1\gamma_2^{-1})^3$. But, since $\gamma_1\gamma_2^{-1}$is fixed by ϱ, then $(\gamma_1\gamma_2^{-1})^3 = (\gamma_1\gamma_2^{-1})^{1+\varrho+\varrho^2}$, so this would place g in the norm again, a contradiction. This completes the proof of the theorem.

This theorem may be extended infinitely many times. The result supporting the previous theorem is based on a geometric version, but essentially by use of the tensor product. But we may also create alternative division rings using central cyclic extensions using either algebraic extensions or transcendental extensions. We will illustrate both methods.

The algebraic extension method: Now, continuing with the setting of the previous result, we may choose an infinite set of elements of $\delta = \delta_1, \delta_2, ..., \delta_k, ...$ such that $x^3 - \delta_i$ is irreducible over $(\mathbb{Q}\varsigma_3)$ and such that $(\mathbb{Q}\varsigma_3))(\sqrt[3]{\delta_i})$ for all $i = 1, 2, ...,$ are mutually non-isomorphic cyclic Galois extensions of $(\mathbb{Q}\varsigma_3)$ and such that δ_i are all non-square in $(\mathbb{Q}\varsigma_3)$. Let $\theta_i = \sqrt[3]{\delta_i}$. The intention is to provide a proof that shows that there is a set of semifield spreads within $PG(3, S(\theta_1, \theta_2, ...\theta_l))$ for all integers $l = 1, 2, ...$ Let σ_i denote the corresponding automorphisms of the series of field extensions. In order to prove that a semifield plane is obtained for the continuing extensions of division rings, we would need to prove at each step that $k^{\sigma_i}k$ is never $\varsigma_3\theta_i$. Notice also that at each step, there is a new center, so that elements of division will always commute with the current θ_i. The proof is by mathematical induction, with step 1 of the induction proof being the previous theorem. Thus, assume true for r, and we consider the step (i) in the previous proof. In the present context, we would assume that

$$k^{\sigma_i} = \varsigma_3\theta_i.$$

Write

$$S = S_0, S_0(\theta_1) = S_1, ..., S_j(\theta_{j+1}) = S_{j+1}, ...$$

Then

$$S_0 \otimes_{Z(S_0)} Z(S_0)(\theta_1), ..., S_{j+1} \otimes_{Z(S_j)} Z(S_j)(\theta_{j+1}).$$

What is occurring is that

$$S = ((\mathbb{Q}\varsigma_3)(\tau)/(\mathbb{Q}\varsigma_3), \rho, g),$$

is being extended to

$$S(\theta) = ((\mathbb{Q}\varsigma_3)(\theta, \tau)/(\mathbb{Q}\varsigma_3)(\theta), \rho, g).$$

By geometric methods or by algebraic methods using tensor products. At stage r, we would have the assumed division ring S_r of degree 3 as

$$S_r = ((\mathbb{Q}\varsigma_3)(\theta_1, \theta_2, ..., \theta_r, \tau)/(\mathbb{Q}\varsigma_3)(\theta_1, \theta_2, ..., \theta_r), \rho, g).$$

To prove that induction stage r implies stage $r+1$, we are trying to prove that the following matrix set

$$x = 0, y = x \begin{bmatrix} u^{\sigma_{r+1}} & \theta_{r+1}t^{\sigma_{r+1}^{-1}} \\ t & u \end{bmatrix}$$

for all t, u in $S_r(\theta_{r+1})$, defined a quasifibration, and hence a semifield spread in $PG(3, S_r(\theta_{r+1}))$.

Refer back to step (i) in the previous theorem, but now in this situation: $k = (\alpha_0 + \alpha_1\theta_{r+1} + \alpha_2\theta_{r+1}^2), \alpha_i \in S_r$, $i = 0, 1, 2, \rho$: so that $k^\sigma = (\alpha_0 + \alpha_1\varsigma_3\theta_{r+1} + \alpha_2\varsigma_3^2\theta_{r+1}^2)$. Follow the previous steps to show that δ_{r+1} is a square in S_r. Then note the next step will show that δ_{r+1} is a square in the associated field $(\mathbb{Q}\varsigma_3)(\theta_1, \theta_2, ..., \theta_r, \tau)$. Note that this is a Galois sequence of dimension 3 cyclic Galois extensions, so that the degree over the base field $(\mathbb{Q}\varsigma_3)$ is 3^{r+1}. But, δ_{r+1} is a non-square in $(\mathbb{Q}\varsigma_3)$ and therefore the generated quadratic extension field is a subfield of $(\mathbb{Q}\varsigma_3)(\theta_1, \theta_2, ..., \theta_r, \tau)$, a contradiction. And, therefore, we have proved the following theorem.

THEOREM 35. *Using the notation established above,*

$$x = 0, y = x \begin{bmatrix} u^{\sigma_k}, & \theta_k t^{\sigma_k^{-1}} \\ t & u \end{bmatrix}$$

for all t, u in $S_{k-1}(\theta_k)$, is a semifield spread with spread components in $PG(3, S_{k-1}(\theta_k)$. And, where

$$S = S_0, S_0(\theta_1) = S_1, ..., S_j(\theta_{j+1}) = S_{j+1}, ...$$

$$\textit{Then } S_0 \otimes_{Z(S_0)} Z(S_0)(\theta_1), ..., S_{k-1} \otimes_{Z(S_{k-1})} Z(S_{k-1})(\theta_k).$$

We now use the transcendental tensor product to basically prove a variation of the same theorem.

Under the assumptions Let $F = (\mathbb{Q}\varsigma_3)$ and $S \otimes_F F(z_1, z_2, z_3)$, which is a division ring with center $F(z_1, z_2, z_3)$, where z_i are commuting in determinants. Let $K_1 = Fix\sigma_1$, where $(z_1 z_2 z_3) = \sigma_1$. Let $K_1 = Fix\sigma_1$, and let $S_1 = S \otimes_F K_1$, and form $S_1 \otimes_{K_1} K_1(z_1)$, and write $K_1(z_1) = K_1(\sqrt[3]{\delta})$, where δ is non-square, which is possible with $F = (\mathbb{Q}\varsigma_3)$. Consider now $F(z_1, z_2, z_3)$, which is Galois over $Fix\sigma_1$. We may repeat this idea with an infinite sequence of commuting indeterminate $z_1, z_2, z_3, ...$, in sequences of of three.

Using the permutation

$$(z_1 z_2 z_3), (z_4 z_5 z_6), ..., (z_{1+n}, z_{2+n}, z_{3+n}),$$

at each 3-interval as defining an automorphism σ_i over $Fix\sigma_i$. Moreover, we obtain an Abelian automorphism group of order 3^n, and a sequence of cyclic Galois extensions over $Fix \oplus_{i=1}^n \{\sigma_i\}$. By changing the associated fields to reflect the location of $Fix\sigma_i$, we see that it is possible to create an infinite class of a sequence of division rings, just as in the previous theorem. Although this is simply a variation of the previous theorem, the idea of using tensor products of transcendental is a useful one.

PROBLEM 36. Work up what the actual theorem would look like using the tensor product of transcendental elements.

REMARK 37. So, in the previous theorems, we established only that the matrix sets define quasifibrations. The covering condition for spreads follows from the fundamental surjective result of quasifibrations. However, in this setting, a more direct result may be obtained. Let (x, y) be any vector such that neither x or y are identically zero. To cover (x, y) by a component $y = x \begin{bmatrix} u^{\sigma} & \theta t^{\sigma^{-1}} \\ t & u \end{bmatrix}$, let $x = (x_1, x_2)$ and $y = (y_1, y_2)$, where x_i and y_i are in $S(\theta)$. Write out the equations indicated by

$$(*)_1 \; : x_1 u^{\sigma} + x_2 t = y_1,$$
$$(*)_2 : x_1 \theta t^{\sigma^{-1}} + x_2 u = y_2.$$

Assume that $x_1 x_2 \neq 0$, otherwise the solutions are immediate. Form

$$x_1^{-1}(*)_1 - (x_2^{-1}(*)_2)^{\sigma},$$

then factor to obtain the following equation:

$$(((x_1^{-1}x_2) - (x_2^{-1}x_1)^{\sigma}\theta^{\sigma})t) = x_1^{-1}y_1 - (x_2^{-1}y_2)^{\sigma}.$$

Letting $(x_1^{-1}x_2) = k$, then, since $k^{\sigma}k \neq \theta^{\sigma}$, we may solve uniquely for t, which then shows that there is a unique solution for u. So, we have the quasifibration property and the surjective property simultaneously. Therefore, we have a semifield plane spread with components in $PG(3, S(\theta))$.

12.1. Quartic extensions

In this section, we show the validity of the construction of semifields with spreads in $PG(3, S(\theta))$, where S is a skewfield and $S(\theta)$ is a central Galois cyclic extension of degree 4 over S.

$$x = 0, y = x \begin{bmatrix} u^{\sigma}, & \theta t^{\sigma^{-1}} \\ t & u \end{bmatrix}; t, u \in S(\theta), \text{ where}$$
$$\sigma \text{ is an induced automorphism of order } n = 4.$$

Using the method of the previous material, we need to show that $k^{\sigma}k \neq \theta^{\sigma}$. So, assume that

$$k^{\sigma}k = \varsigma_4\theta = i\theta.$$

Since, again we have $k^{-1}k^{\sigma}k = k^{\sigma}kk^{-1} = k^{\sigma}$, since $i\theta$ is in the center, which implies that $kk^{\sigma} = k^{\sigma}k$. Letting $k = \alpha_0+\alpha_1\theta+\alpha_2\theta^2+\alpha_3\theta^3$, and $\theta^4 = \delta$, for α_i in S, we have the following equations:

$$\begin{aligned}&(\alpha_0 + \alpha_1\theta + \alpha_2\theta^2 + \alpha_3\theta^3)(\alpha_0 + \alpha_1 i\theta + \alpha_2 i^2\theta^2 + \alpha_3 i^3\theta^3)\\&= (\alpha_0 + \alpha_1\theta + \alpha_2\theta^2 + \alpha_3\theta^3)(\alpha_0 + \alpha_1 i\theta - \alpha_2\theta^2 - \alpha_3 i\theta^3)\\&= (\alpha_0 + \alpha_1 i\theta - \alpha_2\theta^2 - \alpha_3 i\theta^3)(\alpha_0 + \alpha_1\theta + \alpha_2\theta^2 + \alpha_3\theta^3).\end{aligned}$$

The 1-term coefficient is

$$(i)_{kk^\sigma} : (\alpha_0^2 - \alpha_2^2\delta) - \alpha_1\alpha_3 i\delta + \alpha_3\alpha_1 i\delta) = 0$$

also, we then have

$$(i)_{k^\sigma k} : (\alpha_0^2 - \alpha_2^2\delta) + \alpha_1\alpha_3 i\delta - \alpha_3\alpha_1 i\delta = 0.$$

Using these last two equations, we see that $2\alpha_1\alpha_3 = 2\alpha_3\alpha_1$, and as the characteristic is not 2, we have $\alpha_1\alpha_3 = \alpha_3\alpha_1$. Using this identity in $(i)_{kk^\sigma}$ shows that

$$\alpha_0^2 = \alpha_2^2\delta.$$

Recalling that $x^4 - \delta$ is S-irreducible (and $Z(S)$-irreducible), we see that δ is a non-square in S. Therefore, $\alpha_0^2 = 0 = \alpha_2^2$, implying that $\alpha_0 = \alpha_2 = 0$. We now consider that the θ-term of kk^σ must be $i\theta$.

$$(\alpha_1\theta + \alpha_3\theta^3)(\alpha_1 i\theta - \alpha_3 i\theta^3) = (\alpha_1 i\theta - \alpha_3 i\theta^3)(\alpha_1\theta + \alpha_3\theta^3),$$

which is

$$(\alpha_1^2 i - \alpha_3^2\delta)\theta^2,$$

using $\alpha_1\alpha_3 = \alpha_3\alpha_1$.

Note that there are only 1-terms and θ^2-terms in these expressions, a contradiction. Note that the analogous result $kk^{\sigma^{-1}} \neq \theta^3$ also follows. Therefore, we have proved the following theorem.

THEOREM 38. *Let S be a skewfield and let $S(\theta)$ be a central Galois cyclic extension of S of degree 4, and assume that $Z(S)$ is a field of characteristic $\neq 2$, with polynomial $x^4 - \delta$ that is irreducible over $Z(S)$ and S, for $\delta \in Z(S)$. Assume that $Z(S)$ contains the element $i = \sqrt{-1}$. Let σ denote the associated automorphism of order 4 induced by the extension. Then the following are skewfield spreads in $PG(3, S(\theta))$:*

$$x = 0, y = x\begin{bmatrix} u^\sigma & ,\theta^j t^{\sigma^{-1}} \\ t & u \end{bmatrix}; \text{ for all } t, u \in S(\theta) \text{ for } j = 1 \text{ or } 3.$$

PROBLEM 39. In this case, the same fundamental infinite extension of these semifield spreads may be obtained, analogous to the case when $n = 3$. Apply the ideas of the result for extensions of degree 3 in the present setting for degree 4. State the extension theorem as completely as possible.

Try this sort of problem for $n = 5$. Generalize to obtain a corresponding skewfield spread, using a cyclic division ring of degree 5 over a field F.

CHAPTER 13

Central cyclic extensions

In [**34**], we again restate the following theorem.

THEOREM 40. ***(Skew field extension theorem for division rings of degree*** *n.) Let F be a field that contains a primitive n-th root of unity, and so that the characteristic of F does not divide n. Let S be a division ring of degree n with center F that contains a field $F(\theta)$, which is a Galois cyclic field extension of dimension n over F. Assume S induces a subplane covered net on $PG(2n-1, F(\theta,\tau))$, where $F(\theta,\tau)$ is an Abelian central field Galois extension of dimension n^2, the composition of two non-isomorphic cyclic extensions of dimension n, containing $F(\theta)$ and $F(\tau)$: (1) Then S coordinates a matrix skewfield of n by n matrices over $F(\theta)$, defining a left subplane covered net as a suspect subplane covered net relative to the classical (left/right) regulus net over $F(\theta,\tau)$. (2) Also, there is a central division ring extension of S over F, which is a division ring $S(\tau)$ such that $Z(S(\tau)) = F(\tau)$, that coordinates a spread in $PG(2n-1, F(\theta,\tau))$ and a spread in $PG(2n-1, S)$. (3) This $S(\tau)$ central division ring extension defines a Desarguesian translation plane with left spread in $PG(2n-1, F(\theta,\tau))$. Additionally, the dual translation plane may be written in $PG(2n-1, S)$.*

PROPOSITION 41. *Jacobson (1.9.2)* [**18**]*(note in the statement in* [**34**]*, it should be "division algebra" instead of "central division algebra". Let E be a cyclic Galois field extension of F of dimension n with Galois group $\langle\sigma\rangle$. Extend E to $E(z)$, where z is an indeterminate and extend σ to $E(z)$ by defining $\sigma z = z$. Then the cyclic algebra $S = (E(z), \sigma, z)$ is a division algebra.*

In the first volume, the author used the idea of the previous proposition to prove the following result

THEOREM 42. *(*[**34**]*, Theorem 161, and Corollary 50) Given any finite field $F = GF(q)$, let $g(x) = x^2 \pm \alpha x - \beta$, for $\alpha, \beta \in GF(q)$, denote the irreducible quadratic polynomial that extends F to $GF(q^2) = GF(q)(\theta)$. Let σ denote the associated automorphism $x \to x^q$. For uniformity, assume that $\alpha \neq 0$. Form the rational function field $F(z)(\theta)$. Extend σ to $F(\theta)(z)$ by allowing that $\sigma z = z$. Let $f(z) \in F[z]$, where*

DOI: 10.1201/9781003767978-13

the degree of $f(z)$ is an odd integer. Then

$$\begin{bmatrix} u^\sigma & f(z)t^\sigma \\ t & u \end{bmatrix}; \forall t, u \in F(z)(\theta),$$

is a quaternion division ring $(\frac{\beta}{\alpha^2}, f(z))_{GF(q)(z)}^{GF(q^2)(z)}$.

COROLLARY 43. *We note that the result above is not stated in the most general form, as any Galois quadratic extension E of F will provide a quaternion division ring $(\frac{\beta}{\alpha^2}, f(z))_{F(z)}^{E(z)}$, for $f \in F[z]$ of odd dimension and σ, the induced automorphism of E (note that there is a misprint in* [**34**], *stating that $f \in E[z]$, but f is required to be in $F(z)$ to have a division ring).*

More generally, the result of Jacobson is also not stated in the most general form, at least when $n = 2$ or 3. This extension is as follows:

THEOREM 44. *Let F be a field that contains a primitive nth root of unity ς_n, where the characteristic does not divide n. Let E be a cyclic Galois field extension of F of dimension n with Galois group $\langle\sigma\rangle$. Extend E to $E(z)$, where z is an indeterminant, and extend σ to $E(z)$ by defining $\sigma z = z$. Let $f(z) \in F[z]$ such that the degree d of f is relatively prime to n. Then, if $n = 2$, 3, then the cyclic algebra $S = (E(z), \sigma, f(z))$ is a division algebra.*

PROOF. The proof when $f(z) = z$ given in Jacobson [**18**] is not given by using the standard verification of non-norm conditions. Here, we shall use the quasifibration approach when $f(z)$ is a polynomial of positive degree d such that $(n, d) = 1$. First, restrict to the case when $n = 3$. We consider $g = f(z) \in F[z]$ and argue that the associated set of matrices

$$\left\{ \begin{bmatrix} \alpha_0 & \alpha_2^\sigma g & \alpha_1^\sigma g \\ \alpha_1 & \alpha_0^\sigma & \alpha_2^{\sigma^2} g \\ \alpha_2 & \alpha_1^\sigma & \alpha_0^{\sigma^2} \end{bmatrix}; \forall \alpha_0, \alpha_1, \alpha_2 \in E(z) = F(z)(\theta) = F(\theta)(z). \right\}$$

is a quasifibration, which shows that we have a cyclic division ring. This method seems reasonable when n is small but might not work well when n is much larger. □

Since we have an additive set, we need only show that the determinants are all non-zero in the non-trivial case. The determinant is as follows:

$$(*) : \alpha_0^{1+\sigma+\sigma^2} + \alpha_1^{1+\sigma+\sigma^2} g + \alpha_2^{1+\sigma+\sigma^2} g^2 +$$
$$- ((\alpha_0 \alpha_1^\sigma \alpha_2^{\sigma^2}) + (\alpha_0 \alpha_1^\sigma \alpha_2^{\sigma^2})^\sigma + (\alpha_0 \alpha_1^\sigma \alpha_2^{\sigma^2})^{\sigma^2}) g.$$

Since $\alpha_0, \alpha_1, \alpha_2 \in F(z)(\theta)$, we see that all coefficients are in F (we call the negative coefficient of g, the 'middle term'). Write the elements $\alpha_0, \alpha_1, \alpha_2$ as elements of $F(\theta)(z)$ and set the determinant to 0. Form a common denominator and write the resulting equation as a polynomial set to zero. Then all coefficients are equal to zero. Let $\alpha_i = \frac{\delta_i}{\beta_i}$, where δ_i and $\beta_i \in F(\theta)[z]$. Take a common denominator of $(\beta_0\beta_1\beta_2)^{1+\sigma+\sigma^2}$. Note that if $\alpha_{i_0} = 0$ for some i_0 then there is no middle term. Let the degrees of $\delta_0, \delta_1, \delta_2$ be a_0, a_1, a_2 and the degrees of $\beta_0, \beta_1, \beta_2$ be b_0, b_1, b_2, respectively. We assume that all coefficients on the z^{c_i} for $c_i = a_i$ or b_i, $i = 0, 1, 2$ are non-zero.

Let's take the common denominator, clear terms to obtain a polynomial in $F(z)$ equal to zero, which implies that the maximum coefficient is 0. We look at the four resulting terms in $(*) = 0$.

$$(*) : ((\delta_0\beta_1\beta_2)^{1+\sigma+\sigma^2} + (\delta_1\beta_0\beta_2)^{1+\sigma+\sigma^2} g + (\delta_2\beta_0\beta_1)^{1+\sigma+\sigma^2} g^2 +$$
$$- ((\delta_0\delta_1^{\sigma}\delta_2^{\sigma^2}\beta_0^{\sigma+\sigma^2}\beta_1^{1+\sigma^2}\beta_2^{1+\sigma}) + ((\delta_0\delta_1^{\sigma}\delta_2^{\sigma^2}\beta_0^{\sigma+\sigma^2}\beta_1^{1+\sigma^2}\beta_2^{1+\sigma})^{\sigma} +$$
$$(\delta_0\delta_1^{\sigma}\delta_2^{\sigma^2}\beta_0^{\sigma+\sigma^2}\beta_1^{1+\sigma^2}\beta_2^{1+\sigma})^{\sigma^2})g.$$

We note that the degrees of the four terms, in order, left to right, as follows:

$$(3a_0 + 3(b_1 + b_2), 3a_1 + 3(b_0 + b_2) + d, 3a_2 + 3(b_0 + b_1) + 2d,$$
$$a_0 + a_1 + a_2 + 2(b_0 + b_1 + b_2) + d.$$

We may assume that the total middle term is not zero, as otherwise, no pair of the first three terms can be equal. And, no one of these can be largest than all four terms, since the maximum coefficients are not all zero. Therefore, since the sum of the three middle terms is not zero, the associated degree of the middle term must be equal to the degree of exactly one of the first three terms, and the other two terms must have degrees strictly less than the degree of the middle term.

Hence, the sum of the degrees of the first three terms must be strictly less than 3 times the degree of the middle term. We note that the sum of the first three terms is exactly equal to 3 times the degree of the middle term, which implies that $0 < 0$, a contradiction:

$$(3a_0 + 3(b_1 + b_2) + 3a_1 + 3(b_0 + b_2) + d + 3a_2 + 3(b_0 + b_1) + 2d =$$
$$= 3(a_0 + a_1 + a_2 + 2(b_0 + b_1 + b_2) + d).$$

Therefore, the middle term has term zero. But then, the same argument as above shows that one of the α_i has a zero term, but since the terms have at least d degree, we have that two and then three of the α_i are identically zero. This completes the proof of the theorem when $n = 3$.

PROBLEM 45. For $n = 4$, the division ring $S = (E(z), \sigma, z)$ exists. Extend the previous result to replace z by $f(z)$, with the same conditions. It may work for arbitrary $4n$, does it?

13.1. Cyclic division rings

Let S be a cyclic division ring of dimension n over

$$Z(S) = (\mathbb{Q}\varsigma_3)(z), S = ((\mathbb{Q}\varsigma_3)(z)(\tau), \sigma, f(z))$$

of degree 3, where σ is the induced automorphism of order 3. We realize that $x^3 - z$ is irreducible over $Z(S)$, so that we obtain a cyclic Galois extension of $Z(S)$, $Z(S)(\sqrt[3]{z})$. Using the extension of skewfields theorem for cyclic division rings from [**34**], we may extend S to $S(\sqrt[3]{z})$, which is a central Galois extension of degree 3. There is an associated automorphism group $\langle\rho\rangle$, such that $\rho : \sqrt[3]{z} \to \varsigma_3\sqrt[3]{z}$.

PROPOSITION 46. *Since z is non-square, we have a variety of derivable semifield spreads of dimension 2. Therefore, the components are left $S(\sqrt[3]{z})$-subspaces in $PG(3, S(\sqrt[3]{z})$. The set of Baer subspaces are right $S(\sqrt[3]{z})$-subspaces. The derivable nets are type 2 pseudo-regulus nets.*

COROLLARY 47. *There are infinitely many extensions in two or three ways. For example, take any prime number p; then $x^3 - p$ is irreducible over $\mathbb{Q}\varsigma_3$, so*

$$S = (\mathbb{Q}\varsigma_3(z)(\sqrt[3]{p}), \sigma_p, f_p(z)),$$

where σ_p is induced and $f_p(z)$ is any element of $\mathbb{Q}\varsigma_3[z]$ of degree relatively prime to 3. Then one can extend S to $S(z_1 = z, z_2, ..., z_k)$, where $\sigma_p z_i = z_i$, for commuting and independent variables, thus extending the center. Then choose any z_i and extend by $\sqrt[3]{z_i}$ to create a variety of central Galois extensions of dimension 3. Of course, extensions by $\sqrt[3^j]{z_i}$ may also be used to create more variations of division rings that construct classes of semifield planes of dimension 2.

13.2. Iterations-central extensions

There are various examples of cyclic division rings of dimension 3 of the form

$$S = ((\mathbb{Q}\varsigma_3)(\sqrt[3]{b}), \sigma_p, \varsigma_3)$$

constructed in [**34**]. There are many examples of central extensions of the form

$$(\mathbb{Q}\varsigma_3)(\sqrt[3]{b}), \sqrt[3]{c},,, \sqrt[3]{d}), \sigma_b, \varsigma_3)$$

that extend the center using the same automorphism σ_b or change to the automorphism induced from $x^3 - c$. If the b, c, d, etc. are in $\mathbb{Q}$, we can choose a set of mutually non-isomorphic extensions, then all of the cyclic division rings are non-isomorphic, and the extensions are central. Since by the previous lemma, we may assume that all elements b, c, d are always non-square, then there are new examples of derivable semifield planes of dimension 2, where the derivable nets are non-commutative pseudo-regulus nets of type 2.

One can use the results of [**34**] to construct infinitely many cyclic division rings of dimension 3 that have a 3-tower of fields of dimension 3, such that with the skewfield extension theorem, there are infinitely many mutually non-isomorphic central extensions of dimension 3 of cyclic division rings of dimension 3. It is always possible to construct elements $\mathbb{Q}(\varsigma_3)(\sqrt[3]{b_i})$, where b_i is not a square (see [**34**], e.g., Theorem 163). One exceptionally powerful method of Grover and Ling [**16**], also [**44**] constructs cyclic division rings of degree 3, $(\mathbb{Q}(\varsigma_3)(\sqrt[3]{b_i}), \sigma, \varsigma_3)$ is as follows: There are infinitely many primes t_i so that 3 divides $t_i - 1$, but 3^2 does not divide $t_i - 1$ (Dirichlet's Theorem on arithmetic progressions). Form $\mathbb{Q}(\varsigma_3)(\varsigma_{t_i}) = \mathbb{Q}(\varsigma_{3t})$, a Galois extension of $\mathbb{Q}$ of degree $2(t-1)$. There is a cyclic automorphism subgroup $\langle\sigma\rangle$ of order $t-1$, that defines a Galois subfield M_i of degree $(t-1)$ over $\mathbb{Q}$, where $Fix\sigma^3$ fixes $\mathbb{Q}(\varsigma_3)$ pointwise and has index 3 in $\mathbb{Q}(\varsigma_{3t})$. Hence, we have infinitely many cyclic Galois extensions of $\mathbb{Q}(\varsigma_3)$. Since all of these are central extensions, we may couple these with the skewfield extension theorem mentioned previously to create infinitely many central extensions of cyclic division rings of degree 3^2 over $\mathbb{Q}(\varsigma_3)$. Therefore, all of these extensions produce a tremendous variety of new semifield planes of dimension 2 with spreads in 3-dimensional non-commutative projective spaces. But we here we provide a more direct method to construct examples. Let $\varsigma_3 = \frac{-1+\sqrt{3}i}{2}$, a 3rd. root of unity, consider the polynomial $x^3 - \delta$, over $(\mathbb{Q}\varsigma_3)$, where δ is a positive number in $\mathbb{Q}$, for $\mathbb{Q}$ the field of rational numbers. Since $(\mathbb{Q}\varsigma_3)$ has degree 2 as the cyclotomic extension of $\mathbb{Q}$, which has a basis as $\{1, \varsigma_3\}$. We may then assume that the cyclotomic extension of $\mathbb{Q}$ has the form $\alpha + \beta\sqrt{3}i$, where $\alpha, \beta \in \mathbb{Q}$ and

$$(\alpha + \beta\sqrt{3}i)^3 = \alpha^3 + 3\alpha^2\beta\sqrt{3}i - 3\alpha(\beta^2 3) - \beta^3(3^{\frac{3}{2}}i) = \delta,$$

which shows that either $\beta = 0$ or $\alpha = \pm\beta$ and $\delta = \mp 8\beta^3$, where γ is a positive rational number. It is easily seen using prime decomposition that we have a contradiction. Hence, $x^3 - \delta$, for δ a positive prime is irreducible over $(\mathbb{Q}\varsigma_3)$. Therefore, $(\mathbb{Q}\varsigma_3)(\sqrt[3]{\delta})$ is a cyclic Galois field of dimension 3 over $(\mathbb{Q}\varsigma_3)$. Actually, if δ is an integer $\neq (2\gamma)^3$, for any positive integer γ, then $x^3 - \delta$ is irreducible. First, choose δ to be any positive integer not equal to $(2\gamma)^3$, then choose ρ any non-square positive integer such that $(\mathbb{Q}\varsigma_3)(\sqrt[3]{\delta})$ and $(\mathbb{Q}\varsigma_3)(\sqrt[3]{\rho})$ are mutually non-isomorphic cyclic Galois extensions of $\mathbb{Q}\varsigma_3$ of degree 3. Then there are infinitely many non-mutually central cyclic Galois extensions of degree 3. Given any integer δ not equal to $(2\gamma)^3$, then $x^3-\delta$ is irreducible and there is a generator $\sqrt[3]{\delta^*}$ of $(\mathbb{Q}\varsigma_3)(\sqrt[3]{\delta})$, such that δ^* is a non-square. Let $\delta=\rho^2$; then, since $\delta^{2/3}$ is also a generator, we see that $\rho^{1/3}$ is a generator. So, it follows that there is always a generator θ, which is a non-square.

13.3. A few examples of division rings

In [**34**], there are other cyclic division rings of dimension n^2 over their center. In particular, there are several mentioned in Oggier, Belfiorï¿œ, and Viterbo [**45**], of which we consider $n=3$ and 6.

Example:

$$n=3:((\mathbb{Q}\varsigma_3)(\theta)/\mathbb{Q}(\varsigma_3),\sigma,\varsigma_3),$$
$$\text{where } \theta=2\cos(\frac{2\pi}{7})=\varsigma_7+\varsigma_7^{-1},$$
$$\sigma:\varsigma_7+\varsigma_7^{-1}\to\varsigma_7^2+\varsigma_7^{-2}.$$

Note that $\varsigma_7+\varsigma_7^{-1}$ generates the unique
cyclic group of order 3
in the cyclotomic extension $\mathbb{Q}\varsigma_7/\mathbb{Q}$.

Since 3 divides $7-1$, this construction might be related to that of Grover and Ling [**16**], but the next does not.

Example:

$$n=6:((\mathbb{Q}\varsigma_3)(\theta)/\mathbb{Q}\varsigma_3),\sigma,-\varsigma_3),$$
$$\text{where } \theta=2\cos(\frac{\pi}{14})=\varsigma_{28}+\varsigma_{28}^{-1},$$
$$\sigma:\varsigma_{28}+\varsigma_{28}^{-1}\to\varsigma_{28}^2+\varsigma_{28}^{-2}.$$

The Euler ϕ-function $\phi(28)=12$, so that
the extension $\mathbb{Q}\varsigma_{28}+\varsigma_{28}^{-1})/\mathbb{Q}$
is of degree 6.

Since we consider a cyclic subdivision ring of index 2, it is interesting that the use of $-\varsigma_3$ may be considered in the subring but seems distinct from those of the previous constructions. However, note that, by the Lagrange resolvent technique ([**34**] Proposition 10, p. 326), there are always generators of the form $\sqrt[3]{b}$, for which our non-square lemma applies.

Oggier, Belfioré and Viterbo [**45**] also provide the example.

Example:

$$n = 4 : ((\mathbb{Q}i)\,(\theta)/\mathbb{Q}(i), \sigma, i),$$
$$\text{where } \theta = 2\cos(\frac{2\pi}{15}) = \varsigma_{15} + \varsigma_{15}^{-1},$$
$$\sigma : \varsigma_{15} + \varsigma_{15}^{-1} \to \varsigma_{15}^{2} + \varsigma_{15}^{-2}.$$

Recall that $i = \varsigma_4$,

- While it may seem that there must be semifield planes of the form given in this article for perhaps all dimensions n, it appears that one is restricted to small dimensions for the sorts of arguments that are used here.
- Our method used in the cases when $n = 3, 4$ would be useful in general arguments. In the following part, we shall show a very direct argument using nth roots of unity. This idea can also provide a very brief proof of Johnson-Jha lifting skewfield theorem. The original proof of this latter theorem uses the so-called "metro equation" argument, which is important when the forthcoming A_e-addition theorem is proved in part VIII.

THEOREM 48. *Let S be a division ring. Assume that there is a central cyclic division ring $S(\theta)$. If one has a spread of the form:*

$$x = 0, y = x\begin{bmatrix} u^{\sigma} & ,\theta t^{\sigma^{-1}} \\ t & u \end{bmatrix}; t, u \in S(\theta), \textit{ where}$$

σ is an induced automorphism of order n;

choose any other generator of $\langle\theta\rangle$,

$$\Sigma_{i=0}^{n-1}\beta_i\theta^i, where \beta_i \in Z(S), for all i = 0, 1, ..., n-1,$$

the center of S, then, for each generator, there is a spread of the form

$$x = 0, y = x\begin{bmatrix} u^{\sigma}, & \Sigma_{i=0}^{n-1}\beta_i\theta^i t^{\sigma^{-1}} \\ t & u \end{bmatrix}; t, u \in S(\theta), \textit{ where}$$

σ is an induced automorphism of order n.

Therefore, there will be an infinite set of examples of semifields as well.

PROOF. Let $\Sigma_{i=0}^{n-1}\beta_i\theta^i = \lambda$. Given a representation k represented in a θ-basis, then a transformation of this representation to a λ-basis is easily provided. Hence, $kk^{\sigma^{-1}}$ in θ-basis terms is then equal to $kk^{\tau^{-1}}$, in

λ-basis terms, and τ is in the induced form related to λ. Then $kk^{\sigma^{-1}}$(in θ-form)$= kk^{\tau^{-1}}$ (in λ-form). An example, when $n = 2$ will illustrate. Let $k = \theta k_1 + k_2$, and let $\lambda = \theta\delta + \gamma$, for $\delta \neq 0$ and δ, $\gamma \in Z(S)$. Then $k = (\theta\delta + \gamma)\frac{k_1}{\delta} + k_1 - \gamma\frac{k_1}{\delta}$. It is then clear that $kk^{\sigma^{-1}}$ in one form is $kk^{\tau^{-1}}$; in the other form, which shows that $kk^{\sigma^{-1}} \neq \lambda$. Note that the constructed variations of semifield planes may not be isomorphic. This completes the proof. □

Part 5

Lifting skewsfields-degree n

Previously, we have lifted skewfields of degree 2; an analogous procedure may be obtained for any degree n.

CHAPTER 14

General lifting

In this chapter, we continue with the construction of semifield spreads over skewfields D of the following form:

$$x = 0, y = x \begin{bmatrix} u^{\sigma+gt}, & \theta t^{\sigma^{-1}} \\ t & u \end{bmatrix},$$

where g is 0 or 1 for all t, u in D, where θ is in the center K of D, and n is the order of σ. If D is a central cyclic extension of a division ring E, so $E(\theta) = D$, and σ is the induced automorphism, we assume that E contains a primitive nth root of unity and that n does not divide the characteristic of K, and n is a unit.

We give the proof in two parts. First, assume that $g = 0$. Use the $(2,1)$-determinant, and assume for some (t, u), t non-zero, the determinant is 0. This means that:

$$(u^{\sigma}t^{-1}u - \theta t^{\sigma^{-1}})t = 0.$$

Since θ is central, this works out to be the equation for $k = (u^{\sigma}t^{-1})$, that

$$kk^{\sigma^{-1}} = \theta.$$

Since θ is central, then

$$k^{-1}(kk^{\sigma^{-1}}) = (kk^{\sigma^{-1}})k^{-1}$$

which implies that

$$k^{\sigma^{-1}}k = kk^{\sigma^{-1}}.$$

Since $kk^{\sigma^{-1}} = \theta$, then $k^{\sigma}k = \varsigma_n\theta$, where ς_n is a primitive nth. root of unity. Therefore, have

$$k^{\sigma}k\varsigma_n^{-1} = \theta = kk^{\sigma^{-1}} = k^{\sigma^{-1}}k,$$

so either k is zero and we are finished, or $k^{\sigma} = k^{\sigma^{-1}}\varsigma_n$, so we have as ς_n is fixed by σ.

$$k^{\sigma^2} = k\varsigma_n.$$

Now there exists an element δ in K so that $\theta = \sqrt[n]{\delta}$, so letting

$$k = \Sigma_{i=0}^{n-1}\alpha_i\theta^i,$$

DOI: 10.1201/9781003767978-14

where α_i in E, $\theta^{\sigma} = \varsigma_n\theta$, and $\theta^{\sigma^2} = \varsigma_n^2\theta$, and using $k^{\sigma^2} = k\varsigma_n$, we have the following

$$\Sigma_{i=0}^{n-1}\alpha_i\varsigma_n^{2i}\theta^i = \varsigma_n\Sigma_{i=0}^{n-1}\alpha_i\theta^i,$$

using the basis $\{1, \theta, ..., \theta^{n-1}\}$, then

$$\alpha_i\varsigma_n^{2i} = \varsigma_n\alpha_i,$$

for all $i = 0, 1, ..., n-1$. When $n = 2^r n_1$, for n_1 odd, it follows that for $i = 0, 1, 2, ..., n-1, \varsigma_n^{2i}$ can never equal ς_n, so elements α_i are equal to 0. Using the unwrapping procedure, we have a semifield flock spread, as we then have $k = 0$, contrary to our assumptions.

When n is odd, we have exactly one element such that $i = (\frac{n+1}{2})$ that satisfies the possibility that α_i is not zero. Hence, the equation

$$k^{\sigma}k = \varsigma_n\theta$$

can only be given by

$$(\alpha_{\frac{n+1}{2}}\theta^{n+1})^{\sigma}(\alpha_{\frac{n+1}{2}}\theta^{n+1}) = \varsigma_n\theta.$$

Therefore, this becomes:

$$(\alpha_{\frac{n+1}{2}})^2\delta^2\theta = 1,$$

since δ^2 is not zero, we have $(\alpha_{\frac{n+1}{2}}) = 0$, and we have a contradiction.

Therefore, for all positive integers, we have the following result.

THEOREM 49. *If D is a central cyclic extension of a division ring E, so $E(\theta) = D$, and σ is the induced automorphism, assume that E contains a primitive nth root of unity and that n does not divide the characteristic of K, and n is a unit.*

(1) Then for any integer n,

$$x = 0, y = x\begin{bmatrix} u^{\sigma}, & \theta t^{\sigma^{-1}} \\ t & u \end{bmatrix},$$

for all t, u in D, is a semifield flock spread.

(2) All of these flock spreads are simultaneous. The conical flock is linear. The hyperbolic flock spread is linear.

(a) The conical flocks have the following form:

$$\Sigma_t:$$

$$-x_1 0 - x_2 t^{\sigma^{-1}} + x_3\theta t^{\sigma^{-1}} - x_4 = 0$$

for all i in $D(\theta)$.

(b) The hyperbolic flock spreads have the form:

$$x = 0, y = 0, y = x\begin{bmatrix} u^{\sigma} & 0 \\ 0 & u \end{bmatrix},$$

$$y = x\begin{bmatrix} t^{\sigma} & \theta \\ 1 & t \end{bmatrix}\begin{bmatrix} v^{\sigma} & 0 \\ 0 & v \end{bmatrix},$$

for all non-zero u, v and for all t in $D(\theta)$.

(c) The hyperbolic flock is

$$\Pi_t :$$

$$x_1\theta - x_2t + x_3t - x_4 = 0$$

and

$$\rho : x_2 = x_3.$$

(3) All of the transpose matrix sets are conical flock spreads. The conical flocks are nonlinear. And all of these spreads admit the left inversion property. The corresponding hyperbolic flocks are nonlinear.

(a) The transpose matrix set is a spread of the following form:

$$x = 0, y = x\begin{bmatrix} u^{\sigma}, & \theta^{-\sigma}s^{\sigma} \\ s & u \end{bmatrix}$$

for all s, u in $D(\theta)$.

(b) The associated conical flock is

$$\Sigma_s^{trans} :$$

$$-x_1 0 - x_2 s^{\sigma^{-1}} + x_3\theta^{-\sigma}s^{\sigma} - x_4 = 0,$$

for all s in $D(\theta)$, which is nonlinear.

(c) Since

$$\begin{bmatrix} v^{\sigma} & 0 \\ 0 & v \end{bmatrix}\begin{bmatrix} u^{\sigma}, & \theta^{-\sigma}s^{\sigma} \\ s & u \end{bmatrix} = \begin{bmatrix} (vu), & \theta^{-\sigma}(vs) \\ vs & vu \end{bmatrix},$$

the transpose flock spread admits the left-acting pseudo-regulus-inducing homology group. Therefore, using the inverse mapping $(x,y) \to (y, x)$, we obtain the following hyperbolic flock spread:

$$x = 0, y = 0, y = x\begin{bmatrix} u^{\sigma} & 0 \\ 0 & u \end{bmatrix},$$

and

$$y = x\begin{bmatrix} s^{\sigma} & \theta^{-\sigma} \\ 1 & s \end{bmatrix}^{-1}\begin{bmatrix} v^{\sigma} & 0 \\ 0 & v \end{bmatrix},$$

for all non-zero u and v and for all s in $D(\theta)$. Let Δ_s be the (2,1)-determinant. Then, writing out the determinant and converting the hyperbolic flock form, we obtain the following form:

$$y = x\begin{bmatrix} -s & , \Delta_s^{-1}s^{\sigma}\Delta_s^{\sigma^{-1}} - \Delta_s^{\sigma^{-1}} \\ 1 & -\Delta_s^{-1}s^{\sigma}\Delta_s^{\sigma^{-1}} \end{bmatrix}\begin{bmatrix} v^{\sigma} & 0 \\ 0 & v \end{bmatrix},$$

for all s in $D(\theta)$.

(d) The associated hyperbolic flock is

$$\Pi^{trans,left}_{-\Delta_s^{-1}s^\sigma\Delta_s^{\sigma^{-1}}}$$

$$x_1(\Delta_s^{-1}s^\sigma\Delta_s^{\sigma^{-1}}-\Delta_s^{\sigma^{-1}})-x_2(-s^\sigma)+x_3(-\Delta_s^{-1}s^\sigma\Delta_s^{\sigma^{-1}})-x_4=0,$$

and

$$\rho : x_2 = x_3.$$

(4) There are associated right flocks induced from the transposed conical flock spreads. Since θ *is central, there is a conical flock square, where the left and right flocks are the diagonals and are anti-isomorphic. Furthermore, there are associated hyperbolic flock squares with the same properties.*

PROOF. The idea of this proof originates from $n = 3$. That the Kummer theory also works with the given assumptions is given, e.g., in Dummit and Foot [**11**], p. 636, problem # 7. Also, see problem #8; when extensions can be mutually non-isomorphic over the rationals, then use $K = (\mathbb{Q}\varsigma_n)$, the cyclotomic extension. Also, using Cohn and Dicks [**7**]. Also see the treatment of Brauer groups and the idea of "generalized quaternion division rings" in [**34**], p. 57. The central extensions may be realized as central cyclic Galois extensions in terms of tensor products. In the Multiple Replacement Theorem – Redux, the idea of extension by multiple transcendental extension by tensor products is a similar idea.

To see if the transpose is a spread in the original position, use the $(1, 2)$-derivative to show if there is a (t, u),for t non-zero, such that the derivative is 0; one would have

$$t^{-\sigma^{-1}}ut^{-1}u^\sigma = \theta.$$

Letting $t^{-1}u^\sigma$, obtaining:

$$k^{\sigma^{-1}}k = \theta.$$

The argument showing that $k^{\sigma^{-1}}k = kk^{\sigma^{-1}}$ is still valid, and then $kk^\sigma = \varsigma_n\theta = \varsigma_n k^{\sigma^{-1}}k = \varsigma_n kk^{\sigma^{-1}}$. Then, canceling the k, obtains

$$k^{\sigma^2} = \varsigma_n k.$$

Then, the original argument applies to complete the proof. The rest of the results are known results from the simultaneous part. This completes the proof. □

- The procedure in the previous theorem may be used to provide a very short proof of the lifted Skewfield semifield spreads.

THEOREM 50. *Let D be a division ring of dimension n over a field F and let $D(\theta)$ be a central Galois quadratic extension of D. Then*

$$x = 0, y = x \begin{bmatrix} u^\sigma, & (\theta\delta + \gamma)t^\sigma \\ t & u \end{bmatrix}$$

where δ is non-zero and γ is in $F(\theta)$, for all t, u is a semifield spread in $PG(3, D(\theta))$, σ is the induced automorphism of order 2. Note that we are not assuming that the characteristic is not 2.

PROOF. By the previous setup we need only to show that

$$(u^\sigma t^{-1})ut^{-\sigma} = kk^\sigma$$

cannot be $(\theta\delta + \gamma)$. Note that the argument above using the centrality of θ shows that $kk^\sigma = kk^\sigma$. If

$$kk^\sigma = (\theta\delta + \gamma),$$

then

$$k^\sigma k = (\theta\delta + \gamma)^\sigma = k^\sigma k = (\theta\delta + \gamma),$$

a contradiction. □

Part 6

Kantor-Pentilla and CJV σ – Flokki

In [**34**], part 12, chapters 9 and 10, there is a brief history of the legacy of the Thas-Walker flock theory in the finite case to infinite fields. There are considerable differences in the finite field and infinite field theory. But, we are still discussing the same flocks over hyperbolic quadrics and quadratic cones.

Probably the first foray into non-commutative skewfields comes from the work of Jha-Johnson ([**20**],[**21**],[**22**],[**23**]), in a variety of work and, in particular, the "skew-Desarguesian" spreads, which are basically the first known semifield spreads over non-commutative skewfields. The known examples use the non-commutative skewfields that include the quaternion division rings. These semifields are also the first examples of simultaneous spreads, meaning that we have both conical and hyperbolic flocks over generalized hyperbolic quadrics and generalized quadratic cones by a variation of representations of the flock spread. Another direction showing the general representational nature of derivable nets is also due to Jha-Johnson, which was the inclusion of the idea of $\sigma - A$ derivative functions. There is considerable overlap with some of these works and that of Cohn [**6**]. It is shown by Jha-Johnson that $\sigma - A$ derivative functions are inner if the associated division ring is a field, also shown by Cohn. This implies that a basis change provides an isomorphic structure where A is identically zero. For the division ring theory, this might then be thought to somewhat lessen the interest in inner $\sigma - A$ functions in terms of representational value for the theory of spreads. However, we note in these volumes that the geometric value of these "resolutions" is actually quite valuable, as although the spreads contain these two representations of derivable nets that are isomorphic, the associated flocks need not be isomorphic.

The major development of the idea of the connection with derivable nets, spreads and partitions of cones is that of Kantor-Pentilla [**43**], in the finite case, but where the derivable net would be a type 2 representation, and there is now a beginning flock theory for finite fields over $\sigma - cones$. In this case, Kantor-Pentilla uses the Finnish term "flokki" for flocks. Somewhat later, Cherowitzo-Johnson-Vega[**8**], "CJV", extended Kantor-Pentilla to the infinite field case.

In CJV, there arise many of the ideas that are used in the more general theory. In particular, the connection between spreads and bilinear functions, quasifibrations and injective functions.

More recently, the twisted-hyperbolic flock theory over fields was determined by the author [**32**], wherein the idea of the "transform" of an affine point originated.

In [**34**] (4,5,3), there was a result regarding dual partial spreads of flokki spreads.

Now we connect the Kantor-Pentilla/CJV method and our transform method variation of the Thas-Walker method over fields. In [**34**], chapter 8, section 6, p. 160, there is a discussion of what was referenced as an alternative method to the Cherowitzo, Johnson, Vega method of the connection between translation planes and σ-conical flocks. This method follows the Kantor-Pentilla ideas from the finite case $GF(q)$ to the infinite field case K. However, when considered together with the σ-twisted hyperbolic theory of the author, the two methods studied in [**34**] are actually somewhat different.

So, the question is, how do the flocks obtained using the Kantor-Pentilla/Cherowitzo-Johnson-Vega method correspond to those obtained from the same spread using the variation of the Thas-Walker theory, called the transform method?

The following left flock spread is called the generic σ-conical flock spread

$$x = 0, y = x \begin{bmatrix} u^{\sigma} + F(t) & , G(t) \\ t & u \end{bmatrix},$$

for t, u in a field K. Our variation of the Thas-Walker method, the transform method to construct flocks, shows that we obtain a flock with planes, as follows:

$$\Sigma_t^{transform}:$$

$$-a_1 F(t)^{\sigma^{-1}} - a_2 t^{\sigma^{-1}} + a_3 G(t) - a_4 = 0,$$

for all $a_i, i = 1, 2, 3, 4$ in K. When the cone has the following representation: We call the "transform" the mapping from $(x, y) = (x_1, x_2, x_3, x_4)$ that is covered by the conical flock spread after the algebraic elimination of u,u^{σ} terms, and equal to

$$(w_1, w_2, w_3, w_4)$$

such that the cone is defined by

$$x_2 x_3^{((\sigma)^{-1})} = x_1^{-1} + 1$$

with vertex $(0, 0, 0, 1)$.

Whereas the Cherowitzo-Johnson-Vega method uses the spread representation of

$$x = 0, x = 0, y = x \begin{bmatrix} v + F(t) & , G(t) \\ t & v^{\sigma} \end{bmatrix},$$

for all t, v in K, and is equivalent to the conical flock

$$\Sigma_t^{cjv}:$$

$$b_1 t - b_2 F(t)^\sigma + b_3 G(t)^\sigma - b_4 = 0,$$

where the cone is defined by the set of elements (x_1, x_2, x_3, x_4) where

$$x_1^\sigma x_2 = x_3^{\sigma+1},$$

with vertex $(0, 0, 0, 1)$. The conical spreads here are also called flokki spreads

There are two areas of discussion:

(1) how do the planes of the flocks of the two methods compare abstractly? and

(2) how do the intersections of the planes with the σ-cones compare?

Let $K = GF(q)$, and consider flocks of quadratic cones and flocks of hyperbolic quadrics. A hyperbolic quadric has $(q+1)^2$ points, an associated spread with components in $PG(3, GF(q))$ is comprised of $q+1$ regulus nets that share two components. There are then q^2+1 components. But, there are $(q+1)^2$ Baer subspaces, which then are designated as the points of the associated hyperbolic flock. Since the quadratic cone may be embedded in the hyperbolic quadric, the number of non-vertex points for an associated flock of a "quadratic"/σ-cone is $q(q+1)$, it is implicit that the points of the σ-conical flocks always correspond to the set of Baer subspaces of the individual derivable nets. However, the individual flocks produced by one method will not necessarily be isomorphic to those produced by the other method, although as sets, the sets of flocks will be the same.

Some examples, will make these ideas more clear.

THEOREM 51. *When transpose spreads of σ-flokki spreads are σ-flokki spreads, and the methods of Kantor-Pentilla and the transform variation of Thas-Walker (the "transform" method developed here) overlap but do not always produce isomorphic σ-conical flocks from the same spread. However, the conical flock's planes arising from the spread and the transpose spread are isomorphic as a pair.*

In the first (original) volume [**34**], there is a study of σ-cones over fields F due to Cherowitzo, Johnson, Vega using the Kantor-Pentilla ideas. To compare how these compare, we need to organize our notation with the notation of CJV. The CJV spreads use the following notation:

$$x = 0, y = x \begin{bmatrix} u + g(t) & f(t) \\ t & u^\sigma \end{bmatrix}$$

and the associated flocks:

$$\Pi_t^{CJV} : a_1 t - a_2 f(t)^\sigma + a_3 g(t)^\sigma - a_4 = 0$$

for all t in a field F. The spreads and flocks are equivalent. The cone is defined by $w_1^\sigma w_2 = w_3^{\sigma+1}$, with vertex $(0,0,0,1)$.

Using what we shall call the "transform method", the spread has the form:

$$x = 0, y = x \begin{bmatrix} u^\sigma + F(t), & G(t) \\ t & u \end{bmatrix}$$

and the associated flock

$$\Pi_t^{Trans.} : -b_1 F(t)^{\sigma^{-1}} - b_2 t^{\sigma^{-1}} + b_3 G(t) - b_4 = 0.$$

Again the spreads and flocks are equivalent. The cone is defined by $w_2 w_3^{\sigma^{-1}} = w_1^{\sigma^{-1}+1}$, also with vertex $(0,0,0,1)$. Now we make the following changes to the expression for the CJV-spreads and flocks: Let $g(t) = F(t), f(t) = G(t)$, and replace what is called σ in the CJV spread by σ^{-1}. Note that the reason for the switch is to make the following notation common:

$$x = 0, y = x \begin{bmatrix} u^\sigma + F(t) & , G(t) \\ t & u \end{bmatrix}$$

Now return to the spread using the transform method developed in this work to obtain the spread:

$$\begin{bmatrix} u^\sigma & 0 \\ 0 & u \end{bmatrix},$$

letting $u^\sigma = v$, becomes

$$\begin{bmatrix} v & 0 \\ 0 & v^{\sigma^{-1}} \end{bmatrix}.$$

Making these changes then converts the CJV spread and flocks to:

$$x = 0, y = x \begin{bmatrix} u + F(t) & , G(t) \\ t & u^{\sigma^{-1}} \end{bmatrix}.$$

and the associated flock becomes:

$$\Pi_t^{CJV(2)} : a_1 t - a_2 G(t)^{\sigma^{-1}} + a_3 F(t)^{\sigma^{-1}} - a_4 = 0.$$

The cone is now represented by $w_1^{\sigma^{-1}} w_2 = w_3^{\sigma^{-1}+1}$.

Now we try to reconstruct how the spread was formed using the transform method. But, just by looking at the two spreads, they seem identical except for where the exponents are placed and the order of the functions. There are the terms $\{t, G(t)^{\sigma^{-1}}\}$ in the first spread and the analogous terms $\{t^{\sigma^{-1}}, G(t)\}$ in the second spread.

$$\Pi_t^{Trans.} : -b_1 F(t)^{\sigma^{-1}} - b_2 t^{\sigma^{-1}} + b_3 G(t) - b_4 = 0.$$

But first, we explain why the cones are different. Using the transform method, whether the division ring is a field or a non-commutative division ring is ignored. The associated proof might be called the "non-commutative proof". Suppose we now consider the field F case and reprove the theorem using the "commutative proof" using the transform method. Again, the spread:

$$x = 0, y = x \begin{bmatrix} u^\sigma + F(t), & G(t) \\ t & u \end{bmatrix}.$$

Let $x = (x_1, x_2)$ and $y = (x_3, x_4)$. There, this provides the following system of equations:

$$R : x_1(u^\sigma + F(t)) + x_2 t = x_3,$$
$$S : x_1 G(t) + x_2 u = x_4.$$

The intention is to form a relationship between the two equations eliminating the $u^\sigma$and u-terms.

Form

$$R^{\sigma^{-1}}.$$

This creates the term $x_1^{\sigma^{-1}} u$ in the R equation, and we have the term $x_2 u$ in the S equation. Multiply $R^{\sigma^{-1}}$by x_2 and multiply S by $x_1^{\sigma-1}$. The resulting equations are now:

$$x_2 R^{\sigma^{-1}} : x_2 x_1^{\sigma^{-1}}(u + F(t)^{\sigma^{-1}}) + x_2 x_2^{\sigma^{-1}} t^{\sigma^{-1}} = x_2 x_3^{\sigma^{-1}}.$$

And

$$x_1^{\sigma^{-1}} S : x_1^{\sigma^{-1}} x_1 G(t) + x_1^{\sigma^{-1}} x_2 u = x_1^{\sigma^{-1}} x_4.$$

Since we are in a field, we may cancel the terms

$$x_2 x_1^{\sigma^{-1}} u \& x_1^{\sigma^{-1}} x_2 u$$

by subtracting the two equations. We now choose to align the resulting equation using the order $\{G(t), t^{\sigma^{-1}}, F(t)^{\sigma^{-1}}\}$.

We then obtain the following flock expression:

$$x_1^{\sigma^{-1}} x_1 G(t) - x_2 x_2^{\sigma^{-1}} t^{\sigma^{-1}} - x_2 x_1^{\sigma^{-1}} F(t)^{\sigma^{-1}} - (x_2 x_3^{\sigma^{-1}} - x_1^{\sigma^{-1}} x_4) = 0.$$

We are looking for the cone with the equation

$$w_1^{\sigma^{-1}} w_2 = w_3^{\sigma^{-1}+1}.$$

We see that

$$(x_1^{\sigma^{-1}}x_1)^{\sigma^{-1}}(x_2x_2^{\sigma^{-1}}) = (-x_2x_1^{\sigma^{-1}})^{\sigma^{-1}}(-x_2x_1^{\sigma^{-1}}).$$

There is still the question of why the terms $G(t)$ and t are out of place when comparing the commutative transform version and CJV version.

So, take the transpose matrices of the transform matrices to obtain the following spread:

$$x = 0, y = x\begin{bmatrix} u^{\sigma} + F(t) & t \\ G(t) & u \end{bmatrix}.$$

Then the flock becomes

$$\Pi_t^{transpose,trans.} :$$

$$\vdots$$

$$\Pi_t^{CJV(2)} : a_1t - a_2G(t)^{\sigma^{-1}} + a_3F(t)^{\sigma^{-1}} - a_4 = 0.$$

Thus,

$$\Pi_t^{transpose,trans.} = \Pi_t^{CJV(2)}.$$

Here is the formal theorem.

THEOREM 52. *The Kantor-Pentilla/CJV method, determining σ-cone flocks as equivalent to σ-cone spreads over fields, is identical to the commutative transform method on the transpose of the associated σ-cone spread.*

As the transform method developed here produces a σ-conical flock from a σ-conical spread, then the transpose spread produces the flock from the Cherowitzo, Johnson, Vega/Kantor-Pentilla method. Therefore, if the spread and the transpose spread are not isomorphic, the corresponding flocks are not isomorphic.

CHAPTER 15

Transform and CJV-methods

We consider the following examples of spreads

$$x = 0, y = x \begin{bmatrix} u^{\sigma} & \theta t^{\sigma^{-1}} \\ t & u \end{bmatrix},$$

for all t, u in $GF(p^r)$, p odd and s a divisor of r such that r is not s^2. Let θ be a non-square in $GF(p^r)$. Then the Kantor-Pentilla/Cherowitzo, Johnson, Vega construction produces the following σ-conical flock:

$$x_1 t - x_2(\theta t^{\sigma^{-1}})^{\sigma^{-1}} - x_4 = 0$$

which is nonlinear, and the transform method produces the following σ-conical flock:

$$-x_1 0 - x_2 t^{\sigma^{-1}} + x_3 \theta t^{\sigma^{-1}} - x_4 = 0,$$

which is linear, so the flocks cannot be isomorphic, and the transpose spreads are not isomorphic.

Example 53. An infinite field example is as follows: Let Q denote the rational numbers, and let ζ_n denote a primitive nth root of unity, and let $Q(\zeta_n)$ be a cyclotomic extension of Q. Let $Q(\zeta_n)(\sqrt[n]{b})$ be a Galois cyclic extension. Let $\theta = \sqrt[n]{b}$, then the Cherowitzo, Johnson, Vega will construct a nonlinear σ-conical flock, and our transform method will produce a linear flock.

The same proof as in the finite case says that

$$\theta = (u^{-1}s)^{\sigma+1},$$

if there is a zero determinant. Since we have an extension, it remains to show two things for $k = \Sigma_{i=0}^{n-1} t_i \theta^i$, we need to show that $k^{\sigma} k$ cannot be θ. In the section on central cyclic extensions over division rings, there is a more general study of constructing semifield spreads from cyclic division rings of dimension n^2 over their centers, of the form indicated above. There are infinitely many examples of cyclic division rings for all integers n. This is also true for fields, so there are infinitely many non-isomorphic flocks obtained that are constructed from the same spread due to the two different methods under discussion.

DOI: 10.1201/9781003767978-15

In the next part, we consider the question regarding isomorphic spreads and non-isomorphic flocks. But, we pause to look at the finite example above

$$x = 0, y = x \begin{bmatrix} u^{\sigma} & \theta t^{\sigma^{-1}} \\ t & u \end{bmatrix},$$

and the transpose, which is

$$x = 0, y = x \begin{bmatrix} u^{\sigma} & t \\ \theta t^{\sigma^{-1}} & u \end{bmatrix}.$$

Let $\theta t^{\sigma^{-1}} = s$, so that $t = \theta^{-\sigma} s^{\sigma}$, to obtain:

$$x = 0, y = x \begin{bmatrix} u^{\sigma} & \theta^{-\sigma} s^{\sigma} \\ s & u \end{bmatrix}.$$

We point out that the initial spread may be rewritten in the form:

$$y = x \begin{bmatrix} t^{\sigma} & \theta \\ 1 & t \end{bmatrix} \begin{bmatrix} v^{\sigma} & 0 \\ 0 & v \end{bmatrix},$$

for v non-zero and

$$x = 0, y = 0, y = x \begin{bmatrix} u^{\sigma} & 0 \\ 0 & u \end{bmatrix}$$

for u non-zero.

PROBLEM 54. Try the above statement of rewriting the spread as an exercise.

PROBLEM 55. Now show that the first of the above examples admits the right homology group that is regulus-inducing. And the transpose version admits the left homology group that is regulus-inducing.

The right homology group acting defines a generalized hyperbolic flock. The left homology group by taking the mapping $(x, y) \rightarrow (y, x)$ provides another generalized hyperbolic flock. The finite example given above will then produce non-isomorphic hyperbolic flocks from the same hyperbolic spread. We shall explicate this example in the next part.

CHAPTER 16

Choices of representation

In the previous, we saw that the choices of the cone for an σ-conical flock, while not changing the overall set of associated flocks, it is not the case that each conical flock spread will necessarily provide isomorphic flocks. As stated, the choice of the cone and the order of the functions in the representation of the flock planes themselves do change when using the transform version. In addition to σ-conical flocks over fields, there are σ-twisted hyperbolic flocks by the author [**32**], which we also discussed in the original volume [**34**]. In the following parts, we shall also discuss the representation when using non-commutative skewfields.

Here is the field version; refer to pp. 142-143 of [**34**], which states the following:

Let K be a field, and let σ be an automorphism of K.

The following represents a twisted or σ-hyperbolic flock spread:

$$x = 0, y = 0, y = x \begin{bmatrix} u^{\sigma} & 0 \\ 0 & u \end{bmatrix}$$

and

$$y = x \begin{bmatrix} f(t), & g(t) \\ 1 & t \end{bmatrix} \begin{bmatrix} v^{\sigma} & 0 \\ 0 & v \end{bmatrix}$$

for all non-zero u,v and for all t in K. The associated hyperbolic flock is represented as follows:

$$\pi_t : -x_1 g(t)^{\sigma} + x_2 f(t) - x_3 t^{\sigma} + x_4 = 0,$$

for all t in K. Note in volume II, the form is different but equivalent. And ,

$$\rho : x_2 = x_3.$$

And the twisted hyperbolic quadric would look like

$$x_1 x_4^{\sigma} = x_2^{\sigma} x_3,$$

with vertex $(0, 0, 0, 1)$.

The reason for the representation is the method of argument. This went as follows: u and u^{σ} needed to be eliminated. In a general argument, where one might have $\sigma - A$ a σ-derivative function, suppose

DOI: 10.1201/9781003767978-16

one wished to retain the form of A, without applying the automorphism to the equation containing $A(u)$, then alternatively it is possible to apply σ^{-1} to the other equation. Here is what would happen: The representation of the flock would now look like

$$\pi_t^* : -x_1 g(t) + x_2 f(t)^{\sigma^{-1}} - x_3 t + x_4 = 0,$$

And the second twisted hyperbolic quadric would look like

$$x_1^{\sigma-1} x_4 = x_2 x_3^{\sigma^{-1}}.$$

Furthermore, the so-called "transform" of an affine point in the original case is:

$$(x_3 x_1^\sigma, x_4^\sigma x_1, x_3 x_2^\sigma, x_4^\sigma x_2),$$

where the order is important and dictates the order of the functions. Note that this set of points satisfies the equation of the twisted hyperbolic quadric. Note that neither in the Cherowitzo-Johnson-Vega method nor in the previous twisted hyperbolic quadric is there mention of the associated Baer subplanes. Does this mean that the Thas-Walker method is not used? In a finite case, it is relatively easy to check for cover, as one may simply count the number of points in a twisted hyperbolic quadric or a twisted quadratic cone. Therefore, one can think simply of a partition of a σ-cone without the picture associated with a flock of a quadratic cone, and, in this case, Baer subspaces are somewhat superfluous. In the infinite case, there are problems, but using

$$(x_3 x_1^\sigma, x_4^\sigma x_1, x_3 x_2^\sigma, x_4^\sigma x_2),$$

for all affine points, it may be established that the σ-hyperbolic quadric is bijectively obtained as the image set of the set

$$(x_1, x_2, x_3, x_4),$$

which represents the set of affine points. To find the Baer subspace method, take the σ-homogeneous group, defined by the right mappings

$$(x, y) \to (x \begin{bmatrix} u^\sigma & 0 \\ 0 & u \end{bmatrix}, y \begin{bmatrix} u^\sigma & 0 \\ 0 & u \end{bmatrix})$$

for all non-zero u in K. It is not a coincidence that the group fixes each of the Baer subspaces in the derivable net represented by

$$x = 0, y = 0, y = x \begin{bmatrix} u^\sigma & 0 \\ 0 & u \end{bmatrix}.$$

This idea of creating a σ-homogeneous group such that the set of transforms is invariant under that group, then we are able to define the points of the flock to be the 1-dimensional projective representatives of the set of Baer subspaces, which is important, as it is easy to "count" the Baer subspaces, even in the non-commutative cases.

PROBLEM 56. Prove that the set of transforms:

$$(x_3 x_1^\sigma, x_4^\sigma x_1, x_3 x_2^\sigma, x_4^\sigma x_2),$$

is invariant under the σ-homogeneous group. Now find the set of Baer subspaces – this might be difficult at this point, but it will be a good exercise. Go back to the original text to find the representation of the Baer subspaces of each of the σ-regulus nets. This will be a start in understanding the Thas-Walker extension that works for all skewfields finite, infinite fields or non-commutative division rings.

In the general theory, you cannot use commutativity in your arguments. If you never use it, you will be able to find all of the representations of hyperbolic and conical flocks and discover the transforms yourself. The problem will become that part of a σ-hyperbolic quadric.

PROBLEM 57. Do not use commutativity! Here is a derivable net represented by

$$x = 0, y = x \begin{bmatrix} u^\sigma & 0 \\ 0 & u \end{bmatrix},$$

if $(x, y) = (x_1, x_2, x_3, x_4)$ and $x_1 x_2$ is non-zero, use it to define and find the representation of the largest subset of the generalized σ-hyperbolic quadric that is possible. Hint: There will be an inverse that cannot be resolved without commutativity.

Part 7

JPW-hyperbolic flocks

CHAPTER 17

Idea of "left-inversion"

In the text of the author [**34**], σ-flocks of hyperbolic are considered, and it was mentioned that it is possible to construct new examples by use of certain of the j-planes. Here we show how that is possible. Also mentioned in [**34**] is the connection between flocks of quadratic cones and other geometries. In particular, finite translation planes with spreads in $PG(3, GF(q))$ that admit cyclic homology groups of order $q+1$ and such flocks are equivalent; however, there is a many-to-one connection between the planes and the flocks of quadratic cones. The planes, due to the group, admit a set of $q-1$ mutually disjoint regulus nets, which together with the axis and coaxis of the group define the spread. Therefore, any subset of this set of regulus nets may be multiply derived to construct another, potentially non-isomorphic translation plane. However, all of these translation planes correspond to the same flock of a quadratic cone, which does not seem to have any translation planes that admit cyclic homology groups, as the flock spreads of a quadratic cone spread admit elation groups of order q. How this works is considered in the author's text [**31**], Theorem 111, p. 257.

In [**32**], the author shows that it is possible to extend consideration of flocks of hyperbolic quadrics to the so-called σ-hyperbolic flocks. A flock of a hyperbolic quadric is a flock of the projective version of a classical regulus net in generic form. A σ-hyperbolic is a flock of the projective version of a σ-regulus net in generic form. For translation planes with spreads in $PG(3, K)$, that correspond to flocks of hyperbolic quadrics, there is an associated affine homology group that is regulus-inducing in the sense that the orbits union the axis and coaxis define regulus nets whose sets of Baer subspaces define the flocks. In a similar sense, the author shows that for σ-hyperbolic flocks, there is an associated translation plane that admits an affine homology group whose orbits, together with the axis and coaxis, define σ-regulus nets. It was then noticed that the j-planes of Johnson, Pomerada, and Wilke [**42**] admit such an affine homology group. Therefore, these j-planes correspond to flocks of quadratic cones due to the cyclic group of order $q+1$, but also then correspond to flocks of σ-hyperbolic flocks, due

DOI: 10.1201/9781003767978-17

to the existence of a σ-regulus-inducing affine homology group. One of the problems of seeing what the σ-hyperbolic flocks look like is that the associated j-planes are the affine homology groups on the wrong side, as it were. This article explicates the change in construction so that these interesting new σ-hyperbolic flocks may be properly understood. Finally, it might be noted that there are infinitely many classes of these σ-hyperbolic flocks, and there is a procedure analogous to the construction of the Dickson nearfield planes from finite Desarguesian affine planes of odd order that constructs a second infinite class of σ-hyperbolic flocks.

There is another question that comes up with these spreads and flocks. For ordinary hyperbolic flocks, the associated spreads use a regulus-inducing homology group acting on the right of the matrix spread set. In this case, since the kernel of the spread is a mapping

$$(x_1, x_2, x_3, x_4) \to (x_1u, x_2u, x_3u, x_4u)$$

for k non-zero in a field K, and the regulus-inducing homology group with mappings:

$$(x_1, x_2, x_3, x_4) \to (x_1, x_2, x_3v, x_4v),$$

for all v non-zero in K. Letting $u = v^{-1}$, then using these two groups, we also obtain a regulus-inducing homology with mappings

$$(x_1, x_2, x_3, x_4) \to (x_1v^{-1}, x_2v^{-1}, x_3, x_4).$$

So the regulus-inducing homology group with axis $y = 0$ and coaxis $x = 0$ could be interchanged with the regulus-inducing homology group with axis $x = 0$ and coaxis $y = 0$, and the same hyperbolic flocks would be constructed. However, with other regulus-inducing groups or pseudo-regulus-inducing groups that are homology groups, one could not obtain the same form and perhaps not the same flocks. The associated translation planes would be isomorphic, but perhaps one would obtain non-isomorphic hyperbolic flocks. This is the situation with the spreads in question. We often have the regulus-inducing homology group on the wrong side.

There is another connection with j-planes, and this is with flocks of quadratic cones.

In the first volume, we showed how the j-planes of Johnson-Pomareda-Wilke are equivalent to flocks in Part 8, chapter 4 of [**34**]. In particular, there are many-to-one connected to the flocks of quadratic cones. Previously, we have discussed various constructions whereby isomorphic flock spreads may be shown to be in a variety of different types of flocks. Two of these, the Elation double spreads and the $\sigma - A$ additions, may be used to show how the Kantor-Knuth flocks can show up in an assortment of flocks.

In the j-planes, there is an associated cyclic subgroup of order $q+1$

$$\left\{ \begin{bmatrix} 1 & 0 & 0 & 0 \\ 0 & \delta_{u,t}^{-j} & 0 & 0 \\ 0 & 0 & u & t \\ 0 & 0 & ft & ,u+gt \end{bmatrix} ; u,t \text{ in } GF(q) \right\},$$

which is when $\delta_{u,t}$=1. This is a cyclic affine homology group of order $q+1$ with axis $y = 0$ and co-axis $x = 0$. There is also the group of kernel homologies defined by

$$\left\{ \begin{bmatrix} v & 0 & 0 & 0 \\ 0 & v & 0 & 0 \\ 0 & 0 & v & 0 \\ 0 & 0 & 0 & v \end{bmatrix} ; v \text{ non-zero in } GF(q) \right\}.$$

Choose $t = 0$ in the first group, obtaining $\delta_{u,0}^{-j} = u^{-2j}$, then use the second group and multiply by $u^{-1}I_4$, which creates the following group of homologies of order $q-1$ with axis $x = 0$, and coaxis $y = 0$.

$$\left\{ \begin{bmatrix} u^{-1} & 0 & 0 & 0 \\ 0 & u^{-(2j+1)} & 0 & 0 \\ 0 & 0 & 1 & 0 \\ 0 & 0 & 0 & ,1 \end{bmatrix} ; u \text{ non-zero in } GF(q) \right\}.$$

In Johnson, Pomareda, Wilke [**42**], for every finite field of odd order $GF(q)$, there is a class of *j-planes* constructed by choosing an automorphism

$$\sigma = 2j+1,$$
$$j = \frac{\sigma - 1}{2}.$$

Now that we know from the author [**32**] that if the homology group had axis $y = 0$ and coaxis, there would be a set of "twisted" hyperbolic flock spreads and hyperbolic flocks over a finite field. The mapping $(x, y) \to (y, x)$ is an isomorphism of the j-plane but is now realized as a hyperbolic flock spread. However, this was not known for thirty-one years. In between, there were other connecting geometries.

Now looking back at the first-mentioned group of order $q^2 - 1$, we have in the j-planes, a cyclic homology group of order $q + 1$ with axis $y = 0$ and coaxis $x = 0$. In the first volume [**34**], and also in the author's previous work [**31**], chapters 18 and 19, pp. 253-273, there are connections with "hyperbolic fibrations" which are shown by (Pentilla,

Baker, Ebert, see reference in [**31**]) to be equivalent to finite flocks of quadratic cones, and, also by the author, equivalent to finite spreads of dimension 2, that admit a cyclic homology group of order $q+1$.

17.1. Main theorems on JPW-spreads

In this section, we represent the new σ-hyperbolic flocks as well as the second class of σ-hyperbolic flocks constructed from the first class.

Let π be a j-plane that admits an automorphism $\sigma = 2j+1$.

Consider the representation over a field $\begin{bmatrix} u & t \\ ft, & u+gt \end{bmatrix}$ for all u,t in

$GF(q)$, for q odd. Then let $g=0$. The set of j-planes where $2j+1$=p^s, where $q = p^r$ admits an affine homology group that is a σ-regulus-inducing group.

Therefore, there is an associated K^{σ}-hyperbolic flock for $K = GF(q)$. In order to determine the actual form, a coordinate change $(x,y) \to (y,x)$ is required.

Here is the original spread:

$$x = 0, y = 0, y = \begin{bmatrix} u & t \\ \delta^j_{u,t} ft & \delta^j_{u,t} u \end{bmatrix}, \textit{where}$$

$$\delta_{u,t} = u^2 - ft^2.$$

There is a homology group of the following form

$$\begin{bmatrix} \begin{bmatrix} v & 0 \\ 0 & v^{2j+1} \end{bmatrix} & 0_2 \\ 0_2 & I_2 \end{bmatrix} \text{,v in } K\text{^* }.$$

So, in order to realize the K^{σ}-hyperbolic flock, the coordinate change will bring about the following spread

$$x = 0, y = 0, y = \begin{bmatrix} u & t \\ \delta^j_{u,t} ft & \delta^j_{u,t} u \end{bmatrix}^{-1}$$

admitting the group

$$\begin{bmatrix} I_2 & 0_2 \\ 0_2 \begin{bmatrix} w & 0 \\ 0 & w^{2j+1} \end{bmatrix} & 0_2 \end{bmatrix} \text{,w in K^*.}$$

Making the inverse calculations, we obtain the following

$$x=0, y=0, y=\begin{bmatrix} u & t \\ \delta_{u,t}^{j} ft, & \delta_{u,t}^{j} u \end{bmatrix}^{-1}=\begin{bmatrix} \frac{\delta_{u,t}^{j} u}{\delta_{u,t}^{j+1}}, & \frac{-t}{\delta_{u,t}^{j+1}} \\ \frac{-\delta_{u,t}^{j} ft}{\delta_{u,t}^{j+1}}, & \frac{u}{\delta_{u,t}^{j+1}} \end{bmatrix}.$$

When $t=0$, notice that x, y

$$\begin{bmatrix} \frac{\delta_{u,t}^{j} u}{\delta_{u,t}^{j+1}} & \frac{-t}{\delta_{u,t}^{j+1}} \\ \frac{-\delta_{u,t}^{j} ft}{\delta_{u,t}^{j+1}} & \frac{u}{\delta_{u,t}^{j+1}} \end{bmatrix}=\begin{bmatrix} u^{-1} & 0 \\ 0 & u^{-(2j+1)} \end{bmatrix}=\begin{bmatrix} w^{\sigma^{-1}} & 0 \\ 0 & w \end{bmatrix},$$

for $u^{-\sigma}=w$. Then t is not zero, and we use the group with elements for all non-zero w in $GF(q)$,

$$\begin{bmatrix} I_2 & 0_2 \\ 0_2 & \begin{bmatrix} w^{\sigma^{-1}} & 0 \\ 0 & w \end{bmatrix} \end{bmatrix}$$

to place a "1" in the $(2,1)$-entries and create:

$$\begin{bmatrix} -ut^{-1}f^{-1} & t^{-2j}\delta_{u,t}^{j} f^{-(2j+1)} \\ 1 & -ut^{-(2j+1)}\delta_{u,t}^{j} f^{-(2j+1)} \end{bmatrix}\begin{bmatrix} w^{\sigma^{-1}} & 0 \\ 0 & w \end{bmatrix}.$$

The method is to choose $w^{\sigma^{-1}}$ to be $(\frac{-\delta_{u,t}^{j} ft}{\delta_{u,t}^{j+1}})^{-1}$, and adjust to w, and multiply the matrices to obtain the remaining entries in the matrix. Since we have a spread, the combinatorics of finite spreads shows that if one designates the $(2,2)$-entry elements as s, the $(1,1)$-entry elements by $f(s)$ and the $(1,2)$-entry elements by $g(s)$, for functions not co

Note that when $t=0$, we obtain the left generic σ^{-1}-regulus derivable net.

Therefore, we obtain the following results, but before we formally state these, we prepare the reader with what the form should look like by recalling the fundamental theory from the previous results.

THEOREM 58. *Let π be a translation plane in $PG(3,K)$, where K is a field that admits an automorphism σ and admits a σ-regulus-inducing homology group H^{σ}. Then there is a representation for the spread as follows:*

$$x=0, y=0, y=x\begin{bmatrix} u^{\sigma} & 0 \\ 0 & u \end{bmatrix},$$

for all u in K.

$$x=0, y=0, y=x\begin{bmatrix} f(t) & g(t) \\ 1 & t \end{bmatrix}\begin{bmatrix} u^{\sigma} & 0 \\ 0 & u \end{bmatrix},$$

for functions f and g. Then there is a flock of the σ-hyperbolic quadric, defined relative to (x_1,x_2,x_3,x_4) as $x_4^\sigma x_1 = x_3 x_2^\sigma$, when the point is homogeneous relative to the field K^σ, where the multiplicative group is

$$\left\{ \left[\begin{bmatrix} v^\sigma & 0 \\ 0 & v \\ 0 & 0 \\ 0 & 0 \end{bmatrix} \begin{bmatrix} 0 & 0 \\ 0 & 0 \\ v^\sigma & 0 \\ 0 & v \end{bmatrix} \right], v \, in K^{\sigma *}. \right\}$$

Then there is a σ-hyperbolic flock with planes as defined as follows:

$$\Sigma_t : -x_1 g(t)^\sigma - x_2 f(t) + x_3 - x_4 t^\sigma = 0.$$

$$\rho : x_2 = x_3.$$

We now may state our first main theorem.

THEOREM 59. *From the spread*

$$x = 0, y = 0, y = \begin{bmatrix} u & t \\ \delta_{u,t}^j f t & , \delta_{u,t}^j u \end{bmatrix}^{-1},$$

the following flock of the σ-hyperbolic is obtained:

$$\Sigma_{(-ut^{-\sigma}\delta_{u,t}^{(\sigma-1)/2} f^{-\sigma})} : x_1 (t^{-(\sigma-1)} \delta_{u,t}^{(\sigma-1)/2} f^{-\sigma})^{\sigma^{-1}} -$$

$$-x_2(-ut^{-1} f^{-1}) + x_3 - x_4 (-ut^{-\sigma} \delta_{u,t}^{(\sigma-1)/2} f^{-\sigma})^{\sigma^{-1}} = 0,$$

for all (u,t), where t is not zero, and the plane

$$\rho : x_2 = x_3.$$

PROOF. The way to formulate the basic form required for the σ-hyperbolic quadric is to determine the form

$$\begin{bmatrix} f(s) & g(s) \\ 1 & s \end{bmatrix} \begin{bmatrix} w^{\sigma^{-1}} & 0 \\ 0 & w \end{bmatrix}.$$

It is interesting here that the functions f and g involved that are implicitly but not explicitly determined. We obtain:

$$\begin{bmatrix} -ut^{-1} f^{-1} & t^{-2j} \delta_{u,t}^j f^{-(2j+1)} \\ 1 & -ut^{-(2j+1)} \delta_{u,t}^j f^{-(2j+1)} \end{bmatrix} \begin{bmatrix} (-\delta_{u,t}^{-1} f t) & 0 \\ 0 & (-\delta_{u,t}^{-1} f t)^{2j+1} \end{bmatrix}.$$

Therefore, we see that

$$f(-ut^{-(2j+1)} \delta_{u,t}^j f^{-(2j+1)}) = -ut^{-1} f^{-1}$$

and

$$g(-ut^{-(2j+1)} \delta_{u,t}^j f^{-(2j+1)}) = t^{-2j} \delta_{u,t}^j f^{-(2j+1)}.$$

Therefore, we obtain the following flock of the σ-hyperbolic quadric:

$$\Sigma_{(-ut^{-\sigma}\delta_{u,t}^{(\sigma-1)/2}f^{-\sigma})} : x_1(t^{-(\sigma-1)}\delta_{u,t}^{(\sigma-1)/2}f^{-\sigma})^{\sigma^{-1}} -$$

□

$$-x_2(-ut^{-1}f^{-1}) + x_3 - x_4(-ut^{-\sigma}\delta_{u,t}^{(\sigma-1)/2}f^{-\sigma})^{\sigma^{-1}} = 0,$$

for all (u, t), where t is not zero, and the plane

$$\rho : x_2 = x_3.$$

This completes the proof.

- **The Nearfield-Like Class.**

There is actually a class of such flocks, one for each automorphism of $GF(q)$, so for $q = p^r$, every proper divisor s of r, there is an automorphism and therefore a flock of a σ-hyperbolic quadric. The j-planes of order q^2 admit a cyclic homology group of order $q + 1$, and therefore are connected to flocks of quadratic cones (see [**34**]), in this setting, the corresponding cone flock is that of Kantor-Knuth. Since the cyclic group of order $q+1$ creates a set of $q-1$ mutually disjoint regulus nets, it is possible to create a variety of interesting translations by multiple derivation. The orbits of a component under the group of the general form $y = xmM$, where $m = \begin{bmatrix} u & t \\ ft, & u \end{bmatrix}$, where f is a non-square and the determinant of the matrix is 1. This group corresponds to a partial spread of a Desarguesian spread, which is a quadratic extension of $GF(q)$, and which is isomorphic to $GF(\sqrt[2]{f})$, represented by $\sqrt[2]{f}t + u$. The derivation set corresponding to

$$\left\{ y = xmM, where\ m = \begin{bmatrix} u & t \\ ft & u \end{bmatrix}, for \text{ all m of determinant } 1 \right\}$$

$$\left\{ y = x^{\tau}mM, where\ m = \begin{bmatrix} u & t \\ ft & u \end{bmatrix}, for \text{ all m of determinant } 1 \right\}$$

where τ is the inducted automorphism of the corresponding Desarguesian plane. In the classical construction of the Dickson nearfield planes from the Desarguesian planes, the nets of square determinant are multiply derived (this means that since the extension of $GF(q)$ is $GF(q^2)$, then the automorphism is d^q and the determinant has the form m^{q+1}). When the same idea is used here, what happens is that there is a constructed j^*-plane where the new $j^* = \frac{q-1}{2} + j$. Therefore, over $GF(q)$, the corresponding automorphism σ remains the same (since $2j^* + 1) = (q - 1) + 2j + 1)$, whereas the terms change since $\frac{q-1}{2} + j$

is replaced whenever j is seen in the previous flock. Thus, we obtain another class of σ-flocks as follows:

THEOREM 60. The flock corresponding to the nearfield-like construction from the previous flock is as follows:

$$\Sigma_{(-ut^{-\sigma}\delta_{u,t}^{(\sigma-1)/2\frac{+(q-1)}{2}}f^{-\sigma})} : x_1(t^{-(\sigma-1)}\delta_{u,t}^{(\sigma-1)/2+(\frac{q-1)}{2}}f^{-\sigma})^{\sigma^{-1}} -$$

$$-x_2(-ut^{-1}f^{-1}) + x_3 - x_4(-ut^{-\sigma}\delta_{u,t}^{(\sigma-1)/2+(\frac{q-1)}{2}}f^{-\sigma})^{\sigma^{-1}} = 0,$$

$$\varrho : x_2 = x_3$$

.

DEFINITION 61. The first class of σ-hyperbolic flocks will be called the flocks of Johnson-Pomareda-Wilke (see [**34**]) and the second class of flocks constructed using ideas of the general nearfield planes shall be called the nearfield-like flocks of Johnson-Pomareda-Wilke.

So, it is possible to consider the application of a finite Kaleidoscope of derivable nets on the two classes of JPW classes of hyperbolic flocks, for $K = \begin{bmatrix} e & f \\ 0 & h \end{bmatrix}$, where e, f, h eh non-zero in $GF(q)$, for q odd for all such finite fields. Therefore, for each finite field $GF(q)$, there are $(q-1)^2 q$ possible isomorphic hyperbolic flock spreads producing the same number of hyperbolic flocks, where the isomorphism question is more-or-less unresolved. The finite Kaleidoscope of derivable regulus nets is

$$x = 0, y = x \begin{bmatrix} w^{\sigma}, & e^{-1}(w^{\sigma}f - fw) + e^{-1}A(w)h \\ 0 & w \end{bmatrix}.$$

Let

$$e^{-1}(w^{\sigma}f - fw) = A_f(w).$$

THEOREM 62. ***(The Action of the Kaleidoscope on the JPW-Flocks)***

An application of the $K = \begin{bmatrix} e & f \\ 0 & h \end{bmatrix}$ *mappings will show that there are families of the Johnson-Pomareda-Wilke* σ^{-1}*-hyperbolic flocks that produce* $\sigma^{-1} - A_f$*-hyperbolic flocks for all* f *in* $GF(q)$.

PROBLEM 63. Find the forms for the $\sigma - A_c$-hyperbolic flocks of the Johnson-Pomareda-Wilke σ^{-1}-flocks. Show directly (without using a previous theorem) when f is not zero, for an original and the associated isomorphic hyperbolic flock spreads, then the corresponding two hyperbolic flocks are not isomorphic.

- In volume III, the question of changing coordinates will be considered in the following manner: For every hyperbolic flock spread, there is a pseudo-regulus net that contains $x = 0, y = 0, y = x$. Change coordinates by replacing this pseudo-regulus net with one of the other pseudo-regulus nets. It is possible to do this by a mapping, which is an isomorphism of the hyperbolic flock spread. Now there are two different hyperbolic flocks; are they isomorphic? Same question for hyperbolic flock spreads over finite fields or infinite fields.

Part 8

Nonlinear hyperbolic flocks

Previously, we considered the lifted skewfields of Johnson-Jha [**38**]. There are two methods of lifting skew fields, by algebraic extensions as done in the first volume for quaternion division rings and for cyclic division rings. And, using transcendental extensions over division rings D of finite dimension over a field F. In particular, for finite flock spreads, we would need to use only algebraic extensions, recalling that the Hiramine, Matsumoto, Oyama lifting procedure [**17**] is valid for all finite spreads in $PG(3, GF(q))$. When discussing lifting skewfields in the finite case, then we are restricted to lifting finite Desarguesian/Pappian spreads. Let D_1 be a division ring that admits a central Galois quadratic extension. In the Multiple Replacement Theorem – Redux, we have used a tensor product construction for any division ring D that is of finite dimension over a field F, so that in this setting there is an extension division ring D_1 of D that then admits a central Galois quadratic extension $D_1(\theta)$ and this construction is valid for all finite division rings in all characteristics.

Given any division ring D or D_1, where D_1 refers to an extension of D in the transcendental case, we assume that there is a central Galois quadratic extension $E(\theta)$, where E is D or D_1. This extension would have the following matrix form:

$$x = 0, y = x \begin{bmatrix} u^{\sigma} & t^{\sigma}\theta \\ t & u \end{bmatrix}$$

for all t, u in $E(\theta)$ and σ is the induced automorphism.

The actual construction for the transcendental case is as follows. We have considered the details of this construction in other chapters. Consider $D_3 = D \otimes_F F(z_1, z_2)$ where z_1, z_2 are indeterminants. The mapping σ interchanging $z_i, i = 1, 2$ induces an automorphism by defining σ to be 1 on D. $F(z_1, z_2)$ is a Galois quadratic extension over $Fix\sigma = K$. Then, as K and D are subskewfields of D_3, which is a division since D is finite-dimensional, then $D_1 = D \otimes_F K$ is a subdivision ring of $D_1 \otimes_K K(z_1) = D_2$.

CHAPTER 18

Adjoining inner derivation functions

In this part, we use the ideas in the original proof of the lifting skew-field theorem; the metro equation ideas when we work with a similar type of spread. Here we consider the previous putative spreads when σ has order 2 and consider the following problem: When does the matrix set under consideration become a quasifibration? And when is it a quasifibration, and when is it a left semifield spread? For additive quasifibrations, the only method that we have to obtain a spread is the left and right unwrapping result. If we let A be an inner σ-derivation function, assume that $D(\theta)$ is a division ring that is a central Galois 2-dimensional extension of a division ring D, where when the characteristic is not 2, $\theta^{\sigma} = -\theta$, and if the characteristic is 2, $\theta^{\sigma} = \theta + \alpha$, where α is non-zero and in the center. We also assume here that from the chapters on simultaneous spreads, in order to be a simultaneous spread, provided it is a spread, we would need the following: If $\sigma - A$ is a derivation function with respect to a non-trivial outer automorphism σ and $\sigma^2 = 1, A^2 = 0$, and $A(s^{\sigma})^{\sigma} = -A(s)$. And, we would need the left unwrapping property holds. We shall then be interested in four questions:

(i) When is the matrix set a quasifibration? We have seen previously that the answer will be when $k^{\sigma}k - A(k)$ is not θ for all k in D.

(ii) When is the matrix set a spread? Since we only have the unwrapping condition to decide, one possibility is when the constants involved are central elements.

(iii) When is the matrix set a simultaneous quasifibration? When $\sigma^2 = 1, A^2 = 0$, and $A(s^{\sigma})^{\sigma} = -A(s)$.

(iv) When is the matrix set a simultaneous spread? When we have (iii) and the unwrapping property.

The only non-inner $\sigma - A$ function that we have found is when $\sigma = 1$, and the characteristic is 2, which shall be considered in the next section/chapter. Hence, assume that $\sigma^2 = 1$, and non-trivial.

Assume that the derivation function is inner: That $A(u) = (u^{\sigma}c - cu)$, for all u in a division ring D, c a non-zero constant. Then $A(u^{\sigma})^{\sigma} = (u^{\sigma}c^{\sigma} - c^{\sigma}u)$, so if and only if σ has order 2 and $c^{\sigma} = -c$, shows that

DOI: 10.1201/9781003767978-18

$A(s^\sigma)^\sigma = -A(s)$. Now assume that $c^\sigma = -c$, and consider

$$A(A(u)) = A(u^\sigma c - cu) = (u^\sigma c - cu)^\sigma c - c(u^\sigma c - cu).$$

Then, which implies that $A^2(u) = 0$ if and only if c is in $Z(D)$. Therefore, if A is an inner σ-derivation, with σ of order 2 and $c^\sigma = -c$, and c in $Z(D)$, then we will obtain a simultaneous quasifibration provided c is in $Z(D)$. And, since for the unwrapping theorem to hold, we would also require that c is in the center.

It is also possible to define $A(u) = cu - u^\sigma c$, since $dA(u)$ is also a $\sigma - dA(u)$ derivation function, for any non-zero d.

For simplicity, we now take c in $Z(D)$.

We consider $c = c_0 + c_1\theta$, and $D(\theta)$ is a central extension of D, where D is a division ring of finite dimension over a field F, and c_i for $i = 0, 1$ is in $D(\theta)$. We can show that if $c_0c_1 = 0$, and c non-zero, then we may obtain a left flock spread. When $c^\sigma = -c$, then $c_0 - c_1\theta = -(c_0 + c_1\theta)$, when the characteristic is not 2, so that $c_0 = 0$. When the characteristic is 2, then $c^\sigma = c_0 + c_1\alpha + c_1\theta = c_0 + c_1\theta$ shows that $c_1 = 0$. Again, this is what is required for simultaneous flock spreads. The other cases, when $c_1 = 0$ for characteristic not 2 and c_1 not zero for characteristic 2, will also be proved. There are also situations when c_0c_1 is not zero and has an unusual description. Before we prove the main theorem, we consider these situations and whether there are division rings that satisfy the required property.

- The conditions are as follows: First, for characteristic not 2, the division ring D, there cannot be a square k such that

  $$k^2 = \beta(1 - 4c_0c_1),$$

  where

  $$\theta^2 = \beta.$$

  Then this occurs: choose $(1 - 4c_0c_1) = r^2$, where r^2 is in the center. The delicate point is that $k^2r^2 = (kr)^2$, since k and r commute. Then we have

  $$(kr)^2 = \theta^2,$$

  and the idea then shows that

  $$(kr - \theta)(kr + \theta) = 0.$$

  Now we have factored in the standard way, again because θ is central. However, this implies that $kr = \pm\theta$, a contradiction since kr is fixed by σ but θ is not. Therefore, if c_0c_1 is not zero, choose any $c_1$and let $c_0 = \frac{1-4r^2}{c_1}$, then there will be infinitely many solutions involving c_0c_1.

- And, for characteristic 2, when

$$\theta^2 = \theta + \alpha,$$

  for α not equal to 1, we will show that it is always the case that for $k = k_0 + k_1\theta$, then $k_1 = 0$. This will a contradiction, and all elements (c_0, c_1) not equal to $(0, 0)$ will produce conical flocks.
- When $\alpha = 1$, in general, we need to assume that $c_0c_1 = 0$, to obtain conical flocks.

In this part, we consider the form for a lifted skewfield flock spread and then add to the form an inner $\sigma - A_c$ function where $A_c(w) = (w^\sigma - w)c$, for c in the center of the corresponding division ring. We have considered simultaneous spreads previously and have noted that these occur when $c^\sigma = -c$. The question is whether such putative flock spreads exist where c does not have this property. In all cases, we need a central Galois quadratic extension of a skewfield. There are two ways to do this. First of all, there are algebraic extensions, which are studied in the first volume mostly for quaternion division rings ([**34**] p. 294, chapter 3, Galois theory for division rings). Then there are transcendental extensions, using ideas of Jacobson [**18**], which have been considered in this text in part II, in multiple replacement-redux. There is a special situation, which does not quite fit these two categories, where we have a central quadratic extension of a skewfield, but not a Galois extension. We have considered this in volume II, [**35**], 16.1, Theorem 83. This can only occur in characteristic two and can construct very unusual simultaneous flock spreads.

The following theorem involving central Galois quadratic extension is given is a very general form.

THEOREM 64. ***(Lifting Skewfields – Addition of Inner*** $\sigma - A$ ***Derivative Functions)*** *Let D be a division ring. Consider two possible central quadratic extensions with generator θ over D:*

(1) $D(\theta)$ is a central Galois quadratic algebraic extension with automorphism σ fixing D pointwise.

(a) If the characteristic is not 2, assume that the basis is chosen so that the irreducible polynomial is $x^2 - \delta$, where δ is in $Z(D(\theta))$, the center of D.

(b) The characteristic is 2, and a basis is chosen so that the irreducible polynomial is $x^2 + x + \alpha$, where α is in $Z(D(\theta))$.

(2) D is a division ring finite-dimensional over the field center F. Use the tensor product construction to determine a super division ring D_1 of D that has a central Galois quadratic extension $D_1(\theta) = D_2$. σ

is the associated automorphism of order 2 that fixes D_1 pointwise. The basic irreducible polynomial for all characteristics is $x^2-(z_1+z_2)x+z_1z_2$, where z_1 and z_2 are indeterminants with respect to F, where σ: $z_1 \leftrightarrow z_2$.

(a) Hence, for characteristic not 2, noting that $(z_1+z_2)^2-4z_1z_2$ is in $Fix\sigma$, choose the polynomial $x^2-(z_1-z_2)$, with root θ so that $\theta^\sigma=-\theta$.

(b) For characteristic =2, we choose the irreducible polynomial as $x^2+x+(\frac{z_1z_2}{(z_1+z_2)^2}=\alpha)$.

that admits a central Galois quadratic extension, $D_1(\theta)$.

Consider

$$x=0, y=x\begin{bmatrix} u^\sigma+A_c(t), & A_c(u)+\theta t^\sigma \\ t & u \end{bmatrix},$$

for all t,u in $D_1(\theta)$. Let $A(w)=(w^\sigma-w)c$ for c in $Z(D_1(\theta))$. Similarly, the elements could more generally be in $E(\theta)$, and c in $Z(E(\theta))$.

(I) If $c^\sigma=-c$, then the matrix set is a semifield spread that is simultaneous and also is a double cover; there are both conical and hyperbolic spreads.

***(Hyperbolic Flocks and Both Left and Right Pseudo-regulus-Inducing Homology Groups.)** Thus, there are two hyperbolic flocks and one conical flock associated with this spread. There exists a pseudo-regulus-inducing homology group with axis $x=0$ and co-axis $y=0$. This means that the mapping $(x,y)\rightarrow(y,x)$ induces a left hyperbolic spread that induces a corresponding left hyperbolic flock. This result is valid for division rings of all characteristics.*

(II)(a) If c^σ is not equal to c and the characteristic is not 2, then the matrix set is a semifield spread that produces a conical flock and also has a pseudo-regulus-inducing a homology group with an axis $x=0$ and co-axis $y=0$. Again, this means that the mapping $(x,y)\rightarrow(y,x)$ induces a left hyperbolic spread that induces a corresponding left hyperbolic flock. In this case, there is one conical flock and one hyperbolic flock. For $c=c_0+c_1\theta$, either $c_0c_1=0$ or for c_0c_1 not 0, choose any c_1 and then choose $c_0=\frac{1-r^2}{4c_1}$ for any element r in the center.

(II)(b) If c^σ is not c and the characteristic is 2 and α is not 1, then there are two associated flocks: one conical and one hyperbolic, as in (II)(a). Since $\frac{z_1z_2}{(z_1+z_2)^2}$ is not 1, both the algebraic and the transcendental extensions are valid in this case. For $c=c_0+c_1\theta$, or choose any and (c_0,c_1) not $(0,0)$, which defines a conical flock.

(II)(b) If c^σ is not c, the characteristic is 2 and $\alpha=1$, for $c=c_0+c_1\theta$, for c_0,c_1 in D, for the algebraic extension case. Then the algebraic extensions are also valid. In this case, $c_0c_1=0$, not both 0.

(III) Therefore, we use $E(\theta)$ to denote $D(\theta)$ in the algebraic case and $D_1(\theta)$ in the transcendental cases for all characteristics, and for all c in $Z(E(\theta))$, we obtain a set of concal flock spreads and therefore of conical flocks.

(a) In all cases, the conical flocks are linear if $c^\sigma = -c$ and non-linear is c^σ is not $-c$. In the linear case, the flock spread is simultaneous and the hyperbolic flock is linear.

(b) For all cases there is an associated left-acting pseudo-regulus group, and the mapping $(x, y) \to (y, x)$ creates a nonlinear hyperbolic flock. Due to the length of this theorem, this part will be given in a separate chapter.

(c) Comparing (a) and (b), in the hyperbolic linear case, there are isomorphic hyperbolic flock spreads with associated non-isomorphic hyperbolic flocks. This will prove that the hyperbolic isomorphism question may be answered in the negative.

PROOF. **Characteristic not 2.**

We first consider the characteristic not 2 part. For any irreducible quadratic over a field F representing the center of the division ring, we assume that we have a central Galois quadratic extension. Since we have the quadratic equation in this characteristic, if we do not already have the form $x^2 - \beta$, where β is a non-square in F, we may take quadratic $x^2 + bx + c$ and take $b^2 - 4c = \beta$, noting that this must be non-square, then allowing that θ is a root, we see that since $-\theta$ is also a root, it follows that the induced automorphism σ must map θ to $-\theta$.

We have seen that since we may use the left and right unwrapping procedure to determine that we have a conical flock spread provided we have a quasifibration, we therefore assume that we do not have a quasifibration and there is some matrix element such that the $(2, 1)$-determinant is zero. We then have the following: There must be a matrix element with t non-zero such that

$$(u^\sigma + A_c(t))t^{-1}u = A_c(u) + t^\sigma\theta.$$

We have used this idea throughout volumes II and III, and we see that using the properties of $\sigma - A$ derivation sets, this equation can be rewritten as

$$(t^{-1}u)^\sigma(t^{-1}u) - A_c(t^{-1}u) = \theta.$$

Let $k = (t^{-1}u)$. We then have:

$$k^\sigma k - A_c(k) = \theta.$$

Since we have a 2-dimensional central extension, we may write $k = k_0 + k_1\theta$, where k_i are in E, $i = 0, 1$ and σ fixes k_i.

Here is the idea of both this proof and the analogous proof for characteristic 2. When we write out the left part of the equation, we may write it in the form $w_0 + w_1\theta$. This means that $w_0 = 0$ and $w_1 = 1$. Using these two equations, we obtain a contradiction and the proof of the result. However, since commutativity cannot be used, the arguments can be quite complicated. The ideas arise from those given in Johnson and Jha [**38**], where the so-called "metro equation" of Cohn [**18, 6**] is introduced.

We determine

$$-A_c(w) = -(w^\sigma - w)c = -((w_0+w_1\theta)^\sigma - (w_0+w_1\theta))c) = 2w_1(c_0+c_1\theta).$$

So,

$$-A_c(k) = 2k_1\beta c_0 + 2k_1c_1\theta.$$

We obtain:

$$k^\sigma k = (k_0 - k_1\theta)(k_0 + k_1\theta) = (k_0^2 - k_1^2\beta) + (k_0k_1 - k_1k_0)\theta.$$

Combining these two expressions, we obtain the following simultaneous equations:

$$(*)_1 : (k_0^2 - k_1^2\beta) + 2k_1\beta c_0 = 0,$$
$$(*)_2 : (k_0k_1 - k_1k_0) + 2k_1c_1 = 1.$$

Multiply equation $(*)_1$ on the left and then on the right by k_1. Since βc_0 is central, we see that

$$k_1k_0^2 = k_0^2k_1.$$

Now multiply $(*)_2$ on the left and then on the right by k_0, use the above equality, to obtain the following:

$$2(k_0^2k_1 - k_0k_1k_0) = -2c_1(k_0k_1 - k_1k_0),$$

then cancel the 2's and factor k_0 the left to obtain:

$$(k_0 + c_1)(k_0k_1 - k_1k_0) = 0.$$

Since we are working over a skewfield, there are no zero divisors; so one of the two product elements is zero. If $k_0 = -c_1$, then since $c_1$is central, k_0 and $k_1$commute so that

$$(k_0k_1 - k_1k_0) = 0.$$

Since this is then always valid, the conclusion that $k_0 = -c_1$ cannot be substantiated. From $(*)_2$, we have

$$k_1 = \frac{1}{2c_1},$$

or $c_1 = 0$. Since then, c_0 cannot be zero, we see that $c_0c_1 = 0$. And, there is now a contradiction to $(*)_2$. Therefore, c_1 is not zero, and we obtain from $(*)_1$ that

$$(k_0 2c_1)^2 - \beta + \beta 4c_0c_1 = 0.$$

From the preliminary notes, c_0 cannot be zero, or there is a contradiction. Therefore, $c_0c_1{=}0$ always provides a conical flock spread. Assume that c_0c_1 is not zero and

$$1 - 4c_0c_1 = r^2;$$

for r not 0 or 1. That is, choose any c_1 and then choose c_0 equal to

$$\frac{(1-r^2).}{4c_1}$$

Again, by the preliminary notes, we obtain a contradiction. This completes the proof for characteristic not 2.

Characteristic 2.

Therefore, we now assume that the characteristic is 2 and the central extension has an irreducible polynomial of $x^2 + x + \alpha$.

We return to the equation

$$k^\sigma k + A_c(k) = \theta,$$

if we do not have a quasifibration. Then

$$\begin{aligned} A_c(k) = (k^\sigma + k)c = (k_0 + k_1\alpha + k_1\theta) + (k_0 + k_1\theta))c = k_1\alpha(c_0 + c_1\theta) \\ = k_1\alpha c_0 + k_1\alpha c_1\theta. \end{aligned}$$

Also,

$$\begin{gathered} k^\sigma k = (k_0 + k_1\alpha + k_1\theta)(k_0 + k_1\theta) = \\ (k_0^2 + k_1k_0\alpha + k_1^2\alpha) + (k_1^2(\alpha+1) + k_0k_1 + k_1k_0)\theta. \end{gathered}$$

$$k^\sigma k + (k^\sigma + k)c = \theta$$

produces the following two simultaneous equations:

$$(\#)_1 : (k_0^2 + k_1k_0\alpha + k_1^2\alpha + k_1\alpha c_0) = 0,$$

and

$$(\#)_2 : (k_1^2(\alpha+1) + k_0k_1 + k_1k_0 + k_1\alpha c_1) = 1.$$

Now apply the automorphism σ to

$$k^\sigma k + A_c(k) = \theta,$$

to obtain:

$$kk^\sigma + (k + k^\sigma)c^\sigma = \theta^\sigma = \theta + \alpha.$$

Since

$$c^{\sigma} = c_0 + c_1\alpha + c_1\theta$$

and

$$kk^{\sigma} = (k_0^2 + k_0k_1\alpha + k_1^2\alpha) + (k_1^2(\alpha+1) + k_1k_0 + k_0k_1)\theta,$$

$$(\$)_1 : (k_0^2 + k_1k_0\alpha + k_1^2\alpha + k_1\alpha(c_0 + c_1\alpha) = \alpha,$$

and

$$(\$)_2 : (k_1^2(\alpha+1) + k_1k_0 + k_0k_1 + k_1\alpha c_1) = 1.$$

Note that $(\#)_2 = (\$)_2$. Then add $(\#)_1$ and $(\$)_1$to obtain:

$$(k_0k_1 + k_1k_0)\alpha + k_1\alpha^2 c_1 = \alpha.$$

Then we have the following:

$$(k_0k_1 + k_1k_0) + k_1\alpha c_1 = 1,$$

since α is central. Compare this with

$$(\#)_2 : (k_1^2(\alpha+1) + k_0k_1 + k_1k_0 + k_1\alpha c_1) = 1.$$

This shows that

$$k_1^2(\alpha+1) = 0.$$

If α is not 1, then $k_1 = 0$, but if so, this is the contradiction $0 = 1$. Hence, for α not equal to 1, c_0 and c_1 can be any elements of the center of E and are completely independent except that (c_0, c_1) is not $(0,0)$.

$\alpha = \mathbf{1}$.

Finally, assume that $\alpha = 1$. We shall prove that $c_0c_1 = 0$ gives conical flock planes.

$$(\#)_1^{\alpha=1} : (k_0^2 + k_1k_0 + k_1^2 + k_1c_0) = 0,$$

and

$$(\#)_2^{\alpha=1} : (k_0k_1 + k_1k_0 + k_1c_1) = 1.$$

First, consider $\boldsymbol{c_0 = 0}$. For $(\#)_2^{\alpha=1}$, left and right map by k_1, this will show that

$$k_1^2k_0 = k_0k_1^2.$$

This also is equivalent to

Since we now have:

$$C_0 : (k_0^2 + k_1k_0 + k_1^2) = 0.$$

Note that $k_0 = 0$ if and only if $k_1 = 0$. Therefore, we also have

$$k_1^2k_0^{-1} = k_0^{-1}k_1^2.$$

Multiply C_0 on the right by k_0^{-1} to obtain:

$$k_0 + k_1 + k_1^2k_0^{-1} = 0,$$

use the above equality to obtain

$$k_0 + k_1 + k_0^{-1}k_1^2 = 0.$$

Multiply the last equality on the left by k_0 to obtain:

$$k_0^2 + k_0 k_1 + k_1^2 = 0.$$

Comparing with C_0, we see that

$$k_1 k_0 = k_0 k_1.$$

From $(\#)_2^{\alpha=1}$, we see that

$$k_1 c_1 = 1.$$

Then

$$C_0 : (k_0^2 + k_1 k_0 + k_1^2) = 0.$$

Becomes

$$(k_0 c_1)^2 + (k_0 c_1) + 1 = 0.$$

The preliminary material shows that

$$(k_0 c_1 + \theta)(k_0 c_1 + \theta + 1) = 0.$$

But either zero possibility gives a contradiction, $k_0 c_1$ is θ or $\theta + 1$.

Finally, consider

$\boldsymbol{c_1 = 0}$. Hence,

$$(\#)_2^{\alpha=1} : (k_0 k_1 + k_1 k_0) = 1.$$

The procedure of left and right multiplication by k_1 and then by k_0 shows that

$$k_0 k_1^2 = k_1^2 k_0 \& \; k_1 k_0^2 = k_0^2 k_1.$$

Using:

$$(\#)_1^{\alpha=1} : (k_0^2 + k_1 k_0 + k_1^2 + k_1 c_0) = 0,$$

left and right multiply by k_1 and add the two equations together to obtain:

$$k_1 k_0^2 + k_0^2 k_1 + h_1^2 h_0 = h_1 h_0 h_1,$$

using the above identities,we obtain A

$$h_1 h_0 = h_0 h_1,$$

which is a contradiction to $(\#)_2^{\alpha=1}$, and completes the proof to all parts of the theorem except for part III.

Part III. We have seen previously that in the simultaneous case, both the conical and hyperbolic flocks are linear.

(a) The conical flock for any of the conical flock spreads is:

$$-x_1(A_c(t)^\sigma + A_c(t^\sigma)) - x_2 t^\sigma + x_3(t^\sigma \theta) - A_c(A_c(t)^\sigma) - x_4 = 0.$$

Using $A_c(t) = (t^\sigma - t)c$, and c central, we obtain:

$$-x_1(t - t^\sigma)(c^\sigma + c) - x_2 t^\sigma + x_3(t^\sigma \theta - (t - t^\sigma)(c^\sigma + c)c) - x_4 = 0.$$

We see that when $c^\sigma + c = 0$, we have a linear conical flock, and if $c^\sigma + c$ is not zero, the existence of t and t^σshows that the conical flock is nonlinear. To see this, when $t = 1$, we see that $x_2 = x_3\theta$, then we obtain

$$-x_1(t - t^\sigma)(c^\sigma + c) - x_3\theta t^\sigma + x_3(t^\sigma \theta - (t - t^\sigma)(c^\sigma + c)c) - x_4 = 0.$$

Then, when $t - t^\sigma$is not zero, we see that $x_1 = x_3 c$. Therefore, the intersection of the planes of the flock is

$$(x_3 c, x_3\theta, x_3, 0),$$

a point.

Then $c^\sigma + c = 0$, the intersection of the planes of this flock is

$$(x_1, x_3\theta, x_3, 0),$$

a line.

Moving to the hyperbolic flock when $c^\sigma + c = 0$, the associated hyperbolic flock spread is:

$$x = 0, y = x \begin{bmatrix} u^\sigma & A_c(t) \\ 0 & u \end{bmatrix}, y = x \begin{bmatrix} t^\sigma, & \theta + A_c(t) \\ 1 & t \end{bmatrix} \begin{bmatrix} v^\sigma & A_c(t) \\ 0 & v \end{bmatrix}$$

for all u, t in $E(\theta)$, for non-zero v in $E(\theta)$.

The hyperbolic flock is given by

$$\Sigma_t : x_1(\theta + A_c(A_c(t^{\sigma^2}) - x_2 t^{\sigma^2} + x_3 t - x_4 = 0$$

and

$$\varrho : x_2 = x_3.$$

And since this becomes

$$\Sigma_t : x_1\theta - x_2 t + x_3 t - x_4 = 0,$$

we see the intersection of the planes

$$\varrho : x_2 = x_3.$$

is the line

$$(x_1, x_2, x_2, x_1\theta).$$

We now consider part III (b) in a separate theorem. This completes the given theorem. □

In the next few chapters, we look at the resolutions and realize that these conical flock spreads also admit left-inversing hyperbolic flock spreads.

CHAPTER 19

Resolved conical flocks

We now resolve the conical spreads of the previous theorem, the proof of which has been seen in the previous chapter.

COROLLARY 65. *Each of the conical flock spreads of the previous theorem may be resolved using the resolution mappings*

$$(x,y) \to (x\begin{bmatrix} 1 & -e \\ 0 & 1 \end{bmatrix}, y\begin{bmatrix} 1 & -e \\ 0 & 1 \end{bmatrix})$$

to obtain the following set of conical flock spreads:

$$x = 0, y = x\begin{bmatrix} u^{\sigma} + t^{\sigma}(e^{\sigma} + e), & t^{\sigma}(\theta - e^{2}) \\ t & u \end{bmatrix}.$$

For all e in $Z(E(\theta))$.

It is an open question whether the constructed conical spreads in the pre-resolved conical spreads and the constructed conical spreads in the resolved conical spreads are isomorphic.

THEOREM 66. *Each resolved conical flock spread admits a left-acting pseudo-regulus-inducing homology group. All of the associated hyperbolic flock spreads are nonlinear. Hence, we obtain the spread with components*

$$x = 0, y = 0, y = x\begin{bmatrix} u^{\sigma} + t^{\sigma}(e^{\sigma} + e), & t^{\sigma}(\theta - e^{2}) \\ t & u \end{bmatrix}^{-1}.$$

This set of inverses has the following form:

$$x = 0, y = 0, y = x\begin{bmatrix} r^{\sigma} & 0 \\ 0 & r \end{bmatrix}, \textit{and}$$

$$y = x\begin{bmatrix} v^{\sigma} & 0 \\ 0 & v \end{bmatrix}\begin{bmatrix} s^{\sigma} + e^{\sigma} + e, & \theta - e^{2} \\ 1 & s \end{bmatrix},$$

for all non-zero r and v and for all s in $E(\theta)$. Let $\Delta_s = (s^{\sigma} + e^{\sigma} + e)s - (\theta - e^{2}))$, the (2,1)-determinant of the indicated matrix. Applying the $(x,y) \to (y,x)$ left-inversing mapping and modifying the representation

DOI: 10.1201/9781003767978-19

to hyperbolic type to obtain (changing to the use of "t" for general cases (s=t))

$$y = x\begin{bmatrix} -t & ,-(1-t\Delta_t^{-1}(t^\sigma+e^\sigma+e))\Delta_t^\sigma \\ 1, & -\Delta_t^{-1}(t^\sigma+e^\sigma+e)\Delta_t^\sigma \end{bmatrix},$$

for all t in $E(\theta)$.

Therefore, we obtain the following hyperbolic flock:

$$\Pi_{-\Delta_t^{-1}(t^\sigma+e^\sigma+e)\Delta_t^\sigma} :$$

$$b_1(t\Delta_t^{-1}(t^\sigma+e^\sigma+e))\Delta_t^\sigma-\Delta_t^\sigma)+b_2t^\sigma-b_3(\Delta_t^{-1}(t^\sigma+e^\sigma+e)\Delta_t^\sigma)-b_4=0,$$

and

$$\varrho : b_2 = b_3.$$

PROOF. To show that there is a left-acting pseudo-regulus inducing-homology group may be seen by considering the fact that the pre-resolutions admit left-acting pseudo-regulus-inducing homology groups, but with a different group: The group with elements

$$(x,y)\to(x\begin{bmatrix} u^\sigma & A_c(u) \\ 0 & u \end{bmatrix},y),$$

is the homology group acting on the pre-resolution, and the group with elements

$$(x,y)\to(x\begin{bmatrix} u^\sigma & 0 \\ 0 & u \end{bmatrix},y),$$

is the homology group acting on the resolved spread.

Now apply the e-mapping

$$(x,y)\to(x\begin{bmatrix} 1 & -e \\ 0 & 1 \end{bmatrix},y\begin{bmatrix} 1 & -e \\ 0 & 1 \end{bmatrix}),$$

as follows:

$$\begin{bmatrix} 1 & e \\ 0 & 1 \end{bmatrix}\begin{bmatrix} v^\sigma & A_e(v) \\ 0 & v \end{bmatrix}\begin{bmatrix} u^\sigma+A_e(t), & A_e(u)+\theta t^\theta \\ t & u \end{bmatrix}\begin{bmatrix} 1 & -e \\ 0 & 1 \end{bmatrix}$$

is the resolved spread since it admits the left-inversion pseudo-regulus homology group, and since

$$\begin{bmatrix} 1 & e \\ 0 & 1 \end{bmatrix}\begin{bmatrix} v^\sigma & A_e(v) \\ 0 & v \end{bmatrix}\begin{bmatrix} 1 & -e \\ 0 & 1 \end{bmatrix}=\begin{bmatrix} v^\sigma & 0 \\ 0 & v \end{bmatrix},$$

now insert

$$\begin{bmatrix} 1 & -e \\ 0 & 1 \end{bmatrix}\begin{bmatrix} 1 & e \\ 0 & 1 \end{bmatrix}$$

before and after the spread matrix. This will show that the resolved spread admits the form of the pseudo-regulus-inducing a homology group relative to the form of type 2 of the spread.

We use the same analysis as previous when forming the structure of the matrices. This part of the argument will be left to the reader to complete. □

PROBLEM 67. It will be useful to the reader to work through how to obtain the form required for a hyperbolic flock in the situation above.

REMARK 68. (1) It has been assumed that the constant elements in the matrix sets are central or at least are left linear functions when considering a left quasifibration. For right quasifibrations, the constants need to be right linear functions.

(2) In the present situation, the constants are all central elements. This means there must be a method to show that the left semifield spread sets also give rise to right semifield spread sets. Since the right spread sets correspond loosely to the transpose matrices of the left spread sets, as this is true when $E(\theta)$ is a field, if the transpose of the left spreads are also left spreads, all of these left semifield spreads would give rise to right semifield spreads, even in the non-commutative situations.

(3) Previously, it was proved that all transposes are, in fact, spreads. Therefore, all of these transposes provide right conical flocks. All of the transposes of the right conical flocks provide right spreads, which correspond to the original left conical flock spreads.

(4) We also look at similar material, where the elements c are considered in $E(\theta) - Z(E(\theta))$, where now the left matrix sets form quasifibrations but are not spreads. However, their resolutions have the correct form to construct right conical spreads.

(5) There are analogous right conical flock spreads using right lifted skewfield spreads and the analogous versions of the addition of inner σ-derivation functions. That is, all of the left conical spreads have right analogues. Therefore, there are constructions from right quasifibrations that produce left conical spreads.

Again, we have:

$$x = 0, y = x \begin{bmatrix} u^{\sigma} + t^{\sigma}(e^{\sigma} + e), & t^{\sigma}(\theta - e^2) \\ t & u \end{bmatrix}.$$

The resolved left spreads for all e in $Z(E(\theta))$, for all t, u in $E(\theta)$.

CHAPTER 20

The isomorphism questions

In this chapter, given any finite-dimensional division ring D, over a field F, we have constructed the lifted skewfield semifield set of planes as follows: $S_{\delta,\gamma}$ for δ non-zero, where F. Here, σ is an outer automorphism of order 2 and

$$x = 0, y = x \begin{bmatrix} u^{\sigma} & (\theta\delta + \gamma)t^{\sigma} \\ t & u \end{bmatrix}$$

for all t, u in $E(\theta)$, where $E(\theta)$ is the division ring constructed using D, as in the Multiple Replacement Theorem – Redux. When $(\theta\delta + \gamma)^{\sigma} = -\theta$, the spread is simultaneous and is a double homology cover, a double double cover. Using the hyperbolic representation, we have shown that the left-inversing hyperbolic flock is nonlinear, whereas the hyperbolic flock using the standard right-acting pseudo-regulus-inducing homology group is linear. This proves that the hyperbolic isomorphism question may be answered in the negative; two isomorphic hyperbolic flock spreads do not produce isomorphic flocks.

Now we consider the conical representation, and we isolate on using $K = \begin{bmatrix} 1 & f \\ 0 & 1 \end{bmatrix}$ −mappings f not in the center.

20.1. In-depth resolution

In this section, we take another look at the question of whether isomorphic conical flock spreads do not necessarily lead to isomorphic conical flocks. Consider a particular example: Let D be a division ring central Galois quadratic extension of characteristic not 2 of a division ring E, so that $D = E(c)$ where $x^2 - d$ for d in the center of E is a non-square in E. Let $c = \sqrt{d}$ in the formal sense so that $D = E(\sqrt{d}) = E(c)$ is 2-dimensional over E. Define σ as the 2-dimensional automorphism of $Z(E)(c)$, and define σ on E to be trivial. Let $A(u) = (u^{\sigma} - u)c$. Note that $c^{\sigma} = -c$.

The following

$$x = 0, y = x \begin{bmatrix} u^{\sigma} + A(t), & A(u) + t^{\sigma}c \\ t & u \end{bmatrix}$$

DOI: 10.1201/9781003767978-20

for all t, u in D is a semifield conical flock that is equivalent to the following conical flock with center (0,0,0,1):

$$\Sigma_t^1 :$$

$$-a_1(A(t)^{\sigma^{-1}} + A(t^{\sigma^{-1}})) - a_2 t^{\sigma^{-1}} + a_3(t^{\sigma} c - A(A(t)^{\sigma^{-1}}) - a_4 = 0,$$

for all t in D.

Since $\sigma^{-1} = \sigma, A(t)^{\sigma} = ((t^{\sigma} - t)c)^{\sigma} = (t - t^{\sigma})c^{\sigma} = (t^{\sigma} - t)c$, since $c^{\sigma} = -c$. And $A(t^{\sigma^{-1}}) = A(t^{\sigma}) = (t - t^{\sigma})c = -A(t)^{\sigma^{-1}}$. Therefore, the $-a_1$ function is 0. Then we have $A(A(t)^{\sigma^{-1}}) = -A(A(t)) = 0$. Therefore, the a_3 function is $t^{\sigma} c$. Hence, this shows that the conical flock is

$$-a_1 0 - a_2 t^{\sigma} + a_3 t^{\sigma} c - a_4 = 0.$$

If we resolve the above conical flock spread using the mapping

$$(x, y) \to (x \begin{bmatrix} 1 & -c \\ 0 & 1 \end{bmatrix}, y \begin{bmatrix} 1 & -c \\ 0 & 1 \end{bmatrix}),$$

we obtain the following flock spread

$$x = 0, y = x \begin{bmatrix} u^{\sigma}, & (t^{\sigma}(c - c^2) \\ t, & u \end{bmatrix},$$

for all t, u in D, which produces the following conical flock spread:

$$\Sigma_t^2 :$$

$$-b_1 0 - b_2 t^{\sigma} + b_3 t^{\sigma}(c - c^2) - b_4 = 0.$$

These are two linear conical flock spreads that appear to be isomorphic. Let's go back and find out if that could be the case. If these two conical flocks are isomorphic, they should be isomorphic by the mapping

$$(x, y) \to (x \begin{bmatrix} 1 & -c \\ 0 & 1 \end{bmatrix}, y \begin{bmatrix} 1 & -c \\ 0 & 1 \end{bmatrix}).$$

The non-commutative transforms of the first conical flock have the following form: For $x_1(x_2 - A(x_1^{\sigma^{-1}}))$ non-zero, the following point is the "transform" of (x_1, x_2, x_3, x_4).

$$(x_2, x_1 A(x_1^{-\sigma^{-1}} x_2^{\sigma^{-1}}) + x_2 x_1^{-\sigma^{-1}} x_2^{\sigma^{-1}},$$
$$, x_1, x_4 - A(x_3^{\sigma^{-1}}) - (x_2 - A(x_1^{\sigma^{-1}}))x_1^{-\sigma^{-1}} x_3^{\sigma^{-1}}).$$

For the second conical flock, the following set of points represents the transforms:

$$(x_2, x_2 x_1^{-\sigma^{-1}} x_2^{\sigma^{-1}}, x_1, x_4 - (x_2 x_1^{-\sigma^{-1}} x_3^{\sigma^{-1}}).$$

However, we have glossed over something. The previous would be true if, in both the first and the second cases, we use (x_1, x_2, x_3, x_4). But this is not valid. Here is what happens: the first conical flock spread is mapped to:

$$((x,y)\to (x\begin{bmatrix}1 & -c\\ 0 & 1\end{bmatrix}, x\begin{bmatrix}u^\sigma + A(t), & A(u)+t^\sigma c\\ t & u\end{bmatrix}\begin{bmatrix}1 & -c\\ 0 & 1\end{bmatrix}).$$

Let $x^* = x\begin{bmatrix}1 & -c\\ 0 & 1\end{bmatrix}$, and $y^* = x^*\begin{bmatrix}1 & c\\ 0 & 1\end{bmatrix}$. Then we have:

$$y^* = x^*\begin{bmatrix}w^\sigma, & (t^\sigma(c-c^2)\\ t, & w\end{bmatrix}$$

where $u = w + tc$. Now, if we continue to use $(x^*, y^*) = (z_1, z_2, z_3, z_4)$, then it is seen that we have something different: Here are the actual transforms for the 2nd flock:

$$(z_2, z_2 z_1^{-\sigma^{-1}} z_2^{\sigma^{-1}}, z_1, z_4 - (z_2 z_1^{-\sigma^{-1}} z_3^{\sigma^{-1}}).$$

What this is trying to say is this: Does the mapping

$$(x,y)\to (x\begin{bmatrix}1 & -c\\ 0 & 1\end{bmatrix}, y\begin{bmatrix}1 & -c\\ 0 & 1\end{bmatrix}).$$

map the transforms of the first conical flock to the transforms of the second conical flock? What happened is that the mapping resolves the conical flock spread to a related conical flock spread such that, on its own, the transforms appear to look identical to the transforms of the first flock spread by eliminating the $A-\sigma$-derivative function. So, this sort of makes you believe that the two conical flocks might be isomorphic.

There are two ways to see that they are not isomorphic by the induced map: (1) First of all, it would have to be that the transforms of the first conical flock would have to map to the transforms of the second conical flock by the above mappings. (2) Second, if these mappings induce an isomorphism between the two flocks, they would have to map the common line of the first linear flock to the common line of the second linear flock.

Now, since the original conical flock spread is simultaneous, consider the hyperbolic flock spread

$$x = 0, y = x\begin{bmatrix}u^\sigma, & (u^\sigma - u)c = A(u)\\ 0 & u\end{bmatrix},$$

$$y = x\begin{bmatrix}t^\sigma & , A(t)+c\\ 1 & t\end{bmatrix}\begin{bmatrix}v^\sigma & (v^\sigma - v)c\\ 0 & v\end{bmatrix},$$

for all u, t in D and for all v non-zero. Apply the same mapping

$$(x,y)\to (x\begin{bmatrix}1 & -c\\ 0 & 1\end{bmatrix}, y\begin{bmatrix}1 & -c\\ 0 & 1\end{bmatrix}).$$

Then the resolution of this spread is the same as the restructured image of the conical flock spread image. This is

$$x^* = 0, y^* = x^* \begin{bmatrix} u^\sigma & 0 \\ 0 & u \end{bmatrix}, y^* = x^* \begin{bmatrix} s^\sigma, & c - c^2 \\ 1 & s \end{bmatrix} \begin{bmatrix} v^\sigma & 0 \\ 0 & v \end{bmatrix},$$

where $t = s + c$. Here are the two hyperbolic spreads, written in (x^*, y^*).

$$\Pi_t^1 :$$
$$d_1((A(t) + c) - A((t^\sigma)^\sigma) - d_2 t^{\sigma^2} + d_3 t - d_4 = 0,$$

and

$$\varrho^1 : d_2 = d_3.$$

And the second:

$$\Pi_s^2 :$$
$$e_1(c - c^2) - e_2 t + e_3 t - e_4 = 0,$$

and

$$\varrho^2 : e_2 = e_3.$$

PROBLEM 69. Conical Problem: It is left as a problem to verify that neither of these conditions, (1) or (2), can hold. A much harder question would be the following: Let χ be a mapping that has the following properties:

(1) χ maps the cone defined by the pseudo-regulus net

$$x = 0, \begin{bmatrix} u^\sigma, & (u^\sigma - u)c \\ 0 & u \end{bmatrix},$$

for all u in D, as above to the cone defined by the pseudo-regulus net

$$x = 0, \begin{bmatrix} u^\sigma, & 0 \\ 0 & u \end{bmatrix},$$

for all u in D,

(2) χ maps Σ_t^1 to Σ_t^2,

(3) Then χ must be the mapping

$$(x, y) \to (x \begin{bmatrix} 1 & -c \\ 0 & 1 \end{bmatrix}, y \begin{bmatrix} 1 & -c \\ 0 & 1 \end{bmatrix}).$$

If this problem could be solved, this would show that the two conical flocks could not be isomorphic.

Now for the hyperbolic piece:

PROBLEM 70. Now complete the same analysis for the two hyperbolic problems.

CHAPTER 21

Hyperbolic isomorphism question

We continue with the previous chapter and consider the set of the following conical flock spreads:

$$x = 0, y = x \begin{bmatrix} u^{\sigma} + A_c(t) & , A_c(u) + t^{\sigma}\theta \\ t & u \end{bmatrix}$$

for all t, u in $D_1(\theta)$ or $D(\theta)$, use $E(\theta)$ in general. When $c^{\sigma} = -c$. We have seen in other chapters that these flocks are simultaneous flock spreads: a conical and a hyperbolic flock spread, such that **both are linear flocks.**

In general, all of these spreads, whether linear or not, admit left-acting pseudo-regulus homology groups. To see this, consider

$$\begin{bmatrix} v^{\sigma} & A(v) \\ 0 & v \end{bmatrix} \begin{bmatrix} u^{\sigma} + A(t), & A(u) + t^{\sigma}\theta \\ t & u \end{bmatrix}$$

which is equal to:

$$\begin{bmatrix} (vu)^{\sigma} + v^{\sigma}A(t) + A(v)t, & v^{\sigma}A(u) + A(v)u + (vt)^{\sigma} \\ (vt), & vu \end{bmatrix}.$$

Then

$$v^{\sigma}A(t) + A(v)t = A(vt),$$

and

$$v^{\sigma}A(u) + A(v)u = A(vu).$$

Thus, we obtain

$$\begin{bmatrix} (vu)^{\sigma} + A_c(vt), & A_c(w) + (vt)^{\sigma}\theta \\ vt, & w \end{bmatrix},$$

which shows, since v is non-zero, that we have a left-acting pseudo-regulus-inducing homology group. Since we have an additive matrix set and therefore a semifield spread, we may list the set of component matrices separately. Therefore, let $t = 0$, to obtain the partial spread:

$$x = 0, y = 0, y = x \begin{bmatrix} u^{\sigma} & A(u) \\ 0 & u \end{bmatrix},$$

DOI: 10.1201/9781003767978-21

for all non-zero u in $E(\theta)$. Now let $v = t^{-1}$, when t is non-zero, to obtain the matrix

$$\begin{bmatrix} w^{\sigma}, & A_c(w)+\theta \\ 1, & w \end{bmatrix},$$

and, since we still have the left-acting homology group, the spread has the following representation:

$$x = 0, y = 0, y = x\begin{bmatrix} u^{\sigma} & A(u) \\ 0 & u \end{bmatrix},$$

$$y = x\begin{bmatrix} v^{\sigma} & A(v) \\ 0 & v \end{bmatrix}\begin{bmatrix} w^{\sigma}, & A_c(w)+\theta \\ 1, & w \end{bmatrix},$$

for all non-zero u and v, and for all w in $E(\theta)$. We take the following isomorphic spread obtaine by the mapping $(x, y) \to (y, x)$. Note that $x = 0, y = 0$ and are interchanged, and since

$$\begin{bmatrix} r^{\sigma} & A(r) \\ 0 & r \end{bmatrix}; r \text{ in } E(\theta)$$

is a skewfield isomorphic to $E(\theta)$, the set is invariant under the inversing mapping. And, as the inversing of matrices interchanges the order, we have the following spread: (replacing by t for the sake of form and notice that the representation is unchanged as a set, so we continue to use the same notation for u and v)

$$x = 0, y = 0, y = x\begin{bmatrix} u^{\sigma} & A(u) \\ 0 & u \end{bmatrix}, and$$

$$y = x\begin{bmatrix} v^{\sigma} & A(v) \\ 0 & v \end{bmatrix}\begin{bmatrix} t^{\sigma}, & A_c(t)+\theta \\ 1, & t \end{bmatrix}^{-1}\begin{bmatrix} v^{\sigma} & A(v) \\ 0 & v \end{bmatrix},$$

for all non-zero u and v and for all w. In the following, we come back to the analysis of the lifted skewfields adjoined, all of which admit left-acting pseudo-regulus-inducing homology groups.

THEOREM 71. *Let π be any lifted skewfield spread adjoined by $\sigma - A_c$. Applying the isomorphism mapping $(x, y) \to (y, x)$, we consider the hyperbolic spread of the initial spread with the following form:*

$$x = 0, y = 0, y = x\begin{bmatrix} u^{\sigma} & A(u) \\ 0 & u \end{bmatrix},$$

$$y = x\begin{bmatrix} t^{\sigma}, & A_c(t)+\theta \\ 1 & t \end{bmatrix}^{-1}\begin{bmatrix} v^{\sigma} & A(v) \\ 0 & v \end{bmatrix}$$

for all t, u and for all v non-zero in $D(\theta)$, or $D_1(\theta)$, denoted together by $E(\theta)$.

Denote $\Delta_t = (t^\sigma t - (A_c(t) + \theta))$. *Then the representation may be written in the following form:*

$$x = 0, y = 0, y = x \begin{bmatrix} u^\sigma & A(u) \\ 0 & u \end{bmatrix},$$

$$y = x \begin{bmatrix} t\Delta_t^{-1}, & 1 - t\Delta_t^{-1}t^\sigma \\ -\Delta_t^{-1} & \Delta_t^{-1}t^\sigma \end{bmatrix} \begin{bmatrix} v^\sigma & A_c(v) \\ 0 & v \end{bmatrix},$$

for all t *and for all* v *and* u *non-zero. To convert to the hyperbolic spread form, we may write the (2,1)-entries as 1 by first letting an element* $v^\sigma = -\Delta_t$ *and applying the matrix multiplication*

$$\begin{bmatrix} -\Delta_t, & -A_c(\Delta_t^\sigma) \\ 0 & -\Delta_t^\sigma \end{bmatrix}.$$

We, therefore, have the following hyperbolic flock spread:

$$x = 0, y = 0, y = x \begin{bmatrix} w^\sigma & A(w) \\ 0 & w \end{bmatrix}, \text{ and}$$

$$y = x \begin{bmatrix} -t, & -(t\Delta_t^{-1}A_c(\Delta_t^\sigma) + (1 - t\Delta_t^{-1}t^\sigma)\Delta_t^\sigma) \\ 1, & , \Delta_t^{-1}(A_c(\Delta_t^\sigma) - t^\sigma\Delta_t^\sigma) \end{bmatrix} \begin{bmatrix} u^\sigma & A_c(u) \\ 0 & u \end{bmatrix},$$

for all t *and for all non-zero* w *and* u *in* $E(\theta)$.

This produces the following hyperbolic flock:

$$\Pi_{\Delta_t^{-1}(A_c(\Delta_t^\sigma) - t^\sigma\Delta_t^\sigma)}$$

$$x_1(t\Delta_t^{-1}(t^\sigma\Delta_t^\sigma - A_c(\Delta_t^\sigma)) - \Delta_t^\sigma + A_c(t^\sigma)) + x_2t^\sigma +$$

$$+x_3\Delta_t^{-1}(A_c(\Delta_t^\sigma) - t^\sigma\Delta_t^\sigma) - x_4 = 0 \text{ and}$$

$$\varrho : x_2 = x_3.$$

(1) All of these hyperbolic flocks are nonlinear.

(2) If the characteristic is not 2, using the intersection of

$$\varrho \cap \Pi_{\Delta_t^{-1}(A_c(\Delta_t^\sigma) - t^\sigma\Delta_t^\sigma)}; t = 0, 1$$

will show that the planes of the hyperbolic flocks either share a unique point or are disjoint.

The planes are disjoint if and only if using the intersection equation

$$\varrho \cap_t \Pi_{\Delta_t^{-1}(A_c(\Delta_t^\sigma) - t^\sigma\Delta_t^\sigma)}$$

for $R = (1 - \theta)^{-1} = (1 + \theta)(1 - \delta)^{-1}$, *and* $\theta^2 = \delta$, $x_2 = x_3$, *and* $-x_4 = x_1\theta + x_2 2c$.

$$\text{Equation (for } t = 1) : x((2c)(1 - R\theta) + -R(\theta + 1) + 1) = .$$

$$x((2c)R\theta + R(\theta+1) - 1).$$

Not both coefficients for x_1 *and* x_2 *can be zero at the same time. Therefore, the hyperbolic flock is nonlinear.*

(3) When the characteristic is 2, we take both α *is not 1 and equal to 1 at the same time, realizing that the values for* c *are all non-zero values when* α *is not 1 and for* $c = c_0 + c_1\theta$ *and* $c_0c_1 = 0$*, when* $\alpha = 1$.

(4) When $c^\sigma = -c$*, the original hyperbolic flock is linear, whereas the constructed hyperbolic flock from the same spread is nonlinear.*

(5) This answers the hyperbolic isomorphism question in the negative; isomorphic hyperbolic spreads can correspond to non-isomorphic flocks.

Proof. We first assume that the characteristic is not 2. The proof showing that the flock is nonlinear is to consider the partial intersection

$$\varrho \cap \Pi_{\Delta_t^{-1}(A_c(\Delta_t^\sigma)-t^\sigma\Delta_t^\sigma)}; t = 0, 1, -1.$$

This method is useful when $t = 0, 1, -1$, $A_c(t) = 0$. For notational purposes only, let

$$\chi_t = \Pi_{\Delta_t^{-1}(A_c(\Delta_t^\sigma)-t^\sigma\Delta_t^\sigma)}.$$

Therefore, we shall consider $\varrho \cap_{t=0,1,-1} \chi_t$.

When $t = 0$: We have, using the intersection with ϱ,

$$x_1(-\Delta_0^\sigma) + x_2(\Delta_0^{-1}A_c(-\theta^\sigma)) - x_4 = 0.$$

This works out to

$$-x_4 = x_1\theta + x_2 2c.$$

The argument to show nonlinearity is to show that either the plane interesections share either a point or have an empty intersection. This is as far as is needed to consider the proof. However, we shall see that a more complete solution would most likely always show that there is an empty intersection.

When $t = 1$. We shall encounter the term $(1-\theta)^{-1}$, which is $(1+\theta)(1-\delta)^{-1}$. To see this, we see that if $(1-\theta)^{-1} = l_0 + l_1\theta$, where l_i are in a field, $i = 1, 2$, recalling the construction of $E(\theta)$. Then use

$$(l_0 + l_1\theta)(1-\theta) = 1,$$

to obtain

$$l_0 = l_1 \& \ l_1 = (1-\delta)^{-1},$$

where $\theta^2 = \delta$, which is not 1.

Now putting $t = 1$ in the equation for the hyperbolic flock, and using the information that

$$-x_4 = x_1\theta + x_2 2c,$$

we provide a lemma:

Lemma 1: If $R = (1+\theta)(1-\delta)^{-1}$ then

$$x_1((1+2c)R\theta + R - 1) = x_2((1+2c)(1-R\theta) - R).$$

Proof: When $t = 1$, we obtain the following:

$$x_1(\theta + \Delta_1^{-1}(\Delta_1^\sigma - A_c(1-\theta)^\sigma) - \Delta_1^\sigma)+$$
$$+x_2(1+2c+\Delta_1^{-1}(A_c(1-\theta)^\sigma) - \Delta_1^\sigma) = 0.$$

Letting $R = (1-\theta)^{-1} = (1+\theta)(1-\delta)^{-1}$, and working out the above equation, remembering that A_c is additive and $A_c(1) = 0$, we obtain the stated conclusion.

So, if it is not the case that both of the coefficients of $x_1$and x_2 are 0, then the intersection is either a single point or is trivial. But if both of these coefficients are 0, then

Lemma 2: If both coefficients are 0, then $R(1+\theta) = 1$, a contradiction.

Proof: If both coefficients are 0 then solving both coefficient equations for $(1+2c)$, we have the following:

$$(1+2c) = (1-R)(R\theta)^{-1} = R(1-R\theta)^{-1},$$

since $R\theta$ and $1-R\theta$ are both non-zero. Therefore, we have:

$$(1-R)(1-R\theta) = R(R\theta),$$

remembering we are able to use field operations here. This implies that

$$R(1+\theta) = 1,$$

but this says that

$$(1+2\theta+\delta)(1-\delta)^{-1} = 1,$$

a contradiction since

$$2(1-\delta)^{-1}\theta$$

must be 0, a contradiction.

This shows that the hyperbolic flock planes either share one point or have an empty intersection.

So, we have established that the hyperbolic flocks are all nonlinear, what has not been definitely determined is whether the intersection of the points is a point or is empty. This completes the proof of (2) for characteristic not 2.

Now assume that the characteristic is 2. Let $\theta^\sigma = \theta^2 = \theta + \alpha$. In the even characteristic case, then

$$x_1(t\Delta_t^{-1}(t^\sigma\Delta_t^\sigma + A_c(\Delta_t^\sigma)) + \Delta_t^\sigma + A_c(t^\sigma)) + x_2t^\sigma+$$

$$+x_3\Delta_t^{-1}(A_c(\Delta_t^\sigma)+t^\sigma\Delta_t^\sigma)+x_4=0.$$

First, let $t=0$, to obtain (in the intersection of planes $x_2=x_3$),

$$x_1(\theta+\alpha)+x_3(\theta^{-1}\alpha c)+x_4=0.$$

We then obtain, from the expression for x_4 previously,

$$x_1(1+\theta)^{-1}(\alpha c+1+\theta+\alpha)+1)=$$
$$=x_3(\theta^{-1}\alpha c)+(1+\theta)^{-1}(\alpha c+1+\theta+\alpha))+1).$$

Let

$$M=(1+\theta)^{-1}(\alpha c+1+\theta+\alpha)+1.$$

Therefore, we must have a nonlinear hyperbolic flock of c. Since it cannot be determined when one of the coefficients is 0, we consider the possibilities:

$$M=0$$

if and only if $x_1=0$. In this case, the intersection of the planes is

$$\{(0,x_3,x_3,x_3\theta^{-1}\alpha c);x_3\text{ in }E(\theta)\}.$$

When

$$M+\theta^{-1}\alpha c=0,$$

then $x_3=0$, and the intersection is

$$\{(x_1,0,0,x_1(\theta+\alpha);x_1\text{ in }E(\theta)\}.$$

If M is non-zero and $M+\theta^{-1}\alpha c$ is non-zero, then we have the intersection of the planes using

$$x_3=x_1(1+\theta^{-1}\alpha cM^{-1}),$$

as

$$\{(x_1,x_1(1+\theta^{-1}\alpha cM^{-1}),x_1(1+\theta^{-1}\alpha cM^{-1}),x_1(\theta+\alpha+(\theta^{-1}\alpha cM^{-1});$$
$$x_1\text{in }E(\sigma)\}.$$

This completes the proof of the theorem. □

Part 9

The Baer flocks

In volume I [**34**], the theory concentrated on connecting flocks and spreads over skewfields K and considered both left and right flock spreads and left and right flocks. When K is a field, the work of Cherowitzo-Johnson-Vega considered generalizing the theory to σ-conical quasiflocks. We shall extend the theory of flocks and spreads to quasiflocks and quasifibrations over skewfields, although we take a different point of view than when K is a field. Perhaps the main reason for doing this is to show how left and right quasiflocks are generally connected, whereas left and right flocks are not. Although, this is not quite the complete story.

We wish to consider the difference between generic forms of a quasifibration and a spread. However, there are none! They look alike. In the flock theory in volume II, we did not discuss if left flocks and right flocks could be connected. These forms are studied in volume II for left and right hyperbolic and conical flocks. These are exactly the same as the generic forms for quasifibrations. How should these be made distinct? Considering extending the Thas-Walker theory in a different manner. Given a spread matrix set of either hyperbolic or conical generic form, which is a partial spread: The hyperbolic

$$x = 0, y = 0, y = x \begin{bmatrix} u^{\sigma} & A(u) \\ 0 & u \end{bmatrix},$$

$$y = x \begin{bmatrix} f(t) & g(t) \\ 1 & t \end{bmatrix} \begin{bmatrix} v^{\sigma} & A(v) \\ 0 & v \end{bmatrix},$$

for all non-zero u and v and for all t in D, where D is a skewfield, where the indicated functions f and g define a set of pseudo-regulus derivable nets that share two components. We see that if we obtain a partial spread, it must be a "maximal partial spread".

Similarly, for the conical type, we would have a maximal partial spread of the form:

$$x = 0, y = x \begin{bmatrix} u^{\sigma} + F(t), & G(t) + A(u) \\ t & u \end{bmatrix},$$

for all t,u in D, which consists of a set of pseudo-regulus nets sharing one component. Assume that the pseudo-regulus central collineation groups act on the maximal partial spreads, homology group and elation groups, respectively.

DEFINITION 72. Let π be a left maximal partial spread that is of hyperbolic form or conical form that consists of a set of pseudo-regulus nets admitting a central collineation group that is pseudo-regulus-inducing. Define the generalized hyperbolic quadrics and generalized quadratic cones in the same manner as in the previous material on spreads. Define the Baer transforms as the transforms defined previously, but over the Baer subspaces. Define the set of such right Baer transforms, the right Baer hyperbolic and the right Baer cone sets. Consider the left planes as in the spread cases. Then the "Baer flocks" are defined as the intersection with the left planes and the right Baer hyperbolic sets or right Baer cone sets. It is clear that the functions involved in the construction of the maximal partial spreads bear the same relationship to the set $(x_1, x_2, (x_1, x_2)M = (x_3, x_4))$ for all x_1, x_2 in D, as does the spread definitions have for (x_1, x_2, x_3, x_4), for all x_i in D,$i = 1, 2, 3, 4$, where $y = xM$ is a maximal partial spread component. The transforms, with respect to the right Baer sets and then considered as Baer points, produce an associated Baer flock in exactly the same manner and with the same proofs as do the full Baer sets in the flock case. Stated in the manner of the spread cases, the functions on the maximal partial spreads define the required bijective functions relative to the right Baer sets, defined by $(x_1, x_2, (x_1, x_2)M = (x_3, x_4))$, where these functions are only injective on the larger sets (x_1, x_2, x_3, x_4); these are injective due to the required definition of partial spreads.

Therefore, a completely general interpretation of the Thas-Walker theory would include the Baer flocks defined on maximal partial spreads of the type under discussion, as well as flocks defined on spreads. The partial spreads are also the maximal partial spreads.

Stating this another way, the Baer flocks are flocks if the Baer subspaces lie on a spread and otherwise are maximal partial spreads. In this case, we shall call the maximal partial spread quasifibrations and the maximal partial flocks quasiflocks. The reader should note that there are left and right Baer flocks, and the results and ideas for left Baer flocks extend analogously to right Baer flocks. Looking at quasifibrations in this manner shows that the arguments for spreads on transforms and connecting these spreads to flocks are identical to quasifibrations on transforms and connecting quasifibrations to quasiflocks, if we agree not to focus on the covering theorem but just on the form. Looking at quasifibrations from the Baer perspective then provides the following theorem. For emphasis, we state separately for left and right conical Baer flocks and for left and right hyperbolic Baer flocks.

THEOREM 73. *(1) A left generic conical Baer flock partial spread is a left maximal partial spread, which is equivalent to a left maximal partial flock over the left Baer cone, called a left Baer flock. The left*

Baer flock is a flock if and only if the left maximal partial spread is a left spread.

A proper left conical Baer flock is a proper conical quasiflock, and the left maximal partial spread is a left generic proper conical quasifibration.

(2) A right generic conical Baer flock spread is a right maximal partial spread, which is equivalent to a right maximal partial flock over the right Baer cone, called a right conical Baer flock. The right Baer flock is a right flock if and only if the left maximal partial spread is a right spread.

A proper right conical Baer flock is a proper right conical quasiflock, and the right maximal partial spread is a right generic proper conical quasifibration.

(3) A left generic hyperbolic Baer flock spread is a left maximal partial spread, which is equivalent to a left maximal partial flock over the left Baer hyperbolic set called a left Baer hyperbolic flock. The left Baer hyperbolic flock is a hyperbolic flock if and only if the left maximal partial spread is a left spread.

A proper left hyperbolic Baer flock is a proper hyperbolic quasiflock, and the left maximal partial spread is a left generic proper hyperbolic quasifibration.

(4) A right generic hyperbolic Baer flock spread is a right maximal partial spread, which is equivalent to a right maximal partial flock over the right Baer hyperbolic set, called a right hyperbolic Baer flock. The right hyperbolic Baer flock is a right hyperbolic flock if and only if the left maximal partial spread is a right hyperbolic flock spread.

A proper right hyperbolic Baer flock is a proper right hyperbolic quasiflock, and the right maximal partial spread is a right generic proper hyperbolic quasifibration.

- Over a field K, a left Baer flock is either a quasiflock or a flock, and a right Baer flock is identical to the left Baer flock. This would seem to say that over non-commutative skewfields, since left Baer flocks and right Baer flocks are equivalent, it is not true that if a left Baer flock is a flock, then right Baer flock is also a flock. One could be a maximal partial flock, and the other could be a flock.
- Over a non-commutative skewfield when the left Baer flocks and the right Baer flocks are both flocks. The code-argument in volume II, part IV, chapter 6, proposition 26, shows how to construct the transforms of a right flock by the form of the

transforms of a left transpose flock. This is as follows: The specific code transfers the algebraic terms as follows:

- (a) Left to right: $ab \to ba, abc \to cba$.
- (b) The automorphisms: $\sigma^{-1} \to \tau$.
- (c) The derivative functions: $A(u^{\sigma^{-1}}) \to A^*(u)$.
- (d) The functions: $\{f(t), g(t)\} \to \{r(t), s(t)\}$.

The right generic conical spread form is designed to be isomorphic to the original spread when the skewfield is a field, considering the column vectors' structure. Suppose that we have a left conical flock spread over D, a skewfield such that the transpose is also a left spread. Then it follows that the transforms of the original left conical flock spread construct the right transforms of the associated right spread obtained from

$$\{f(t), g(t)\} = \{r(t), s(t)\}.$$

Then the right transforms may be constructed from the left transforms by the above code. This says that the left flocks have points that correspond to the Baer subspaces as 1-dimensional right homogeneous subspaces, and the right flocks have points that correspond to the Baer subplanes as 1-dimensional left homogeneous subspaces. That is, if the two Baer flocks are flocks and the transpose of the two Baer flocks are flocks, then there are two pairs of Baer flocks that are anti-isomorphic, since the transforms are anti-isomorphic and the transforms of the Baer subplanes now will cover the set of homogeneous points as 1-dimensional subspaces over the skewfield corresponding to the particular pseudo-regulus net in question.

This will become clear after we consider Draxl's theorem in the following chapter.

CHAPTER 22

Draxl's theorem

Although the primary interest in the connection of flocks and planes is in the flock and the flock spread situation, the emphasis is really on the so-called quasiflock, quasifibrations left and right cases. Generally stated, left quasiflock, quasifibrations are equivalent to a right transposed quasiflock, quasifibrations context. Proper quasifibrations are interesting geometries and seem to be quite rare in the field cases. There are infinite chains of such geometries arising from a small set of examples, as seen in other parts of this text and, in particular, in [**34**], part 7, chapter 9. If a left flock spread exists, it does not automatically correspond to a right flock spread, simply to the same right quasiflock. The question of an adequate covering criterion is unanswered in the non-semifield cases and appears for semifields only in the unwrapping procedures.

Since left quasiflock quasifibrations are equivalent to right quasiflock transposed quasifibrations, it would seem to say that the left transposed matrix sets are also quasifibrations. However, this is not the case.

In the work of Cherowitzo-Johnson-Vega [**8**], the σ-conical flock theory over fields, the conical flock spreads and the conical quasiflock quasifibrations were given simultaneously. All of the work on left and right flock spreads could be generalized to left and right quasiflock quasifibrations. It was decided to split the work over skewfields to first emphasize the spread-flock theory and then to bring back the more general situations.

A "quasiflock" will look exactly like a flock but has the exception that the generalized hyperbolic partial quadric or generalized quadratic partial cone will not be completely covered. However, the part that is covered is "maximal" in the sense that additional "points" may not be added as they arise from Baer subspaces that sit in a maximal partial spread; the quasifibration. Since we now consider the more general situation, we revisit the definitions. However, the reader will need to consult the original text for the required background. This is the only place in this text where the more general subplane covered nets are

DOI: 10.1201/9781003767978-22

mentioned. The connection between left and right quasifibration in the more general subplane-covered nets seems appropriate here.

Definition 74. Let D be any division ring, and let R be any subplane covered net; it is recalled from [**34**], part 1, chapter 2 p. 17, also part 11, that R may be embedded in a projective space $PG(V-1,D)$ with a designated co-dimension 2 subspace N. In the finite-dimensional case, the projective space is $PG(n+1,D)$, where n is at least 2. There are exactly $2n-1$ types $i=0$,1,2..., $2n-1$. The analysis of the general subplane covered net is more difficult than that of the derivable net when $n=2$, and there are four types: $0,1,2,3$. The standard examples for type 0 when $n=2$ are the quaternion division ring planes. When $n>2$, the standard examples of type 0 are the cyclic division ring planes. Recall the contracted net from a subplane covered net of degree n over a skewfield D describes both a left and right subplane covered net written over a D-vector space V_{2n}, which is a left-right D-module of dimension $2n$. The left classical pseudo-regulus subplane covered has the form for a contraction as

$$x=0,y=x\delta,\text{for all }\delta\text{ in }D.$$

If D is a field, we use the term regulus net, since it has the same form as a classical regulus net when $n=2$. The right-left classical pseudo-regulus subplane covered net has the form

$$x=0,y=\delta x,\text{for all }\delta\text{ in }D.$$

The logical extension of left and right spreads or quasifibrations in $PG(3,D)$ containing a generic pseudo-regulus derivable net are left and right spreads or quasifibrations in $PG(2n-1,D)$ containing a generic pseudo-regulus subplane covered net. In this case, here is the more general definition.

(1) A left quasifibration of degree n is a matrix set of n by n matrices over D, which is a set λ of left n-dimensional left subspaces whose distinct differences are also left n-dimensional subspaces. Furthermore, we may choose row n as $(t_1,t_2,...,t_n)$ for all $t_1,t_2,...,t_n$ in D. A left quasifibration is a maximal partial spread.

(2) If the quasifibration admits the "covering condition", for all (x,y), where x,y are n-vectors, there is a unique cover by elements of λ together with $x=0$, then the quasifibration is a spread.

(3) Let R be any subplane covered net of type i,$i=1,2,...2n-1$. And assume that the quasifibration contains R. The quasifibration now has the following form:

$$x=0,y=xM,$$

$M \supset R$, such that M is contained in χ, a set of left D subspaces, where D is a division ring. Consider then the associated right matrix set:

$$x^T = 0^T, y^T = Mx^T,$$

for all M in χ. We shall show below that this is also a quasifibration fundamental connections, we look at some examples.

- **Draxl's Theorem: States that for an n by n matrix M over D, the following are equivalent: (1) there is a right inverse if and only if the rows are left linearly independent, and (2) there is a left inverse if and only if the columns are right linearly independent,** (see also [**34**] p. 41 Theorem 3). **When $n = 2$, we may add (3) if and only if there is a non-zero (i, j)-determinant.**

THEOREM 75. *Draxl's Theorem relative to quasifibrations: Left quasifibrations and right quasifibrations are equivalent: Let R be any subplane covered net of type i,$i = 1, 2, ...2n - 1$. And assume that the quasifibration contains R. The quasifibration now has the following form:*

$$x = 0, y = xM,$$

$M \supset R$, such that M is contained in χ, a set of left D subspaces, where D is a division ring. Consider then the associated right matrix set:

$$x^T = 0^T, y^T = Mx^T,$$

for all M in χ. We shall show below that this is also a quasifibration.

- The import of Draxl's theorem in reference to our study of left and right spreads over skewfields D is as follows: The left uses rows to represent vectors as in

$$(x_1, x_2)\begin{bmatrix} e & f \\ g & h \end{bmatrix} = x_1\begin{bmatrix} e \\ f \end{bmatrix} + x_2\begin{bmatrix} g \\ h \end{bmatrix} = 0.$$

A left linear independence question of the rows: Therefore, we have

$$x_1e + x_2g = 0,$$
$$x_1f + x_2h = 0.$$

Assume that g is non-zero, then we would obtain:

$$x_1(eg^{-1}h - f) = 0,$$

The $(2, 1)$-*row* determinant is

$$(eg^{-1}h - f)g,$$

so we have linear independence if and only if the $(2,1)$-*row* determinant is non-zero, if and only if we have a right inverse.

$$y^T = \begin{bmatrix} e & f \\ g & h \end{bmatrix} x^T,$$

the right type of structure. We test the columns for right linear dependence:

$$\begin{bmatrix} e & f \\ g & h \end{bmatrix} x^T = \begin{bmatrix} e \\ g \end{bmatrix} x_1 + \begin{bmatrix} f \\ h \end{bmatrix} x_2$$

and obtain the same dominant: $eg^{-1}h - f$ is non-zero if and only if there is a left inverse. Along with the upper form theory for left quasiflock and the "opposite homology pseudo-regulus-inducing group, this will then prove the following fundamental result.

- A note on matrix inverses: When $g = 1$ is non-zero, the determinant $(2,1)$-determinant and inverse of

$$\begin{bmatrix} e & g \\ 1 & h \end{bmatrix}$$

are

$$eh - g = \Delta,$$

and

$$\begin{bmatrix} h\Delta^{-1}, & 1 - h\Delta^{-1}e \\ -\Delta^{-1}, & \Delta^{-1}e \end{bmatrix}.$$

Then the determinant (2,1)-determinant is also $eh - g$, and the inverse of

$$\begin{bmatrix} h & 1 \\ g & e \end{bmatrix}$$

is

$$\begin{bmatrix} \Delta^{-1}e, & -\Delta^{-1} \\ 1 - h\Delta^{-1}e, & h\Delta^{-1} \end{bmatrix},$$

using the double-transpose mapping.

DEFINITION 76. Using Draxl's result for a left derivable net, then the left derivable net applying the double-transpose isomorphism gives the following connection between left and right generic derivable nets: The left derivable net maps to the right derivable net as follows:

$$x = 0, y = x \begin{bmatrix} u^{\sigma} & A(u) \\ 0 & u \end{bmatrix}$$

for all u in a division ring D,maps to

$$x^T = 0^T, y^T = \begin{bmatrix} u & 0 \\ A(u), & u^\sigma \end{bmatrix} x^T,$$

for all u in D. Since given a non-commutative division ring D,there is an anti-isomorphism between the left and right multiplications. We may this be in terms of the so-called original space and the "o_1−opp space" with multiplication $D = \{D, +, juxtaposition\}$, and $\{D, +, o_1\}$ where

$$ao_1b = ba.$$

Given any non-commutative division ring D, there are two associated pseudo-regulus nets in $PG(3, D)$, where the left classical pseudo-regulus net is

$$x = 0, y = x\delta$$

for all δ in D and the right classical pseudo-regulus net, which is

$$x = 0, y = \delta x = xo_1\delta$$

for all δ in D. We have chosen in this text to use the left/right notation instead of the "opp-space" notation. However, the opp-space notation shows the connection between the left and right structures as being "anti-isomorphic". For the study of Baer flocks, we have chosen the generic derivable nets, both left and right. We have also defined generalized hyperbolic quadrics and generalized quadratic cones using derivable nets.

When we use

$$xou = x \begin{bmatrix} u^\sigma & A(u) \\ 0 & u \end{bmatrix},$$

to create another scalar multiplication and then use this to define the left hyperbolic quadric; the right hyperbolic form uses the right pseudo-regulus net written in terms of column vectors. Since the transforms of the left Baer subplanes as 1-dimensional right homogeneous subspaces satisfy the left generalized Baer hyperbolic sets or generalized Baer cones, and the transforms of the right Baer subspaces as 1-dimensional left homogeneous satisfy the right generalized Baer hyperbolic sets or generalized Baer cones, we see that the connection between the right Baer flocks and the left Baer flocks is connected by an anti-isomorphism. We then define the right Baer flock that is connected to the left Baer flock using Draxl's theorem to be "anti-isomorphic" Baer flocks.

THEOREM 77. (***The Fundamental Theorem on Quasifibrations, Part 1.***) *A left generic hyperbolic quasiflock quasifibration exists if and only if a generic right hyperbolic quasiflock quasifibration exists.*

(1) If there is a left generic hyperbolic quasifibration over a division right D exists, then there is an associated generic right hyperbolic quasifibration over D.

$$x = 0, y = 0, y = x \begin{bmatrix} u^\sigma & A(u) \\ 0 & u \end{bmatrix},$$

$$y = x \begin{bmatrix} f(t) & g(t) \\ 1 & t \end{bmatrix} \begin{bmatrix} v^\sigma & A(v) \\ 0 & v \end{bmatrix},$$

for all non-zero u and v and for all t in D. With associated left hyperbolic Baer flock, written over the left Baer hyperbolic set: (corresponding to the Baer transforms of the affine points of the Baer cover).

$$\Pi_t^{B,L} : x_1(g(t) - A(f(t)^{\sigma^{-1}}) - x_2 f(t)^{\sigma^{-1}} + x_3 t - x_4 = 0,$$

and

$$\varrho : x_2 = x_3.$$

Then

$$x^T = 0^T, y^T = 0^T, y^T = \begin{bmatrix} u^\sigma & A(u) \\ 0 & u \end{bmatrix} x^T,$$

$$y^T = \begin{bmatrix} f(t) & g(t) \\ 1 & t \end{bmatrix} \begin{bmatrix} v^\sigma & A(v) \\ 0 & v \end{bmatrix} x^T,$$

for all non-zero u and v and for all t in D, it is a right quasifibration. The double transpose mapping changes the form to a right-acting hyperbolic right-quasifibration of the following form:

$$x^T = 0^T, y^T = 0^T, y^T = \begin{bmatrix} u & 0 \\ A(u) & u^\sigma \end{bmatrix} x^T,$$

$$y^T = \begin{bmatrix} f^{-1}(s) & 1 \\ g(f^{-1}(s)), & s \end{bmatrix} \begin{bmatrix} v, & 0 \\ A(v), & v^\sigma \end{bmatrix} x^T,$$

Following this mapping by the right "inverse mapping",

$$(x^T, y^T)^T \to (y^T, x^T)^T,$$

constructs the following generic right hyperbolic quasiflock quasifibration:

$$x^T = 0^T, y^T = 0^T, y^T = \begin{bmatrix} u & 0 \\ A(u) & , u^\sigma \end{bmatrix} x^T,$$

$$y^T = \begin{bmatrix} v, & 0 \\ A(v), & v^\sigma \end{bmatrix} \begin{bmatrix} f^{-1}(s) & 1 \\ g(f^{-1}(s)), & s \end{bmatrix}^{-1} x^T.$$

for all non-zero u, v and all s in D (the notation should be read setwise on the u and u). Letting the determinant be the (1,2)-determinant, which is

$$\Phi_s = sf^{-1}(s) - g(f^{-1}(s)),$$

the inverse is

$$\begin{bmatrix} \Phi_s^{-1}s & -\Phi_s^{-1} \\ 1 - f^{-1}(s)\Phi_s^{-1}s & f^{-1}(s)\Phi^{-1} \end{bmatrix}.$$

To obtain the proper form to determine the associated quasiflock, we need a "1" in the (1,2)-entry. Therefore, we apply the left-acting homology group action, letting

$$v = -\Phi_s.$$

The resultant right generic hyperbolic quasiflock quasifibration. For the second part, it is then:

$$y^T = \begin{bmatrix} v, & 0 \\ A(v), & v^\sigma \end{bmatrix} Wx^T,$$

where W is

$$\begin{bmatrix} -s, & 1 \\ \Phi_s^\sigma - (A(\Phi_s) + \Phi_s^\sigma f^{-1}(s))\Phi_s^{-1}s, & (A(\Phi_s) + \Phi_s^\sigma f^{-1}(s))\Phi_s^{-1} \end{bmatrix}.$$

The associated quasiflock, noting that $A^* = A$, in this situation, is the generic right hyperbolic Baer flock (corresponding to the right Baer transforms of the right Baer hyperbolic set).

$$\Pi^{B,R}_{(A(\Phi_s)+\Phi_s^\sigma f^{-1}(s))\Phi_s^{-1}} :$$

$$((\Phi_s^\sigma - (A(\Phi_s) + \Phi_s^\sigma f^{-1}(s))\Phi_s^{-1}s) + A(s))a_1 -$$

$$+ s^\sigma a_2 + (A(\Phi_s) + \Phi_s^\sigma f^{-1}(s))\Phi_s^{-1})a_3 - a_4 = 0,$$

$$\varrho^{B,R} : a_2 = a_3.$$

(2) If there is a right generic hyperbolic quasiflock quasifibration, then there is a corresponding left generic hyperbolic quasiflock quasifibration.

$$x^T = 0^T, y^T = 0^T, y^T = \begin{bmatrix} u & 0 \\ A^*(u), & u^\tau \end{bmatrix} x^T,$$

$$y^T = \begin{bmatrix} v & 0 \\ A^*(v), & v^\tau \end{bmatrix} \begin{bmatrix} r(t) & 1 \\ s(t), & t \end{bmatrix} x^T,$$

for all non-zero u,v and for all t in D,and with corresponding right hyperbolic Baer flock ((corresponding to the right Baer transforms of the right Baer hyperbolic set):

$$\Pi_t^{B,R} :$$

$$(s(t) - A^*(r(t))x_1 - r(t)^\tau x_2 + tx_3 - x_4 = 0,$$

$$\varrho^{B,R} : x_2 = x_3.$$

Then there is a left hyperbolic quasiflock quasifibration, inducing and then there is also a left hyperbolic generic quasiflock quasifibration as follows:

$$x = 0, y = 0, y = x \begin{bmatrix} u^{\tau} & A^*(u) \\ 0 & u \end{bmatrix},$$

$$y = x \begin{bmatrix} r^{-1}(k), & s(r^{-1}(k)) \\ 1 & k \end{bmatrix}^{-1} \begin{bmatrix} v^{\tau} & A^*(v) \\ 0 & v \end{bmatrix},$$

for all non-zero u, v and for all k in D. Using the analogous procedure to convert the inverse term to generic form, the last expression is:

$$y = x \begin{bmatrix} -k, & -\Delta_k^{\tau^{-1}} - k\Delta_k^{-1}((A^*(\Delta_k^{\tau^{-1}}) - r^{-1}(k)\Delta_k^{\tau^{-1}}) \\ 1, & \Delta_k^{-1}(A^*(\Delta_k^{\tau^{-1}}) - r^{-1}(k)\Delta_k^{\tau^{-1}}) \end{bmatrix} \begin{bmatrix} v^{\tau} & A^*(v) \\ 0 & v \end{bmatrix},$$

where

$$\Delta_k = r^{-1}(k)k - s(r^{-1}(k)).$$

The corresponding left generic hyperbolic Baer flock (corresponding to the left Baer transforms of the left Baer hyperbolic set) is given by

$$\Pi^{B,L}_{s=\Delta_k^{-1}(A^*(\Delta_k^{\tau^{-1}})-r^{-1}(k)\Delta_k^{\tau^{-1}})} :$$

$$a_1((-\Delta_k^{\tau^{-1}} - k\Delta_k^{-1}((A^*(\Delta_k^{\tau^{-1}}) - r^{-1}(k)\Delta_k^{\tau^{-1}}) - A^*(-k^{\tau^{-1}}) -$$
$$-a_2(-k^{\tau^{-1}}) + a_3(\Delta_k^{-1}(A^*(\Delta_k^{\tau^{-1}}) - r^{-1}(k)\Delta_k^{\tau^{-1}}) - a_4 = 0.$$
$$\varrho^{B,L} : a_2 = a_3.$$

(3) If there is a left generic conical quasiflock quasifibration then there is also a right generic conical quasiflock quasifibration.

If

$$x = 0, y = x \begin{bmatrix} u^{\sigma} + F(t), & A(u) + G(t) \\ t & u \end{bmatrix},$$

for all t, u in D, if is a left conical quasifibration with an associated left conical Baer flock (corresponding to the Baer transforms of the Baer cone):

$$\Sigma_t^{B,L} :$$
$$-x_1(F(t)^{\sigma^{-1}} + A(t^{\sigma^{-1}})) - x_2 t^{\sigma^{-1}} + x_3(G(t) - A(F(t)^{\sigma^{-1}}) - x_4 = 0.$$

Then there is a right generic conical quasiflock quasifibration as follows:

$$x^T = 0^T, y^T = \begin{bmatrix} w - F(t)^{\sigma^{-1}}, & t \\ G(t) - A(F(t)^{\sigma^{-1}}) + A(w), & w^{\sigma} \end{bmatrix},$$

for all t, w in D. This form begins with the same form arising from Draxl's theorem and then using the double-transpose mapping, and letting $u^{\sigma} + F(t) = w^{\sigma}$. Notationally, $A(w) = A^{*}(w)$ and $\tau = \sigma$, in the standard right form.

The corresponding right conical Baer flock is given by

$$\Sigma_t^{B,R}:$$

$$-(-F(t) + A(t))x_1 - t^{\sigma}x_2 + (G(t) - A(F(t)^{\sigma^{-1}})x_3 - x_4 = 0.$$

(4) If there is a right generic conical quasiflock quasifibration, then there is a left generic conical quasiflock quasifibration.

If

$$x^T = 0^T, y^T = \begin{bmatrix} u + R(t) & t \\ S(t) + A^{*}(u), & u^{\tau} \end{bmatrix} x^T,$$

for t,u in D, with right conical Baer flock

$$\Sigma_t^{B,R}:$$

$$-(R(t)^{\tau} + A^{*}(t))x_1 - t^{\tau}x_2 + (S(t) - A^{*}(R(t))x_3 - x_4 = 0.$$

Then a right generic form then

$$x = 0, y = x\begin{bmatrix} v^{\tau} - R(t)^{\tau}, & A^{*}(v) - A^{*}(R(t)) \\ t & v \end{bmatrix},$$

for all t, v in D, is a left generic conical quasiflock quasifibration. The associated left conical Baer flock is:

$$\Sigma_t^{B,L}:$$

$$-x_1(-R(t) + A^{*}(t^{\tau^{-1}}) - x_2 t^{\tau^{-1}} + x_3(-A^{*}(R(t)) - A^{*}(R(t)) - x_4 = 0.$$

PROOF. In volume II, all of the left and right hyperbolic and conic flock theory is covered. It should be clear that when we consider Baer flock theory, all of the connecting arguments using the lower form of theory in chapter 4 of part III of volume II also apply to the more general situation when we use left Baer flock coverings. Similarly, the use of upper form theory applies when we consider right Baer flock coverings. We relist all of how the use of this material will change the general forms, and what this implies using the upper and lower form theory. Furthermore, the left inversing theory for left flock spreads and the right inversing theory for right flock spreads also may be applied more generally for left and right Baer flock spreads (the quasiflock quasifibrations), respectively. This latter material is in part VI in volume II. The details of all these general applications are complicated, but the reader should be able to work through them. □

In the original text and also in this text, the set of known quasifibrations over fields is mentioned. The material on strings of quasiflocks and flocks shows that even though left and right quasiflocks are equivalent, left and right flocks are not or may not be. We have also mentioned attempts to prove that what looks to be a quasifibration may not be. So, the related unsolved problem for non-commutative skewfields is

PROBLEM 78. Given a non-commutative division ring D. There is no known example of a proper quasifibration of any finite dimension. So, the specific problem is to prove that what appears to be a proper quasifibration in the examples on strings (to follow) is really proper.

CHAPTER 23

Transposed Baer flocks

Definition 79. In the work on transposition in volume II, we showed that there are conical flock spreads whose transpose matrix sets are also spreads and that there are transformations that produce isomorphic generic flock spreads. This means that there is a fundamental theorem on quasifibrations using the transpose spreads in generic form. When this occurs in general cases over quasifibrations, the transpose quasifibrations produce conical Baer flocks called "transposed conical Baer flocks".

Theorem 80. *The Fundamental Theorem of Quasifibrations (part 2 – the transpose – for conical quasifibrations).*

Given any left or right generic conical quasiflock quasifibration whose matrix transpose set is also a quasifibration. Then there is a generic quasiflock quasifibration isomorphic to the transpose quasifibration. Hence, there is a set of Baer flocks corresponding to these matrix quasiflock quasifibrations called the "transposed conical Baer flocks".

(1) There is a "square of four conical Baer flocks" constructed as follows: On the Northwest point, there is a left conical Baer flock. On the Northeast point, there is a right conical Baer flock, obtained using the Fundamental Theorem, part 1. On the Southwest point is the transposed left Baer flock, and on the Southeast point is the right conical Baer flock obtained from the transposed left conical Baer flock by the transposition of the Northeast right conical Baer flock.

To travel east or west on the square, use the Fundamental Theorem applied to the corresponding generic quasiflock quasifibrations. To travel north or south on the square, use the transpose on the corresponding generic quasiflock quasifibrations.

Here is the proof of the Conical Flock Square using the generic forms. Everything will follow provided that starting on the NW point and moving south and then east to the SE point, the NE point is the transpose of the SEpoint.

We know that the NW point is represented by

$$x = 0, y = x \begin{bmatrix} u^{\sigma} + F(t), & A(u) + G(t) \\ t & u \end{bmatrix},$$

DOI: 10.1201/9781003767978-23

and the NE point is represented by

$$x^T = 0^T, y^T = \begin{bmatrix} w - F(t)^{\sigma^{-1}}, & t \\ G(t) - A(F(t)^{\sigma^{-1}}) + A(w), & w^\sigma \end{bmatrix}.$$

Then transpose NW followed by the double-transpose to obtain:

$$(*) : x = 0, y = x \begin{bmatrix} u, & A(u) + G(t) \\ t & u^\sigma + F(t) \end{bmatrix},$$

letting $v{=}u^\sigma + F(t)$*,so that* $u{=}v^{\sigma^{-1}} - F(t)^{\sigma^{-1}}$*,and*

$$A(u) = A(v^{\sigma^{-1}}) - A(F(t)^{\sigma^{-1}}).$$

Then $A^*(r) = A(r^{\sigma^{-1}})$*,produces*

$$x = 0, y = x \begin{bmatrix} v^{\sigma^{-1}} - F(t)^{\sigma^{-1}}, & A^*(v) - A^*(F(t)) + G(t) \\ t & v \end{bmatrix},$$

as representing the Southwest point.

Now go back to the left transpose $(*)$*, traveling east, and applying the double-transpose mapping, we have:*

$$x^T = 0^T, y^T = \begin{bmatrix} u, & A(u) + G(t) \\ t, & u^\sigma + F(t) \end{bmatrix} x^T.$$

Before changing to the generic form, take the transpose of the Southeast point and let $w^\sigma = u^\sigma + F(t)$ *to obtain the representation of the Northeast point*

$$x^T = 0^T, y^T = \begin{bmatrix} w - F(t)^{\sigma^{-1}}, & t \\ G(t) - A(F(t)^{\sigma^{-1}}) + A(w), & w^\sigma \end{bmatrix}.$$

This proves the assertion of the two transposed conical Baer flocks.

To make

$$x^T = 0^T, y^T = \begin{bmatrix} u, & A(u) + G(t) \\ t, & u^\sigma + F(t) \end{bmatrix} x^T$$

into generic form, take the double-transpose and let $u = v^{\sigma^{-1}}$*, to obtain the Southeast point represented as*

$$x^T = 0^T, y^T = \begin{bmatrix} v + F(t), & t \\ A^*(v) + G(t), & v^{\sigma^{-1}} \end{bmatrix} x^T,$$

where $A^*(v) = A(v^{\sigma^{-1}})$*.*

(2) Here are the points of the conical Baer flock square.

- $NW : \Sigma_t^{B,L}$: Cone for the non-commutative part is

$$-x_1(F(t)^{\sigma^{-1}} + A(t^{\sigma^{-1}})) - x_2 t^{\sigma^{-1}} + x_3(G(t) - A(F(t)^{\sigma^{-1}}) - x_4 = 0.$$

$NE : \Sigma_t^{B,R}$:

$$-(-F(t) + A(t))x_1 - t^{\sigma} x_2 + (G(t) - A(F(t)^{\sigma^{-1}})x_3 - x_4 = 0.$$

$SW : \Sigma_t^{B,trans,L}$:

$$-x_1(-F(t) + A(t)) - x_2 t^{\sigma} + x_3 G(t) - x_4 = 0.$$

$SE : \Sigma_t^{B,trans,R}$:

$$-(-F(t) + A(t^{\sigma^{-1}}))x_1 - t^{\sigma^{-1}} x_2 + (G(t) - A(F(t)^{\sigma^{-1}})x_3 - x_4 = 0.$$

Noting that the cone associated with the NW point is defined using the pseudo-regulus net

$$x = 0, y = x \begin{bmatrix} u^{\sigma} & A(u) \\ 0 & u \end{bmatrix},$$

for all u in D, defines the cone associated with the SW point (transpose) is applying the transpose and then the double-transpose (Baer flock-inducing isomorphism mapping) is

$$x = 0, y = x \begin{bmatrix} v^{\sigma^{-1}} & A^*(v) \\ 0 & v \end{bmatrix},$$

where $A^*(v) = A(v^{\sigma^{-1}})$, for all v in D. Therefore, the cone associated with the SW point is not generally isomorphic to the cone associated with the NW point.

- The NW left Baer flock and the NE right Baer flock are anti-isomorphic.
- The SW left transformed Baer flock and the SE right transposed Baer flock are anti-isomorphic.
- It was remarked in the section of type (0) flocks that every left quaternion division ring plane provides a left type (0) flock, which is anti-isomorphic to the associated right quaternion division ring plane.

Part 10

Anti-isomorphic flocks

In this part, we discuss when we may construct anti-isomorphic conical flocks. By the methods considered in this text, we determine additive quasiflock quasifibrations such that we have both the left and right unwrapping procedures by choosing all constants as central elements over the associated skewfield D.

Now assume that we have a left semifield flock spread over a division ring that admits both the left and right unwrapping theorems. Furthermore, assume that the transpose matrix sets also produce left and right flocks. Then, by the fundamental theorem of quasifibrations, part 1, all Baer flocks are flocks, so that we have connected left flocks and right flocks as anti-isomorphic geometries. There are also examples of simultaneous flocks studied in part III, which shows that there are both left and right conical anti-isomorphic flocks and left and right hyperbolic anti-isomorphic flocks.

In all of the examples, we have the left and right unwrapping procedure.

THEOREM 81. *Let π be a left simultaneous flock spread over a skewfield D that admits both the left and right unwrapping procedures. This means we have a skewfield conical square flock procedure. Assume that the conical transpose of the left simultaneous flock is also a left spread.*

Then the left flock transpose spread constructs also a right flock spread, where the right flock spread is written in the upper form. Using the fact that the right upper form provides isomorphic geometries to the generic/lower right flock form, we realize the left generic flock as anti-isomorphic to the right generic flock arising from the transpose of the left generic flock. And the left generic transpose flock is anti-isomorphic to the right generic flock obtained from the left generic conical flock.

Hence the NW left conical flock is anti-isomorphic to the SE right conical flock, and the SW left conical flock is anti-isomorphic to the NE right conical flock.

It is instructive to use two examples to illustrate how this works.

EXAMPLE 82. Consider

$$x = 0, y = x \begin{bmatrix} u^{\sigma} & \theta t^{\sigma^{-1}} \\ t & u \end{bmatrix},$$

for all t, u in $D = E(\theta)$, where E is a division ring, and D is a central cyclic Galois extension of E, of degree $n>2$. We have seen that when we have a quasifibration, and since θ is central, we have the left and right unwrapping procedure. Assume then that this is a left simultaneous flock spread. More over, the transpose of this is

$$x=0, y=x\begin{bmatrix} u^{\sigma} & t \\ \theta t^{\sigma^{-1}} & u \end{bmatrix},$$

assume that there is a singular matrix, so that using the (1,2)-determinant, we see that, for some u, t, t non-zero.

$$ut^{-1}u^{\sigma}=\theta t^{\sigma^{-1}}.$$

This is true if and only if

$$(ut^{-1})(ut^{-1})^{\sigma^{-1}}=\theta.$$

Let $k=(ut^{-1})$, and since θ is central, it follows that

$$k^{-1}$$

commutes with

$$kk^{\sigma^{-1}},$$

therefore

$$k^{-1}(kk^{\sigma^{-1}})=k^{\sigma^{-1}}=kk^{\sigma^{-1}}k^{-1}.$$

This then shows that

$$k^{\sigma^{-1}}k=kk^{\sigma^{-1}}.$$

Now look at the original matrix set, which is assumed to be a spread. Therefore, the following

$$\begin{bmatrix} u^{\sigma} & \theta t^{\sigma^{-1}} \\ t & u \end{bmatrix}$$

is non-singular for all t, u. Choosing the same t, u as above, we see that

$$u^{\sigma}t^{-1}u=\theta t^{\sigma^{-1}}$$

cannot be true. But note that this is equivalent to

$$k^{\sigma^{-1}}k=\theta.$$

This then shows that the transpose matrix set is non-singular and θ is central, so by the unwrapping procedure, the transpose is also a spread. This implies that we have a left conical flock, which is anti-isomorphic to a right conical flock.

Now consider any division ring D of finite dimension over a field F. Then, by the Multiple Replacement Theorem – Redux, we obtain the set

$$S_{\delta,\gamma}$$

for all δ, γ, δ non-zero of F of lifted skewfield spreads of the following form

$$x=0, y=x\begin{bmatrix} u^{\sigma}, & (\theta\delta+\gamma)t^{\sigma} \\ t & u \end{bmatrix},$$

for all t, u in a central Galois quadratic extension $D(\theta)$ of D. Since σ has order 2, we see that the transpose is also a spread.

Here is the main general theorem, part of which is repeated from part IV.

- We note that once a flock using a generator θ of the central group is used in the expression as a constant, then any generator will produce another example. For example, when $n = 2$, we obtain the lifted skewfield valid for any finite division ring D (see section VI). The main idea in this case is that "lifting" can be changed by choosing various pre-skewfields of the same skewfield. This is also valid in the finite field case, where such isomorphisms may be completely determined as in Cardinalli, Polverino, and Trombetti [**4**]. Although these ideas are valid in infinite fields or finite division rings, there have been no other such studies. However, it seems reasonable to assume that there are a variety of mutually non-isomorphic flock spreads in the infinite cases, as well.

THEOREM 83. ***Theorem of 8 flocks.***

(1) Let S be a division ring. Assume that there is a central cyclic division ring extension $S(\theta)$ of S. If one has a spread of the form:

$$x = 0, y = x\begin{bmatrix} u^{\sigma} & , \theta t^{\sigma^{-1}} \\ t & u \end{bmatrix}; t, u \in S(\theta),\ where$$

σ is an induced automorphism of order n.

and choose any other generator of $\langle\theta\rangle$,

$$\Sigma_{i=0}^{n-1}\beta_i\theta^i, where \beta_i \in Z(S), for all i = 0, 1, ..., n-1,$$

the center of S, then, for each generator, there is a spread of the form

$$x = 0, y = x\begin{bmatrix} u^{\sigma}, & \Sigma_{i=0}^{n-1}\beta_i\theta^i t^{\sigma^{-1}} \\ t & u \end{bmatrix}; t, u \in S(\theta),$$

σ is an induced automorphism of order n.

$for\ the\ cases$: $where\, n > 2\, then$

Therefore, there will be an infinite set of examples of semifields, as well.

(2) Each of these spreads is simultaneous with the hyperbolic flock spreads and the conical spreads are both linear.

(3) The transpose of this matrix set is isomorphic to the semifield flock spread of the form

$$x = 0, y = x\begin{bmatrix} u^{\sigma}, & (\Sigma_{i=0}^{n-1}\beta_i\theta^i)^{-\sigma} t^{\sigma} \\ t & u \end{bmatrix},$$

for all t, u in $S(\theta)$. Use the notation for this semifield flock spread as $S^{trans}_{(\Sigma_{i=0}^{n-1}\beta_i\theta^i)^{-\sigma}}$.

(4) If σ is not of order 2, the conical flock obtained from the original flock is linear, whereas the conical flock obtained from the transpose is nonlinear.

(5) Noting that we have left an inversing situation, so there is also an inverse type of hyperbolic flock obtained from the transpose. This will not be a semifield flock.

(6) Therefore, the transpose of the left semifield spread $S_{\Sigma_{i=0}^{n-1}\beta_i\theta^i}$ is isomorphic to the left semifield spread $S^{trans}_{(\Sigma_{i=0}^{n-1}\beta_i\theta^i)^{-\sigma}}$.

(7) Further, the left semifield flock obtained from the transpose of the left semifield spread $S_{\Sigma_{i=0}^{n-1}\beta_i\theta^i}$ is

(a) isomorphic to the left semifield flock obtained from the left semifield spread $S^{trans}_{(\Sigma_{i=0}^{n-1}\beta_i\theta^i)^{-\sigma}}$, and

(b) is anti-isomorphic with the right semifield flock obtained from the left semifield spread $S^{trans}_{(\Sigma_{i=0}^{n-1}\beta_i\theta^i)^{-\sigma}}$, by the Fundamental Theorem of quasifibrations, part 2; the conical flock square.

(c) The left semifield flock arising from $S^{trans}_{(\Sigma_{i=0}^{n-1}\beta_i\theta^i)^{-\sigma}}$, is anti-isomorphic with the right semifield spread obtained from the original semifield spread $S_{\Sigma_{i=0}^{n-1}\beta_i\theta^i}$ that lies in the conical flock square.

(8) Therefore, there are eight flocks involved in the sequence, which we shall call a "string of 8 flocks". Four conical flocks and four hyperbolic flocks.

There are four conical flocks in the conical square: two left conical flocks and two right conical flocks. The two left conical flocks are transposes of each other and the two right conical flocks are transposes of each other. Traveling on the diagonals of the conical square, the left and right sides are anti-isomorphic.

There are four hyperbolic flocks, extra to the conical square: two left hyperbolic and two right hyperbolic squares. The original left conical flock is also simultaneous, and the left transpose conical flock is left-inversive. The fundamental theorem on quasifibrations produces the other two. When $n > 2$, the original hyperbolic flock is linear, and the left-inversive flock is nonlinear.

***The theorem of twelve**: When $n = 2$, the left lifted skewfield conical flock admits both left and right pseudo-regulus-inducing homology groups. The left transpose is also a different form of a lifted skewfield conical, so the transpose also admits both left and right pseudo-regulus-inducing homology groups. This means there are four extra hyperbolic flocks; two left and two right. Therefore, there are a total of four conical* *flocks and eight hyperbolic flocks. The complete isomorphism questions are mostly.*

- Note that the above results extend the same result of
- Cherowitzo-Johnson, Vega (see Theorem 47 in [**34**]) over fields and when $A = 0$, as anti-isomorphisms are isomorphisms in the field cases.
- The idea that various constructions produce connected flocks is studied in the next part, where they set are said to be "strings of flocks".

CHAPTER 24

The hyperbolic flock square

In the non-commutative case, there is no direct way to construct a transpose of a left hyperbolic flock square. But, if the original left flock spread is simultaneous, there is a possibility. In a left simultaneous flock spread, the matrices may be stated in the conical flock form, but the elements can be rearranged to obtain the form for a hyperbolic flock form. If the conical transpose of a simultaneous flock spread exists and admits the left-inversing procedure, then the conical transpose spread points may be rearranged using the left-acting pseudo-regulus-inducing homology. Using the Fundamental Theorem of Quasifibrations, we may directly take the generic form of the right Baer flock spread. If we have a central unwrapping principle also, then we obtain a right hyperbolic flock spread in generic form, which has the same points as the transpose conical flock spread. This says that the original left hyperbolic flock spread and the constructed right hyperbolic flock spread produce transform points that are anti-isomorphic. Here is the theorem in the basic forms. We shall include when the transpose of a general simultaneous spread exists and when it admits the left-acting pseudo-regulus-inducing homology group.

THEOREM 84. *(**When the Transpose of the Left Simultaneous Spread Exists.**)*

Let

$$x = 0, y = \begin{bmatrix} u^{\sigma} + F(t) & , A(u) + G(t) \\ t & u \end{bmatrix},$$

for all t,u in D, where D is a division ring, and F,G are functions on D. Be a simultaneous spread. Then

(1) $F(t) = ft - A(t^{\sigma^{-1}})^{\sigma}$,

(2) $G(t) = gt^{\sigma^{-1}} + fA(t^{\sigma^{-1}}) - A^2(t^{\sigma^{-1}})$, *and*

The representation for the left hyperbolic flock spread is:

$$y = x \begin{bmatrix} t^{\sigma} + f & , A(t) + g \\ 1 & t \end{bmatrix} \begin{bmatrix} u^{\sigma} & A(u) \\ 0 & u \end{bmatrix},$$

DOI: 10.1201/9781003767978-24

and,

$$x = 0, y = x \begin{bmatrix} v^{\sigma} & A(v) \\ 0 & v \end{bmatrix},$$

for all v, t in D and for all non-zero u in D, where $f = F(1)$ and $g = G(1)$.

(3) The conical transpose, in generic form, is:

$$x = 0, y = x \begin{bmatrix} w^{\sigma^{-1}} + H(t), & A^*(w) + J(t) \\ t & u \end{bmatrix},$$

where

$$H(t) = A^*(t) - f^{\sigma^{-1}} t^{\sigma^{-1}},$$
$$J(t) = A^*(ft) - 2A^{*2}(t) + gt^{\sigma^{-1}} + fA^*(t),$$

and

$$A^*(r) = A(r^{\sigma^{-1}}).$$

(4) ***The conical transpose, in generic form, admits the left-acting pseudo-regulus-generating homology group if and only if***

(a) $f^{\sigma^{-1}}$ and f arein $Z(D)$, the center of D,&

(b) $J(t) = t^{\sigma^{-1}}(A^*(f)+g) = A^*(ft) - 2A^{*2}(t) + gt^{\sigma^{-1}} + fA^*(t)$. Then g is in $Z(D)$. Therefore, $A^*(ft) - 2A^{*2}(t) + fA^*(t) = 0$, for all t.

So, let $J(t) = t^{\sigma^{-1}} j$, where $j = A^*(f) + g$, and $H(t) = A^*(t) - f^{\sigma^{-1}} t^{\sigma^{-1}}$.

(5) Since we obtain semifield spreads, if spreads are obtained, then under the fundamental theorem of quasifibrations, the associated right and left Baer flocks are semifield flocks if the automorphisms are outer or 1, and all constants are central.

(6) Hence, under these conditions, we obtain both the conical square whose diagonals are anti-isomorphic and the hyperbolic square whose diagonals are anti-isomorphic.

PROOF. Take the transpose and then apply the double-transpose mapping. The latter mapping preserves or discovers the flock, assuming a spread is obtained. The reader is invited to finish this proof as a sort of homework. The following problem should be attempted: Determine all possible solutions to:

$$A^*(ft) - 2A^{*2}(t) + fA^*(t) = 0, \textit{for all } t \text{ in } D.$$

Where D is a division ring. Here is a start: (i) $f = 0$, characteristic 2. (ii) $f = 0$ and $A^{*2}(t) = 0$. When does this occur? For all of our examples, we had $f = 0$. We have no examples of part (3) that do not satisfy part (4). □

Part 11

Elation group double covers

In the first volume, there was a discussion on double covers using sets of regulus nets with two lines (i.e., over fields). In Jha and Johnson [**21**], there is a similar work on double covers using cones but over pseudo-regulus nets (that is, the right classical pseudo-regulus nets) with one line. Here we generalize the situations to the variety of different types of pseudo-regulus nets and then recall the specific instances just mentioned.

In particular, left derivable nets are classified relative to a related classical right pseudo-regulus net. Every derivable net may be embedded in a 3-dimensional projective space and then re-represented in affine form in a 4-dimensional vector space over a skewfield K. There are two possible forms, depending on whether the representation is chosen to be a right space or a left space. We are initially interested in considering the spreads as sets of left 2-dimensional K-subspaces and the related classical right pseudo-regulus net as a reference net. The "type" of the associated "imposter" derivable net that drives the spreads related to the flocks is defined as the number of Baer subspaces (incident with the zero vector) that the two nets share, with three shared Baer subspaces to have the meaning "at least three" shared Baer subspaces. The meaning of a derivable net "driving the spread", means that there is a central collineation group acting on the spread that is defined using that particular derivable net; we call these central collineation groups pseudo-regulus-inducing groups. The elation groups will then generate the form for a conical (generalized) flock, and the homology groups will generate the form for a hyperbolic (generalized) flock. When both elation groups and homology groups exist in the same spread, the spread form is said to be "simultaneous", as we have seen. We shall define double covers relative to the types of derivable nets and carry the term over to the associated spread and always use the conical form initially. We have mentioned that the "generic forms" of derivable nets may be used when we embed a derivable net in a translation plane.

In the following definition, we shall use regular double covers with two axes only when K is a field. Generally, this means that there are two pseudo-regulus nets: one is a classical pseudo-regulus as a driver, and one is a courier net with two components. With an elation double cover, there are always two elation groups. One group can be relative to a generic pseudo-regulus-inducing group as a driver, but the other is not. The second elation group is necessarily of a different form, as it is relative to a second representation of the general pseudo-regulus-inducing group, one that is not defined using a generic pseudo-regulus net For example, consider

$$x = 0, y = x \begin{bmatrix} \rho t & \gamma t \\ t & 0 \end{bmatrix}, \text{ for all } u \text{ in } K, \text{ contains}$$

$x = 0$ and $y = 0$ but not $y = x$. Change bases by $\begin{bmatrix} \begin{bmatrix} \varrho & \gamma \\ 1 & 0 \end{bmatrix} & 0_2 \\ 0_2 & I_2 \end{bmatrix}$.

This basis change will bring this pseudo-regulus net into a generic representation of the same type 3. If a conical spread of type 3 would be a regular double cover with one axis, with the two regulus-inducing elation groups, one of which is generic and one of which is the additive group of the previous derivable net.

When the generic derivable net is of type 2, it is what might be called a regular double cover with a single axis. This would be a conical spread admitting two regulus-inducing elation groups of type 2, one of which is relative to the generic type 2 pseudo-regulus net with the automorphism σ, and the second regulus-inducing elation group of type 2 would be with the automorphism δ equal to α or α^{-1} and not generic.

DEFINITION 85. Let D be a derivable net of the specified form in $PG(3, K)$, where K is a skewfield.

$$x = 0, y = x \begin{bmatrix} u^\sigma & A(u) \\ 0 & u \end{bmatrix} \text{ for all } u \text{ in } K.$$

σ is an automorphism, possibly trivial, and A is a σ-derivative function, possibly identically zero.

(1) Assume that there is a set of P_1 of pseudo-regulus derivable nets that are invariant under a pseudo-regulus-inducing elation group, where the elation group E_1 has the axis $x = 0$ and acts transitively on the components of D other than $x = 0$. Choose any component of D, $y = 0$, and assume there is a second pseudo-regulus-inducing elation group E_2 with axis $y = 0$, and a corresponding set P_2 of pseudo-regulus derivable nets that are invariant under the group E_2. Assume that the two groups leave D invariant. These two sets of pseudo-reguli and two pseudo-regulus-inducing groups are said to be a "regular double cover with two axes", if the axes for the two groups are not equal. One of these elation groups can be chosen to be generic, and one is not. For type 2, the two automorphisms could be either equal or paired as $\{\alpha, \alpha^{-1}\}$.

(2) Under essentially the same conditions, when the two axes are equal, the two groups do not leave D invariant, and the two sets of pseudo-regulus nets and two elation groups have the same axis, which is called an "elation double cover with one axis". In both cases, one elation group is relative to a generic pseudo-regulus group, and one is

relative to another type of pseudo-regulus group but is not a generic representation.

When K is a field and the pseudo-regulus net D is a classical regulus net over the field K, this is the situation studied in the first volume, where K is an ordered field and conical bilinear flocks were constructed and classified as regular double covers with two axes. In this case, $SL(2,K)$ is constructed as a collineation group of the associated bilinear flock plane and is completely classified (see [**34**], p. 223, sections 3,4,5).

When K is a skewfield and the pseudo-regulus net D is a classical left pseudo-regulus net over K, the regular double covers with a single axis were studied in Jha and Johnson [**21**] to prove the following result on conical flocks over K. The actual theorem is rephrased into the current language. In that noted article, the study was not on pseudo-regulus-inducing elation groups instead of normal sets of pseudo-regulus nets, which then gave pseudo-regulus-inducing elations. That is, the Jha-Johnson study was a stronger work. Here the emphasis is on regulus-inducing central collineation groups, thereby possibly bypassing certain interesting spreads.

THEOREM 86. *Let $\pi_{\varrho,\gamma}$ be a conical spread of type 3 of primitive type (1, 0) with $PG(3,K)$,*

where K is a skewfield.

(1) Then the following matrix set

$$x=0, y=x\begin{bmatrix} u+\varrho t, & \gamma t \\ t & u \end{bmatrix}$$

for all t,u in K is a semifield spread if and only if the $(2,1)$-determinant is non-zero for non-zero matrices: This is true if and only if

$$k^2+\varrho k-\gamma \text{ is non-zero,}$$

for all k in K.

(2) The spread is Desarguesian if and only if ϱ and γ are both in $Z(K)$, the center of K.

(3) The spread is a regular elation double cover of type 3 with one axis. The spread is said to be a "skew-Desarguesian" spread.

(3) Every conical spread, which is a regular double cover with one axis and with respect to a pseudo-regulus net of type 3, is a skew-Desarguesian spread.

(4) This spread is a simultaneous spread and admits an affine homology pseudo-regulus-inducing a homology group; a double double cover.

Let π be a conical spread in $PG(3,K)$, for K a skewfield that is a regular double cover of type 2 with one axis. Choose one elation group relative to a generic type 2 pseudo-regulus with automorphism σ and the second elation group of type 2 with automorphism δ, where $\delta = \sigma$ or σ^{-1}.

(1) If $\delta = \sigma$, then consider the following putative conical spread:

$$x = 0, y = x\begin{bmatrix} u^{\sigma} & t^{\sigma}\gamma \\ t & u \end{bmatrix}.$$

This is a quasifibration if and only if

$$k^{\sigma}k \text{ is never } \gamma,$$

for all k in K. Note this is a lifted spread if and only if γ is a generator of a central quadratic extension of Z(E) of a skewfield E, such that K is that extension. The quasifibration is an elation double cover of type 2.

Furthermore, the quasifibration is a spread and is a simultaneous spread if and only if γ is in $Z(K)$ and σ has order 2.

(2) If $\delta = \sigma^{-1}$ then $x = 0, y = x\begin{bmatrix} u^{\sigma} & \gamma t^{\sigma^{-1}} \\ t & u \end{bmatrix}$, is a spread if and only if $kk^{\sigma^{-1}}$is never γ.

The spread is a regular elation double cover of type 2. Furthermore, the spread is a simultaneous spread.

(3) If $\delta = \sigma$, σ of order 2, then $x = 0, y = x\begin{bmatrix} u^{\sigma} + \varrho t & , \gamma t^{\sigma} \\ t & u \end{bmatrix}$ is a conical spread if and only if

$$kk^{\sigma} + \rho k^{\sigma} \text{is never } \gamma.$$

The spread is a regular double cover of type 2. And the spread is a simultaneous spread.

Now the type 1 case.

THEOREM 87. *Consider a type 1 (or fake type 1) conical flock spread, which is a type 1 (or fake type 1) double cover, where both elation groups have the same automorphism σ.*

Then assume that γ is in the center of the division ring D.

$$x = 0, y = x\begin{bmatrix} u^{\sigma} + A(t) & , A(u) + \gamma t^{\sigma} \\ t & u \end{bmatrix}$$

for all t,u in a skewfield K is a type 1 elation double cover if and only if $k^{\sigma}k - A(k)$ is never γ.

The spread is a simultaneous spread if and only if σ has order 1 or 2. Note that the two pseudo-regulus nets are

$$x = 0, y = x \begin{bmatrix} u^{\sigma} & A(u) \\ 0 & u \end{bmatrix}$$

for all u in D, and there is an elation group with elements of the form

$$\begin{bmatrix} I_2 & \begin{bmatrix} u^{\sigma} & A(u) \\ 0 & u \end{bmatrix} \\ 0_2 & I_2 \end{bmatrix}$$

for all u in D. The second pseudo-regulus net is

$$x = 0, y = x \begin{bmatrix} A(t), & A(u) + \gamma t^{\sigma} \\ t & 0 \end{bmatrix}$$

for all t in D. And, there is an elation group with elements of the form

$$\begin{bmatrix} I_2 & \begin{bmatrix} A(t) & , A(u) + \gamma t^{\sigma} \\ t & 0 \end{bmatrix} \\ 0_2 & I_2 \end{bmatrix}$$

for all t in D. When separating the full elation group into subgroups, the identity mapping is doubled, viewed from the spread.

Then, transforming the spread to use the second elation group as the primary driver derivable net defining a conical spread, we consider that the second derivable net shares $x = 0$ and $y = 0$. Using the idea of carrier nets, select one other component of the spread M of the second derivable net, other than the two components mentioned, and use the mapping $(x, y) \to (xM, y)$ to construct the primary derivable net as

$$x = 0, y = xM^{-1} \begin{bmatrix} A(t) & A(u) + \gamma t^{\sigma} \\ t & 0 \end{bmatrix}.$$

Use the mapping mentioned to transform the spread into a related conical spread.

Similarly, use the mapping $(x, y) \to (x, yM^{-1})$ to transform the second derivable net to the primary net as

$$x = 0, y = x \begin{bmatrix} A(t), & A(u) + \gamma t^{\sigma} \\ t & 0 \end{bmatrix} M^{-1}.$$

Then use the second mapping to transform the spread into another related conical spread. In general, the three spreads will construct three conical flocks that are not necessarily isomorphic.

These two theorems are proved elsewhere in volume II. Here, the point is to note the similarity of the ideas. We shall find examples of conical flocks of these types by various means. The set of such conical spreads are also hyperbolic spreads and will admit examples of the various categories listed in volume III.

There is one example that was given by [**21**], which showed that all quaternion division rings of all characteristics will provide examples of the type 3 example above, exactly when there is a non-square not in the center.

CHAPTER 25

Three spreads of a double cover

We have shown previously how to construct three spreads from an elation double cover. This method is completely general and can, for example, construct conical spreads that may not satisfy either the left or right unwrapping theorem. In the following, we shall study various elation double covers that have this property, in particular, some interesting examples of inner flocks, where the automorphisms are inner.

When there is an elation double cover, there are actually three conical flocks from the two covers. One of the spreads is represented using the pseudo-regulus net represented by one of the classical forms, from which we have given a conical flock, using the pseudo-regulus-inducing elation group. The points being associated with the Baer subspaces of the cover of pseudo-regulus nets associated with the action of the elation group, the drivers. Since there are two covers, there is another elation group that is pseudo-regulus-inducing and another pseudo-regulus associated with containing the axis of the elation group and $y = 0$, but not $y = x$. We have seen in the material of direct summation of derivable nets that the double cover spreads are semifield spreads that are a form of direct sum of two derivable nets, except for the component $x = 0$. But there is a duplication of $x = 0$ and $y = 0$ here. Some of the double covers are direct sums of the same type of derivable net, but some are not. If we recall how a conical flock is obtained, we initially have a spread of the form

$$x = 0, y = x \begin{bmatrix} u^{\sigma} + F(t) & , A(u) + G(t) \\ t & u \end{bmatrix}$$

for all t, u in a division ring D, where A is either 0 or a $\sigma - A$ derivation, where σ is either 1 or an outer automorphism. We have mentioned that theoretically, σ could also be an inner automorphism, and our study could include these types of flocks as well. We will offer a few remarks on these "inner flocks" in due course. The derivation or σ-derivations could be inner σ-derivations, however. For double cover spreads, we

DOI: 10.1201/9781003767978-25

have:

$$x = 0, y = x \left[\left[\begin{array}{cc} u^{\sigma} & A(u) \\ 0 & u \end{array} \right] + \left[\begin{array}{cc} F(t) & G(t) \\ t & 0 \end{array} \right] \right],$$

where the partitioned matrix set, as indicated, refers to the following two derivable nets:

$$x = 0, y = x \left[\begin{array}{cc} u^{\sigma} & A(u) \\ 0 & u \end{array} \right]$$

for all u in D and

$$x = 0, y = x \left[\begin{array}{cc} F(t) & ,G(t) \\ t & 0 \end{array} \right],$$

for all t in D. Note that we are indicating a union to form the spread; when we focus on one of the derivable nets, we consider $x = 0$ and $y = 0$, when we focus on the other, we refocus on that net, including $x = 0, y = 0$. Using the ideas of the derivable net of "carrier nets" of type 2 for the non-standard derivable net of [**34**], we may view the representation of the type of derivable net by either post-multiplication or pre-multiplication to create a component $y = x$. Therefore, there are three quasifibrations; the initial is a spread by definition. If we have a left spread, then to obtain a spread from this spread, we transform the spread to use the second derivable net in standard form, where it will have two components $x = 0, y = 0$ in common. This derivable net is a 2-carrier net. There using post- and pre-multiplication produce a spread and quasifibration.

Therefore, a convenient construction would be to take any two derivable nets and form a direct summation, using the ideas above, and proceed in certain cases to show that a quasifibration is obtained, which then provides a cover, thereby constructing a spread of conical type. Once the types of derivable nets are known, we then know what the set of Baer subspaces is that are considered "points" of one of the two conical flocks. Thus, in particular, the points of two conical flocks cannot be the same, and it would be possible to determine if the two flocks are isomorphic, in theory. Furthermore, since we may pre- or post-multiply to determine the second derivable net, when we do this, and if we would do this for the entire spread, the spread would be changed so that the focus is on the second derivable net.

Method. We note that since we are merely trying to express a unit that is missing in the skewfield in question, we may either post- or pre-multiply by the same matrix. However, in the division ring situation, this will produce much different-looking spreads, even though they are technically isomorphic. There may be two non-isomorphic conical flocks associated in this way. If we push the multiplication through to the entire spread, we have then changed positions from u to t viewpoints.

It might be pointed out that all of this work is original, and little theory has been completed.

PROBLEM 88. Give additional examples of elation double covers and determine the associated conical flocks, and work toward some way of determining the isomorphism of flocks.

As an example, we consider when

$$x=0, y=x\begin{bmatrix} F(t), & G(t) \\ t & 0 \end{bmatrix},$$

is

$$x=0, y=x\begin{bmatrix} A(t) & ,\theta t^{\sigma} \\ t & 0 \end{bmatrix}.$$

All of the known examples have the condition θ in $Z(D)$, the center of the associated division ring, and σ has order 2 or $\sigma=1$. For this example, simply assume the center condition only. We will start with method 1, and apply the same sort of argument to determine the conical flock associated with the classical pseudo-regulus net (the u-net). Since to determine a unit element in the t-net, we may post multiply by $\begin{bmatrix} 0 & 1 \\ \theta^{-1} & 0 \end{bmatrix}$. To obtain $\begin{bmatrix} A(t), & \theta t^{\sigma} \\ t & 0 \end{bmatrix}\begin{bmatrix} 0 & 1 \\ \theta^{-1} & 0 \end{bmatrix}=\begin{bmatrix} t^{\sigma} & A(t) \\ 0 & t \end{bmatrix}$. Therefore, in this example, we have used the same pseudo-regulus net in two ways and taken the direct sum. Since the set of "points" now arises from the set of Baer subplanes of the standard pseudo-regulus net, we obtain an isomorphic spread.

If we try to resolve the form of the derivable net by pre-multiplication (on the left), this leads to the associated right spread form as follows:

$$\begin{bmatrix} 0 & 1 \\ \theta^{-1} & 0 \end{bmatrix}\begin{bmatrix} A(t) & ,\theta t^{\sigma} \\ t & 0 \end{bmatrix}=\begin{bmatrix} t & 0 \\ \theta^{-1}A(t), & t^{\sigma} \end{bmatrix},$$

and

$$\begin{bmatrix} t & 0 \\ \theta^{-1}A(t), & t^{\sigma} \end{bmatrix}+\begin{bmatrix} 0 & u \\ \theta^{-1}u^{\sigma}, \theta^{-1}A(u) & \end{bmatrix}=$$

$$=\begin{bmatrix} t & u \\ \theta^{-1}(u^{\sigma}+ & A(t), \theta^{-1}A(u)+t^{\sigma} \end{bmatrix}.$$

This is a spread that also produces a right conical flock spread. Since θ is central, it is also a left spread, but we have not yet proved that it is also a left conical flock spread.

THEOREM 89. *The spread*

$$x = 0, y = x \begin{bmatrix} t & u \\ \theta^{-1}(u^{\sigma}+ & A(t), \theta^{-1}A(u) + t^{\sigma} \end{bmatrix}$$

produces both a left spread and a right conical flock spread. For a standard representation of conical flock spreads, we change symbols (u,t)↔ (t, u) and conjugate by the matrix $\begin{bmatrix} 0 & 1 \\ 1 & 0 \end{bmatrix}$. *Letting* $B(t) = \theta^{-1}A(t)$, *then B is also a σ-B derivation function, and we obtain the left conical flock spread:*

$$x = 0, y = x \begin{bmatrix} u^{\sigma} + B(t) & , B(u) + \theta^{-1}t^{\sigma} \\ t & u \end{bmatrix},$$

for all t,u in a division ring D, where θ is in the center of D.

The two conical flocks are:

$$x_1(G(t) - A(F(t)^{\sigma^{-1}})) - x_2(F(t)^{\sigma^{-1}} + A(t^{\sigma^{-1}}) + x_3 - x_4 t^{\sigma^{-1}} = 0,$$

where the functions $F(t)$ and $G(t)$ are

(1) $F(t) = A(t)$, $G(t) = \theta t^{\sigma}$, *and*

(2) $F(t) = B(t), G(t) = \theta^{-1}t^{\sigma}$.

PROOF. The form with θ central will provide the covering principle. The set of left determinants is non-zero provided the following condition cannot hold.

$$(B(u) + \theta^{-1}t^{\sigma})u^{-1}t = u^{\sigma} + B(t).$$

It is easily checked that this is the same as the original right determinant condition. This completes the theorem. □

- In the part on simultaneous spreads of type 2, elation double covers of the spread

$$x = 0, y = x \begin{bmatrix} u^{\sigma} + gt & , \theta t^{\sigma^{-1}} \\ t & u \end{bmatrix}$$

for all t, u in a division ring $D(\theta)$,a Galois cyclic extension of a division ring D, is discussed. The reader is invited to review this material. We shall then move to another situation involving elation double covers.

CHAPTER 26

Skew-Desarguesian spreads

This chapter will consider the skew-Desarguesian spreads (Jha, Johnson [**21**]), which are defined as follows:

$$x = 0, y = x \begin{bmatrix} u + \delta t & , \gamma t \\ t & u \end{bmatrix},$$

for all t, u in a division ring D, and δ, γ constants in D. Since the $(1, 2)$-determinant is

$$k^2 + \delta k - \gamma$$

is not 0 for all k in D; $x^2 + \delta x - \gamma$ is irreducible over D. Now let δ be in $Z(D)$ and the characteristic not 2. At this point, we may complete the square. Then $x^2 - (\frac{4\gamma}{\delta^2}+1)$ is irreducible if and only if $x^2 + \delta x - \gamma$ is irreducible. Let $(\frac{4\gamma}{\delta^2} + 1) = \gamma^*$, and assume γ^* is not in $Z(D)$ and is a non-square. Since we have the left unwrapping principle, we see that this is a Desarguesian plane if and only if δ and γ are in the center of D and a semifield spread if at least one of the constants is not in the center. Therefore, γ cannot be in the center as γ^* is not in the center. The associated pseudo-regulus net

$$x = 0, y = x \begin{bmatrix} u & 0 \\ 0 & u \end{bmatrix}$$

for all u in D is a type 3 derivable net, as is the net

$$x = 0, y = x \begin{bmatrix} \delta t, & \gamma t \\ t & 0 \end{bmatrix}.$$

We determine the two spreads arising from the latter net. The initial conical flock is:

$$\Sigma_t^1 : -x_1\delta t - x_2 t + x_3\gamma t - x_4 = 0,$$

which is a linear flock such that

$$\cap_t \Sigma_t^1 = \{(x_1, -x_1\delta + x_3\gamma, x_3, 0)\}.$$

The right inverse of $\begin{bmatrix} \delta & \gamma \\ 1 & 0 \end{bmatrix}$, is $\begin{bmatrix} 0 & 1 \\ \gamma^{-1} & -\gamma^{-1}\delta \end{bmatrix}$.

DOI: 10.1201/9781003767978-26

After applying the right inverse on the right side to the existing spread, then let $s = u\gamma^{-1}$ and $v = t - u\gamma^{-1}\delta$, define $\gamma w\gamma^{-1} = w^{\sigma}$ and $\delta r - r^{\sigma}\delta = A(r)$, then the matrix set may be transformed into the following form:

$$x = 0, y = x \begin{bmatrix} v^{\sigma} + s^{\sigma}\delta^{\sigma}, & s\gamma + A(v) + A(s\delta) \\ s & v \end{bmatrix}$$

for all s, v in D. Note that σ and $\sigma - A$ are inner automorphisms and inner σ-derivation functions, respectively. Therefore, even as we have an isomorphic spread, this is not clear for the flocks, as both are linear with distinct lines. Note that the second spread does not satisfy the left unwrapping principle but is still a spread. Moreover, it also does not satisfy the right unwrapping principle, although the notation hides this. Therefore, we obtain a left spread, but perhaps not a right spread.

Assume that δ is not zero and γ is non-zero and not in the center of D. The associated conical flock Σ_s^2 has the form:

$$-a_1((s^{\sigma}\delta^{\sigma})^{\sigma^{-1}} + A(s^{\sigma^{-1}})) - a_2 s^{\sigma^{-1}} + a_3(s\gamma + A(s\delta) - A((s^{\sigma}\delta^{\sigma})^{\sigma^{-1}}) - a_4 = 0.$$

Rewriting, we have:

$$-a_1(s\delta + A(s^{\sigma^{-1}})) - a_2 s^{\sigma^{-1}} + a_3(s\gamma) - a_4 = 0.$$

This reduces to

$$-a_1(\delta\gamma^{-1}s\gamma) - a_2(\gamma^{-1}s\gamma) + a_3(s\gamma) - a_4 = 0.$$

Therefore, the spread is linear, and the planes share

$$\left\{(a_1, a_2, a_1\delta\gamma^{-1} + a_2\gamma^{-1}, 0); a_i \text{ in } D, \text{i=1,2}\right\}.$$

To see that when in quaternion division rings there are non-squares, see Jha and Johnson [**19**]. This completes the proof of all parts of the following theorem.

THEOREM 90. *(1) The first left conical spread is linear with common line:*

$$\{(x_1, -x_1\delta + x_3\gamma, x_3, 0)\}.$$

(2) The second left conical spread is also linear with common line:

$$\left\{(a_1, a_2, a_1\delta\gamma^{-1} + a_2\gamma^{-1}, 0); a_i \textit{ in } D, \textit{i=1,2}\right\}.$$

(3) The mapping establishing the possible isomorphism between the conical flock spreads does not map the line of (1) to the line of (2), nor does it leaves invariant the common vertex of the cones. This suggests that the two flocks cannot be isomorphic.

(4) Assume that δ is not zero, $k^2 + \delta k - \gamma$ is not zero for all k in a skewfield D, and γ is not in $Z(D)$, the center of D. For example, let D be any quaternion division ring of characteristic not 2. Let δ be non-zero in $Z(D)$ and $\gamma + (\frac{\delta}{2})^2 = \gamma^*$. Then if there exist non-squares $\gamma^* =$ not in $Z(D)$. Or in general characteristic, if $\delta = 0$, then we have a skew-Desarguesian spread.

PROBLEM 91. We now consider the left mapping construction using $\begin{bmatrix} 0 & 1 \\ \gamma^{-1} & -\gamma^{-1}\delta \end{bmatrix}$. convert to standard form, conjugate by $\begin{bmatrix} 0 & 1 \\ 1 & 0 \end{bmatrix}$. Then let $u = s$ and $t = v$.

(5) Show that the spread has the following form:

$$x = 0, y = x \begin{bmatrix} v+ & \gamma^{-1}\delta\gamma s, \gamma s \\ s & v \end{bmatrix}$$

for all s, v in D. Note that this is a type 3 left spread and does satisfy the left unwrapping principle.

(6) Find the associated conical flock. Is this flock isomorphic to the original conical flock?

In later chapters, we shall have a discussion on the "inner flocks" and on a related quasifibration of the form:

$$x = 0, y = x \begin{bmatrix} u, & t\gamma \\ t & u \end{bmatrix}$$

where γ is a non-square in D that is not in the center $Z(D)$ of D.

PROPOSITION 92. *Consider any Hamiltonian division ring*

$$(-1, -1)_K,$$

where K is a subfield of the field of real numbers. To determine the set of non-squares not in K, we determine the squares

$$(\alpha_0 + \alpha_1 U + \alpha_2 V + \alpha_3 UV)^2$$

as

$$\alpha_0^2 - \alpha_1^2 - \alpha_2^2 - \alpha_3^2 +$$
$$2\alpha_0\alpha_1 U + 2\alpha_0\alpha_2 V + 2\alpha_0\alpha_3 UV.$$

We wish to find non-squares γ not in K such that $\gamma + (\frac{\delta}{2})^2 = \gamma^*$ is also a non-square not in K, for some δ. For simplicity, assume that K is the field of rational numbers.

Here are some examples: c_1U, c_2V, c_3UV, where c_i is an odd non-square in K, for $i = 1, 2, 3$. And, $1+c_1U, 1 + c_2V, 1 + c_3UV$ are also non-squares where c_i is chosen so that $\frac{1+\sqrt{1+c_i^2}}{2}$ is a non-square. There are infinitely many ways to do this.

Another way to find examples is to extend the quaternion division ring by tensor product over $K(z)$, the rational function field. Then let $c_i{=}z^{2k_i+1}$, to obtain non-squares.

Proof. Consider c_1U and assume that it is a square. Then $\alpha_2 = \alpha_3 = 0$, which shows that $\alpha_0^2 - \alpha_1^2 = 0$ and $2\alpha_0\alpha_1 = c_1$. Therefore, $\alpha_0 = \pm\alpha_1$, which implies that $\pm 2\alpha_0^2 = c_1$. Since -1 and 2 are both non-square, choose c_1 to be an $\pm q$ (the $\pm$ terms are independent), where q is an odd positive non-square, if $c_1 = -q$ or q, then $\pm 2\alpha_0^2$ can never be c_1.

Now consider

$$c_1U + 1$$

and assume that this element is a square. Then

$$\alpha_0^2 - \alpha_1^2 = 1 \text{ and } 2\alpha_0\alpha_1 = c_1.$$

Then we obtain:

$$4\alpha_0^4 - 4\alpha_0^2 - c_1^2 = 0.$$

This implies that

$$\frac{1 + \sqrt{1 + c_1^2}}{2}$$

is a square, since α_0 is a rational number.

If we extend any field K to $K(z)$, consider zU. If this element is a square, then we must have

$$\alpha_0^2 - \alpha_1^2 = 1 \text{ and } 2\alpha_0\alpha_1 = z.$$

This implies that $z = \pm 2\alpha_0^2$, a contradiction.

Therefore, at least for Hamiltonian division rings, we obtain skew-Desarguesian spreads for either δ zero or non-zero. This completes the proof.

We will discuss the hyperbolic flocks after (19.6). □

CHAPTER 27

Right skew-Desarguesian spreads

When considering what might be called "right" skew-Desarguesian spreads, we place the constants on the right and essentially define these right spreads in $PG(3, D)$, for D a division ring, as follows: Here, u^T denotes the transpose of a $row2$-vector over D.

The right spread is then

$$x^T = 0^T, y^T = \begin{bmatrix} u+ & t\delta, t \\ t\gamma & u \end{bmatrix} x^T,$$

for all t, u in D. We use the $(1,2)$-(row) column determinant, for t non-zero to obtain:

$$ut^{-1}(u + t\delta) - t\gamma = 0,$$

then shows that, and if zero, we would obtain:

$$(t^{-1}u)^2 + (t^{-1}u)\delta - \gamma = 0.$$

and now the constants are on the right; using the unwrapping procedure, we see that we obtain a right conical flock spread, provided

$$k^2 + k\delta - \gamma$$

is non-zero for all k in D. Since this right; spread is a right elation double spread, we may choose to interchange these two pseudo-regulus nets by post-multiplication or pre-multiplication by $\begin{bmatrix} \delta & ,1 \\ \gamma, & 0 \end{bmatrix}^{-1}$. Using the mapping

$$(x^T, y^T)^T \to (x^T, \begin{bmatrix} \delta & ,1 \\ \gamma, & 0 \end{bmatrix}^{-1} y^T)^T,$$

it is seen that we have isomorphic spreads, and the image spread will be

$$x^T = 0^T, y^T = (\begin{bmatrix} \delta & ,1 \\ \gamma, & 0 \end{bmatrix}^{-1} \begin{bmatrix} u+ & t\delta, t \\ t\gamma, & u \end{bmatrix})x^T.$$

The inverse matrix is

$$\begin{bmatrix} 0, & \gamma^{-1} \\ 1, & -\delta\gamma^{-1} \end{bmatrix}.$$

Therefore, we obtain:

DOI: 10.1201/9781003767978-27

$$x^T = 0^T, y^T = \begin{bmatrix} w + (\delta s)^{\tau^{-1}}, & s \\ \gamma s + B^*(\delta s) + A^*(w), & w^{\tau} \end{bmatrix} x^T,$$

for all s, w in D. Where $u^{\tau} = \gamma u \gamma^{-1}, A^*(u) = u^{\tau}\delta - \delta u, B^*(u) = u\delta - \delta u^{\tau^{-1}}$.

Using Theorem 51, the following right conical flock is obtained:

$$\Sigma_s^{2,right} :$$

$$-((\delta s)^{\tau^{-1}})^{\tau} + A^*(s))x_1 - s^{\tau}x_2 + (\gamma s + B^*(s\delta) - A((\delta s)^{\tau^{-1}})x_3 - x_4 = 0,$$

for all s in D. Reducing, the flock is

$$-(\gamma s \gamma^{-1}\delta)x_1 - \gamma s \gamma^{-1} x_2 + (\gamma s)x_3 - x_4 = 0,$$

which is a linear flock with a common left line

$$\left\{(x_1, x_2, (\gamma^{-1}\delta)x_1 + \gamma^{-1}x_2, 0); x_i \text{ in } D, \text{i=1,2.}\right\}$$

The original conical flock is

$$\Sigma_t^{1,right} :$$

$$-t\delta x_1 - tx_2 + t\gamma x_3 - x_4 = 0.$$

This is a linear flock with a common line

$$\left\{(x_1, x_2, (\gamma^{-1}\delta)x_1 + \gamma^{-1}x_2, 0); x_i \text{ in } D, \text{i=1,2.}\right\},$$

that is, the "same" common line.

However, the mapping

$$(x^T, y^T)^T \to (x^T, \begin{bmatrix} \delta & ,1 \\ \gamma, & 0 \end{bmatrix}^{-1} y^T)^T,$$

cannot act as an isomorphism of flocks even as it acts as an isomorphism of spreads.

- So, we have isomorphic conical flock spreads corresponding to two linear conical flocks sharing the same line, but are not isomorphic by the spread isomorphism. In particular, the common vertex is not invariant.

Problem 93. Write $x^T = 0^T, y^T = \begin{bmatrix} u+ & t\delta, t \\ t\gamma & u \end{bmatrix} x^T$, in hyperbolic flock form:

$$x^T = 0^T, \ y^T = \begin{bmatrix} u & 0 \\ A^*(u), & u^{\tau} \end{bmatrix} x^T, y^T = \begin{bmatrix} v & 0 \\ A^*(v) & v^{\tau} \end{bmatrix} \begin{bmatrix} r(t) & 1 \\ s(t) & t \end{bmatrix} x^T.$$

So, the form is for right skew-Desarguesian flock spreads:

$$x^T = 0^T, y^T = 0^T, y^T = \begin{bmatrix} u, & 0 \\ 0, & u \end{bmatrix} x^T, y^T = \begin{bmatrix} v & 0 \\ 0, & v \end{bmatrix} \begin{bmatrix} t+\delta, & 1 \\ \gamma & t \end{bmatrix} x^T,$$

for all non-zero u, v and for all t in a non-commutative division ring D.

(1) For any non-zero t_0 in D, construct the right t_0-interchange, hyperbolic flock spread.

(2) Construct the right hyperbolic flock and show it is nonlinear, whereas the original right hyperbolic spread is linear.

(3) After your construction, decide if you can solve this by using the left t_0-interchange. Explain how to go about this.

Part 12

Strings

In this part, we consider when there are left Baer flocks whose maximal partial spreads do not appear to be spreads. In particular, the left matrices satisfy the right unwrapping theorem but not the left. We have mentioned previously that we do not have proof that such Baer flocks are not flocks, as this is an open question. However, shall such maximal partial spreads be "quasifibrations" without using the term "proper quasifibrations".

CHAPTER 28

Strings of quasfibrations and spreads

It has been mentioned that it is possible to consider isomorphic flock spreads to construct non-isomorphic flocks. This seems counter-intuitive, but there are many instances of such many-to-one connected geometries. But, with non-commutative skewfields, we have yet to determine any two isomorphic flock spreads defined by different types of generic pseudo-regulus derivable nets for whose two flocks can be shown to be isomorphic. Then the fact that there are right and left flocks that cannot be isomorphic brings another dimension to the possible open problems. It has also been mentioned that there are quasifibrations that produce what are called quasiflocks, which do not appear to be proper/are not flocks, but again resist a complete understanding. Also, we have listed four different determinants of 2 by 2 matrices written over non-commutative skewfields. Given any matrix, if any of the determinants is 0, they are all zero. If anyone is non-zero, then the matrix is non-singular. To understand how certain affine matrix sets correspond to sets of left or right planes in the projective case, we distinguish "left generic derivable nets" where the "right generic derivable" resemble the associated transpose matrix sets. This is chosen in the form due to the fact that when the skewfield is a field, the right form would turn out to be the transposed form. Therefore, we have chosen the so-called "upper form" for the left and "lower form" for the right.

There is exactly one class of left conical spread sets for which knowing one of the determinants is non-zero also implies that the covering condition is also satisfied, and this is the class of "lifted skewfield flock spreads". Since these flocks were constructed prior to the understanding of how unwrapping works, we again see that the covering condition is a central unwrapping. Also, the lifted skewfield flock spreads admit left- and right-acting pseudo-regulus-inducing homology groups, as was pointed out in the theorem of 12. For additive matrix sets, the only general covering condition known is the left and right unwrapping principle.

Furthermore, for conical quasifibrations, those corresponding to a conical quasiflock, given a left quasifibration where the pseudo-regulus

DOI: 10.1201/9781003767978-28

is in generic form, there is a corresponding right quasifibration, whose form may be adjusted so that it is also generic without destroying the isomorphism of the associated quasiflock.

Let

$$x = 0, y = x \begin{bmatrix} u^{\sigma} + F(t), & A(u) + G(t) \\ t & u \end{bmatrix}$$

for all t, u in a non-commutative division ring D, be a left conical flock spread, which admits the left but not the right unwrapping principle. Without changing the matrix spread set, consider the following right structure:

$$x^T = 0^T, y^T = \begin{bmatrix} u^{\sigma} + F(t), & A(u) + G(t) \\ t & u \end{bmatrix} x^T.$$

Now use the double transpose mapping, which turns this structure into:

$$y = x \begin{bmatrix} w - F(t)^{\sigma^{-1}}, & A(u) + G(t) \\ A^*(w) + G(t) - A^*(F(t)^{\sigma^{-1}}), & w^{\sigma} \end{bmatrix},$$

where $A^* = A, w^{\sigma} = u^{\sigma} + F(t)$. This is a generic quasiflock, which is isomorphic to the quasiflock associated with the original left flock. But here is the question: we see that we have a quasiflock; why would this be a flock? In particular, if the original matrix spread satisfied the left unwrapping principle but not the right, then we have no particular way to determine that we would have a right flock.

In this part on strings, these ideas will be primary, and in the next material, we will provide a method to decide whether certain quasiflocks could possibly be flocks.

What we will do in this part is to look at very primitive-appearing left quasifibrations that do not satisfy the left unwrapping procedures but do admit the right unwrapping procedures and show what sort of "strings" of left quasiflocks and right flocks can be obtained, or the reverse, such a study of strings of left flocks and corresponding right quasiflocks.

We begin with simply the matrix set and do not initially make an attempt to describe whether it should be connected to a right system or a left system.

It turns out that there are infinitely many flocks in a string. The isomorphism of the flocks is a completely open question. There will be infinite strings of left quasiflocks and infinite strings of right flocks, but the ideas are easily reversed to obtain strings of left flocks and strings of right quasiflocks. These ideas are more generally discussed in the material on left and right "skew-Desarguesian spreads".

Let D be a division ring of characteristic 2 such that $D(z)(\sqrt{z})$ is a division ring. In particular, consider any quaternion division ring of characteristic 2. Form $D \otimes_F F(z)$, where $F(z)$ is the rational function field over F. Since z is a non-square, $x^2 + z$ is an irreducible polynomial, which leads to a quadratic but not Galois extension field over $F(z), F(z)(\sqrt{z})$. By the skewfield extension theorem ([**34**]), if $D = (a,b)_F$, then $D_1 = D \otimes_F F(z)$, then $D_1 \otimes_{F(z)} F(z)(\sqrt{z})$ is a quaternion division ring $(a,b)_{F(z)(\sqrt{z})}$. There are also quaternion division rings that begin at $(a,b)_{F(z)}$. In any case, we will show that, in all cases, there are elements e not in $F(z)$ such that $\sqrt{z} + e^2$ is a non-square.

So, consider the following four matrix sets over $D_1(\sqrt{z})$:

$$S_1 : \begin{bmatrix} u + te = w, & t(e^2 + \sqrt{z}) \\ t & , u + te = w \end{bmatrix}$$

and

$$S_1^{trans} : \begin{bmatrix} u + te = w, & t \\ t(e^2 + \sqrt{z}) & , u + te = w \end{bmatrix},$$

$$S_2 : \begin{bmatrix} u + et = w, & (e^2 + \sqrt{z})t \\ t & , u + et = w \end{bmatrix},$$

$$S_2^{trans} : \begin{bmatrix} u + et = w, & t \\ (e^2 + \sqrt{z})t & , u + et = w \end{bmatrix}.$$

We will show that all of these sets are quasifibrations; that is, they consist of non-singular matrices over $(a,b)_{F(z)\sqrt{z}} = D_1(\sqrt{z})$, a two-dimensional central non-Galois quadric extension of $D_1 = (a,b)_{F(z)}$. These are quite similar; the first two are transposes, as are the two pairs. The second pair has the constants on the left, and the first two have the constants on the right. $S_1$and S_1^{trans} may not be isomorphic, and $S_2$and S_2^{trans} may not be isomorphic.

Consider S_1 and S_1^{trans} as defining left quasifibrations. Then, since they do not admit the left unwrapping procedure, they are probably not spreads. However, consider them as defining right quasifibrations; then they both are right semifield spreads since they are quasifibrations that satisfy the right unwrapping theorem.

The same sort of analysis holds for S_2 and S_2^{trans}; both of these can define left semifield spreads, and both define right quasifibrations that are probably not spreads.

What we can then do is apply various constructions to obtain strings of left quasiflocks and connecting right spreads, and left spreads and right quasiflocks.

THEOREM 94. *Under any of the above assumptions, for any quaternion division ring of characteristic 2, choose an element e in* $(a,b)_{F(z)(\sqrt{z})} - F(z)(\sqrt{z})$*, such that* $e^2 + \sqrt{z}$ *is a non-square.*

(1) Then all four matrix sets are quasifibrations.

(2) The quasifibration is a semifield if e^2 *is in* $F(z)$*. In this case, the semifield spread is a Desarguesian spread.*

(3) For example, when writing out $(a,b)_{F(z)(\sqrt{z})}$ *as a 4-dimensional vector space over* $F(z)(\sqrt{z})$*, with basis*

$$\{1, U, V, W\},$$

with defining relations:

$$U^2 = \alpha U + a, V^2 = b, W^2 = ab.$$

$$UV = W, \; VU = UV + \alpha V.$$

Let $V = e$ *then e is a non-square, but* $e^2 = b$ *is in* $F \subseteq F(z)(\sqrt{z})$*. But, if* $\delta U + \gamma = e$*, where* δ *non-zero and* γ *are in* $F(z)$*, then* $e^2 = \delta^2(\alpha U + a) + \gamma^2$*. Then* $e^2 + \sqrt{z}$ *is a non-square. In all cases,* $F(z)(e^2)$ *is a field.*

(4) There are four sets of quasifibrations and flocks as left and right matrix systems. These first sets are all conical quasiflocks or flocks. Therefore, we have left Baer flocks that are not flocks and right Baer flocks that are flocks. All of these sets may be rewritten in hyperbolic form. To avoid confusion, we shall list the hyperbolic forms in the following theorem.

(i) For S_1*:*

$$x = 0, y = x\begin{bmatrix} u + te = w, & t(e^2 + \sqrt{z}) \\ t & , u + te = w \end{bmatrix},$$

for all t, u *in* $D_1(\sqrt{z})$*; a left additive quasiflock quasifibration.*

(ii) For S_2*:*

$$x = 0, y = x\begin{bmatrix} u + et = w, & (e^2 + \sqrt{z})t \\ t & , u + et = w \end{bmatrix},$$

for all t, u *in* $D_1(\sqrt{z})$*; a left semifield conical flock spread.*

(iii) For S_1*:*

$$x^T = 0^T, y^T = \begin{bmatrix} u + te = w, & t(e^2 + \sqrt{z}) \\ t & , u + te = w \end{bmatrix} x^T,$$

for all t, u *in* $D_1(\sqrt{z})$*; a right semifield flock spread.*

(iv) For S_2*:*

$$x^T = 0^T, y^T = \begin{bmatrix} u + et = w, & (e^2 + \sqrt{z})t \\ t & , u + et = w \end{bmatrix} x^T,$$

for all t, u *in* $D_1(\sqrt{z})$; *a right additive quasiflock quasifibration.*

(v) For S_1^{trans}:

$$x = 0, y = x\begin{bmatrix} u + te = w, & t \\ t(e^2 + \sqrt{z}) & , u + te = w \end{bmatrix},$$

for all t, u *in* $D_1(\sqrt{z})$; *a left additive quasiflock quasifibration.*

(vi) For S_2^{trans}:

$$x = 0, y = x\begin{bmatrix} u + et = w, & t \\ (e^2 + \sqrt{z})t & , u + et = w \end{bmatrix},$$

for all t, u *in* $D_1(\sqrt{z})$; *a left semifield conical flock spread.*

(vii) For S_1^{trans}:

$$x^T = 0^T, y^T = \begin{bmatrix} u + te = w, & t(e^2 + \sqrt{z})^{-1} \\ t & , u + te = w \end{bmatrix} x^T,$$

for all t, u *in* $D_1(\sqrt{z})$; *a right semifield flock spread (linear). Note the transpose becomes similar to* S_1 *but with* $(e^2 + \sqrt{z})$ *becoming* $(e^2 + \sqrt{z})^{-1}$.

(viii) For S_2^{trans}:

$$x^T = 0^T, y^T = \begin{bmatrix} u + et = w, & (e^2 + \sqrt{z})^{-1}t \\ t & , u + et = w \end{bmatrix} x^T,$$

for all t, u *in* $D_1(\sqrt{z})$; *a right additive quasiflock quasifibration.*

PROOF. Thus, writing out $(a, b)_{F(z)(\sqrt{z})}$ as a 4-dimensional vector space over $F(z)(\sqrt{z})$,with basis

$$\{1, U, V, W\},$$

with defining relations:

$$U^2 = \alpha U + a, V^2 = b, W^2 = ab.$$

$$UV = W,\ VU = UV + \alpha V,$$

we see that if a typical element is

$$a_1 + a_2U + a_3V + a_4W,$$

for a_i in $F(z)(\sqrt{z})$, i $= 1, 2, 3, 4$.

Start with S_1. Now with the assumptions on e, take the $(2, 1)$-determinant of any non-zero matrix. If the determinant is 0, then the t must be non-zero. Let $k = t^{-1}w$ and write out the determinant to obtain the following condition:

$$k^2 = \sqrt{z} + e^2.$$

Let $k = k_0 + k_1\sqrt{z}$, where k_0, k_1 are in $(a, b)_{F(z)}$. Then

$$k^2 = (k_0^2 + k_1^2 z) + (k_0k_1 + k_1k_0)\sqrt{z}.$$

Since $e^2$is also in $(a,b)_{F(z)}$, it follows that

$$(*):(k_0^2+k_1^2z)=e^2,$$

and

$$(**):(k_0k_1+k_1k_0)=1.$$

Therefore, k_0k_1 cannot be zero, and k_0 and k_1 cannot commute. Multiply the second equation $(**)$ by k_0 on the left side and then on the right side to obtain that

$$k_0^2k_1=k_1k_0^2.$$

Similarly, the same argument will show that

$$k_1^2k_0=k_0k_1^2.$$

Now take equation $(*)$, use the previous identities, and multiply $(*)$ on the left side and then the right side to by k_0 to conclude that k_0 commutes with e^2. Analogously, k_1 commutes with e^2. Since e^2 is not in $F(z), k_0$ and k_1 are in the centralizer of e^2 and must be of the form

$$\{c_1+c_2e^2;c_1,c_2\text{ in }F(z)\}=F(z)(e^2).$$

The centralizer of this previous set is the set itself. The proof just verifies this. Let

$$r=(b_1+b_2U+b_3V+b_4W)$$

be an element in the centralizer of $F(z)(e^2)$. If $e^2=\delta^2(\alpha U+a)+\gamma^2, write$ out r

$$re^2+e^2r=0=b_3(UV+VU)+b_4(UW+WU),$$

after canceling $\delta^2\alpha$. $UV+VU=\alpha V$ and $UW+WU=U(UV+VU)=\alpha W$. Hence, we have

$$\alpha(b_3V+b_4W)=0,$$

so that $b_3=b_4=0$. This completes the proof. This is a quadratic field extension of $F(z)$ by e^2. Hence, since k_0,k_1 are elements of the centralizer, they commute, which is a contradiction. This completes the proof of the first part. Now, assume that e^2 is in $F(z)$. Using the $(1,2)$-determinant and the w-notation, then we see that we have a quasifibration with constant terms in the center, which means the left and right unwrapping procedures work, proving both a left and right flock spread. This completes the proof that S_1 is a quasifibration. Now consider S_1^{trans}: Now use the $(1,2)$-row determinant, and realize that the same proof applies. For S_2 and S_2^{trans}, using the $(2,1)$ and $(1,2)$-row determinants, we obtain $(t^{-1}w)^2=e^2+\sqrt{z}$, whereas for S_1 and S_1^{trans}, we obtain $(wt^{-1})^2=e^2+\sqrt{z}$, all contradictions to the determinants equal to 0. This completes the proof of all parts. □

THEOREM 95. *Now we may realize that the S_1 and S_1^{trans} left quasifibrations and the right semifield flock spread admit left-acting pseudo-regulus-inducing homology groups. This provides a right associated hyperbolic flock spread and a left-inverse left hyperbolic quasiflock quasifibration. To obtain a right acting pseudo-regulus homology group requires the mapping $(x,y) \rightarrow (y,x)$ and constructs a new quasiflock. The right hyperbolic flock spread admits the left-acting group acting in the natural manner for right hyperbolic flock spreads. The constructed left hyperbolic quasiflock and right hyperbolic flock are very different structures, even as they were very similar when considering the left conical quasiflock and right conical flock.*

Similarly, S_2 and S_2^{trans} right quasifibrations and left semifield flock spreads admit right-acting pseudo-regulus-inducing a homology group. Analogously, we obtain a left hyperbolic flock spread and a right inversing quasifibration. There are then normal acting left hyperbolic flocks and the right inversing quasiflock.

Here are the associated hyperbolic structures: There will be four linear flocks and four nonlinear quasiflocks (proof for nonlinearity given elsewhere).

(i) For S_1

$$x = 0, y = 0, y = x\begin{bmatrix} v & 0 \\ 0 & v \end{bmatrix},$$

$$y = x\begin{bmatrix} u & 0 \\ 0 & u \end{bmatrix}\begin{bmatrix} t & e^2 + \sqrt{z} \\ 1 & t \end{bmatrix},$$

becomes

$$y = x\begin{bmatrix} t, & e^2 + \sqrt{z} \\ 1 & t \end{bmatrix}^{-1}\begin{bmatrix} u & 0 \\ 0 & u \end{bmatrix},$$

and then for $\Delta_t = t^2 + e^2 + \sqrt{z}$, we have

$$y = x\begin{bmatrix} t\Delta_t^{-1} & ,1 + t\Delta_t^{-1}t \\ \Delta_t^{-1} & \Delta_t^{-1}t \end{bmatrix}\begin{bmatrix} u & 0 \\ 0 & u \end{bmatrix},$$

for all non-zero v,u and for all t in $D_1(\sqrt{z})$. Note that, as a set

$$\left\{\begin{bmatrix} u & 0 \\ 0 & u \end{bmatrix}^{-1}; u \text{non-zero}\right\} = \left\{\begin{bmatrix} u & 0 \\ 0 & u \end{bmatrix}; u \text{non-zero}\right\}.$$

The following is a left hyperbolic quasiflock quasifibration (nonlinear hyperbolic quasiflock).

(ii) S_2 :

$$x = 0, y = 0, y = x\begin{bmatrix} v & 0 \\ 0 & v \end{bmatrix},$$

$$y = x\begin{bmatrix} t, & e^2+\sqrt{z} \\ 1 & t \end{bmatrix}\begin{bmatrix} u & 0 \\ 0 & u \end{bmatrix}.$$

For all non-zero v, u *and for all* t *in* $D_1(\sqrt{z})$*. This is a left semifield hyperbolic flock spread (linear hyperbolic flock).*

(iii) S_1:

$$x^T = 0^T, y^T = 0^T, y^T = \begin{bmatrix} v & 0 \\ 0 & v \end{bmatrix} x^T,$$

$$y^T = \begin{bmatrix} u & 0 \\ 0 & u \end{bmatrix}\begin{bmatrix} t, & e^2+\sqrt{z} \\ 1 & t \end{bmatrix},$$

for all non-zero v,u *and for all* t *in* $D_1(\sqrt{z})$*. A right hyperbolic semifield flock spread (linear hyperbolic flock).*

(iv) S_2:

$$x^T = 0^T, y^T = 0^T, y^T = \begin{bmatrix} u & 0 \\ 0 & u \end{bmatrix},$$

$$y^T = \begin{bmatrix} t & , e^2+\sqrt{z} \\ 1 & t \end{bmatrix}\begin{bmatrix} v & 0 \\ 0 & v \end{bmatrix} x^T,$$

becomes

$$y^T = \begin{bmatrix} v & 0 \\ 0 & v \end{bmatrix}\begin{bmatrix} t & , e^2+\sqrt{z} \\ 1 & t \end{bmatrix}^{-1} x^T$$

then expressed as

$$y^T = \begin{bmatrix} v & 0 \\ 0 & v \end{bmatrix}\begin{bmatrix} t\Delta_t^{-1} & , 1+t\Delta_t^{-1}t \\ \Delta_t^{-1} & \Delta_t^{-1}t \end{bmatrix} x^T,$$

for all non-zero v*,* u *and for all* t *in* $D_1(\sqrt{z})$*. A right non-additive hyperbolic quasiflock quasifibration.*

(v) S_1^{trans}:

$$x = 0, y = 0, y = x\begin{bmatrix} v & 0 \\ 0 & v \end{bmatrix},$$

$$y = x\begin{bmatrix} u & 0 \\ 0 & u \end{bmatrix}\begin{bmatrix} t, & (e^2+\sqrt{z})^{-1} \\ 1 & t \end{bmatrix},$$

becomes

$$y = x\begin{bmatrix} t, & (e^2+\sqrt{z})^{-1} \\ 1 & t \end{bmatrix}^{-1}\begin{bmatrix} u & 0 \\ 0 & u \end{bmatrix}$$

and letting $\Delta_t^* = t^2 + (e^2+\sqrt{z})^{-1}$*, this form is:*

$$y^T = \begin{bmatrix} t\Delta_t^{*-1} & , 1+t\Delta_t^{*-1}t \\ \Delta_t^{*-1}, & \Delta_t^{*-1}t \end{bmatrix}\begin{bmatrix} u & 0 \\ 0 & u \end{bmatrix} x^T,$$

for all non-zero v, u *and for all* t *in* $D_1(\sqrt{z})$*. A left non-additive hyperbolic quasiflock quasifibration.*

(vi) S_2^{trans}:

$$x = 0, y = 0, y = x \begin{bmatrix} v & 0 \\ 0 & v \end{bmatrix},$$

$$y = x \begin{bmatrix} t, & (e^2 + \sqrt{z})^{-1} \\ 1 & t \end{bmatrix} \begin{bmatrix} u & 0 \\ 0 & u \end{bmatrix},$$

for all non-zero v, u *and for all* t *in* $D_1(\sqrt{z})$. *A left semifield hyperbolic flock spread (linear hyperbolic flock).*

(vii) S_1^{trans}:

$$x^T = 0^T, y^T = 0^T, y^T = \begin{bmatrix} u & 0 \\ 0 & u \end{bmatrix} x^T,$$

$$y^T = \begin{bmatrix} v & 0 \\ 0 & v \end{bmatrix} \begin{bmatrix} t & , (e^2 + \sqrt{z})^{-1} \\ 1 & t \end{bmatrix} x^T,$$

for all non-zero u, v *and for all* t *in* $D_1(\sqrt{z})$. *A right semifield hyperbolic flock spread (right linear hyperbolic flock).*

(viii) S_2^{trans}:

$$x^T = 0^T, y^T = 0^T, y^T = \begin{bmatrix} u & 0 \\ 0 & u \end{bmatrix} x^T,$$

$$y^T = \begin{bmatrix} t & , (e^2 + \sqrt{z})^{-1} \\ 1 & t \end{bmatrix} \begin{bmatrix} v & 0 \\ 0 & v \end{bmatrix} x^T,$$

becomes

$$y^T = \begin{bmatrix} v & 0 \\ 0 & v \end{bmatrix} \begin{bmatrix} t & , (e^2 + \sqrt{z})^{-1} \\ 1 & t \end{bmatrix}^{-1} x^T,$$

which is

$$y^T = \begin{bmatrix} v & 0 \\ 0 & v \end{bmatrix} \begin{bmatrix} t\Delta_t^{*^{-1}} & 1 + t\Delta_t^{*^{-1}} t \\ \Delta_t^{*^{-1}} & \Delta_t^{*^{-1}} t \end{bmatrix} x^T,$$

for all non-zero u, v *and for all* t *in* $D_1(\sqrt{z})$. *A right non-additive hyperbolic quasiflock quasifibration (nonlinear hyperbolic quasiflock).*

So, we could begin four strings of quasiflocks and flocks.

The four strings are as follows:

(1) The S_1 string of left quasifibrations and right semifields.
(2) The S_2 string of left semifields and right quasifibrations.
(3) The S_1^{trans} string of left quasifibrations and right semifields.
(4) The S_2^{trans} string of left semifields and right quasifibrations.

We are trying to show how to connect the strings when they are considered as joined. At the end of the exposition, starting with S_1, one could trace through the results to see the string of left quasiflock quasifibrations and then go back and trace through to see the string of right flock spreads. There are then a total of four quasiflock strings and four flock strings using this sort of analysis. While it might seem obvious that the strings to S_1 are isomorphic to those from S_1^{trans}. Try to prove this! Also, there is no particular order to the strings. For example, the conical quasiflocks and hyperbolic quasiflocks are loosely connected, but there is no particular reason to begin with a conical quasiflock and then consider a corresponding hyperbolic.

To connect conical-type quasiflocks:

Here is an ad-hock quasiflock string from S_1. We follow a different string order to provide something of an idea of the richness of the quasiflocks.

The **left quasifibrations** are described briefly:

1. Original quasifibration.
2. Noting that the original quasifibration is a direct sum of two pseudo-regulus nets, elation double net, idea is to interchange the two nets so a different derivable net is directing the connection to a quasiflock. We look for a way to find a different unity element; for example, for left quasifibrations, multiply on the right "post" or the left "pre" mappings two different choices, and for the right, the reverse, post would be on the left. Why this is more interesting is that it will create inner automorphisms.
3. Post map: In our case, then we would consider

$$x=0, y=x\begin{bmatrix} u+et=w, & (e^2+\sqrt{z})t \\ t & ,u+et=w \end{bmatrix}\begin{bmatrix} 0 & ,1 \\ (e^2+\sqrt{z})^{-1}, & 0 \end{bmatrix},$$

which becomes

$$x=0, y=x\begin{bmatrix} t^{\sigma}, & s(e^2+\sqrt{z}) \\ s & t \end{bmatrix},$$

where $t^{\sigma}=(e^2+\sqrt{z})t(e^2+\sqrt{z})^{-1}$,for all s,t in $D_1(\sqrt{z})$.

4. Re-resolved quasifibration; we use the mapping

$$(x,y)\to(x\begin{bmatrix} 1 & e \\ 0 & 1 \end{bmatrix}, y\begin{bmatrix} 1 & e \\ 0 & 1 \end{bmatrix})$$

This transforms the original left quasifibration of S_1 to the following:

$$x=0, y=x\begin{bmatrix} r^{\sigma}-(es+s^{\sigma}e^{\sigma}), & A_e(r)-s^{\sigma}e^{\sigma}e+s(e^2+\sqrt{z}) \\ s & r \end{bmatrix},$$

for all s,r in $D_1(\sqrt{z})$.

4(c). The quasifibration obtained could also use the mapping $(x,y)\to(x\begin{bmatrix} 1 & c \\ 0 & 1 \end{bmatrix}, y\begin{bmatrix} 1 & c \\ 0 & 1 \end{bmatrix})$, for any element c not in

$Z(D_1(\sqrt{z}))$ (the center). There are infinitely many possibilities here. Try this as a problem.

5. The left hyperbolic homology group leads to the left inverse type hyperbolic quasifibration/spread of the original quasifibration.
6. The left hyperbolic homology group of the resolved quasifibration/spread leading to the left inverse type hyperbolic quasifibration.
7. The transpose of the resolved quasifibration is a quasifibration.
8. The left hyperbolic homology group of the transpose of the resolved quasifibration/spread leading to the left inverse type hyperbolic quasifibration.
9. The resolved quasifibration is an elation double quasifibration/spread, leading to quasifibrations #9 and #11.
10. The pre-map.
11. The pre-map quasifibration/spread admits a left homology group of pseudo-regulus type, leading to the left inverse by hyperbolic quasifibration.
12. The post-map.
13. The post-map quasifibration/spread admits a left homology group of pseudo-regulus type, leading to a left inverse type hyperbolic type quasifibration.
14. The transpose of the resolved quasifibration is an elation double quasifibration/spread, leading to quasifibrations.
15. The post-map of #13.
16. The resulting post-map quasifibration admits a left homology group of pseudo-regulus type, leading to a left inverse type hyperbolic type quasifibration.
17. The pre-map of #13.
18. The resulting pre-map quasifibration admits a left homology group of pseudo-regulus type, leading to a left inverse type hyperbolic type quasifibration/spread.

CHAPTER 29

Corresponding right "flocks"

The **right** flock spreads:

- Since the resolved left quasifibration satisfies the right unwrapping principle, we obtain a right conical spread, which is simultaneous, so we obtain a right conical and a right hyperbolic spread. The right spread of #1 is an elation double spread producing right spreads (post-map and pre-map types). There are four right spreads here.

Now we consider how to induce the same family of right flock "spreads". Take the transpose of quasifibration #1 to obtain the following right conical flock spread:

$$x = 0, x = 0, y = x \begin{bmatrix} u + te = w, & t \\ t(e^2 + \sqrt{z}) & , u + te = w \end{bmatrix}.$$

Now note that we can turn the left quasifibration into a right semifield flock as

Right semifield conical flock spread #1:

$$x^T = 0^T, y^T = \begin{bmatrix} u + te = w, & t \\ t(e^2 + \sqrt{z}) & , u + te = w \end{bmatrix} x^T,$$

since we now have the right unwrapping principle holding. Since the transpose of these conical flock spreads gets back to the beginning and is also a right conical flock spread, we could have started the right semifield conical flock string with the right version of the left quasifibration. There are many variations how to construct left quasiflocks and right semifield conical flocks.

Right semifield conical flock spread #2: σ-Conical flock spread.

$$x^T = 0^T, y^T = \begin{bmatrix} r, & s \\ s(e^2 + \sqrt{z}) & , r^{\sigma^{-1}} \end{bmatrix} x^T,$$

where $r^{\sigma^{-1}} = (e^2 + \sqrt{z})^{-1} r (e^2 + \sqrt{z})$.

Now apply the following mapping to right conical flock spread #2.

DOI: 10.1201/9781003767978-29

$$(x^T, y^T)^T \to (\begin{bmatrix} 1 & 0 \\ c & 1 \end{bmatrix} x^T, \begin{bmatrix} 1 & 0 \\ c & 1 \end{bmatrix} y^T)^T.$$

This resulting right flock spread is:

Right conical flock #3(c), for all c not in $F(z)$.

$$x = 0, y = x \begin{bmatrix} m + (sc)^\sigma + cs, & s \\ A_c^*(m) + A_c^*(sc) + s(e^\sigma + \sqrt{z}), & m^{\sigma^{-1}} \end{bmatrix},$$

where $A_c^*(u) = uc^{\sigma^{-1}} + cu$.

PROBLEM 96. Note that one could have also taken the transpose of the first right conical flock. Then start the string from this new spot. This is completely different from the above. Try this as a problem.

THEOREM 97. *What this means is that we obtain the/an infinite string of right semifield flock spreads by applying the corresponding analogous mappings to this initial flock semifield spread (which corresponds to the transpose of the original left quasifibration) or to the original quasifibration. Although all quasifibrations are isomorphic, the sorts of quasiflocks will be very different.*

PROBLEM 98. Show that the previously fully represented quasifibration #1 is a spread if and only if each$k$ in $D(z)(\sqrt{z})$, the function

$$\phi_k(t) = k^2 t + t(c^2 + \sqrt{z})$$

is bijective, for all k. We shall revisit this problem when we discuss "skew-Desarguesian flock spreads".

To start this problem, ask if, given any vector (x, y), there is a unique component $y = xM$ such that try to solve for the unique pair $\{t, u\}$. Once this is established, try to show there is a k such that ϕ_k is not bijective.

The reader will need to follow all of the arguments for the right spreads to then see that there are left quasifibrations, with an analogous argument. The first quasifibration is a left spread if and only if the second quasifibration is a left spread. However, looking at the matrices, it is a quasifibration that satisfies the right unwrapping principle, and there is a corresponding right spread.

Returning to the discussion on the string of right flock spreads, we also obtain a right conical flock and a right hyperbolic flock of the following form:

$$x^T = 0^T, y^T = \begin{bmatrix} r & 0 \\ 0 & r \end{bmatrix} x^T,$$

for all r in $D(\sqrt{z})$.

$$x^T = 0^T, y^T = \begin{bmatrix} v & 0 \\ 0 & v \end{bmatrix} \begin{bmatrix} t^{-1}u + c = s, & \sqrt{z} + c^2 \\ 1 & t^{-1}u + c = s \end{bmatrix} x^T.$$

for all s and non-zero r in $D_1(\sqrt{z})$. This defines a right hyperbolic spread form.

Right Elation Double Cover. And Right Simultaneous.

Also, now we see that we have an elation double cover, with the first derivable net as

$$x^T = 0^T, y^T = \begin{bmatrix} u, & 0 \\ 0 & ,u \end{bmatrix} x^T,$$

for all u in $D(z)(\sqrt{z})$ and the second derivable net is

$$x^T = 0^T, y^T = \begin{bmatrix} 0, & t \\ t(c^2 + \sqrt{z}), & 0 \end{bmatrix} x^T$$

for all t in $D(z)(\sqrt{z})$. Note that this spread is right simultaneous. Therefore, there is a hyperbolic spread/flock and a conical spread/flock.

Since the spread is an elation double spread, we may choose two mappings, a left and a right, by

$$\begin{bmatrix} 0, & (c^2 + \sqrt{z}) \\ 1, & 0 \end{bmatrix}^{-1} = \begin{bmatrix} 0 & 1 \\ (c^2 + \sqrt{z})^{-1} & 0 \end{bmatrix} \}.$$

that shows that this is a pseudo-regulus net. The right mapping preserves the unwrapping property and thus produces a right spread. The left mapping does not preserve the spread and is therefore only a quasifibration, unless the above problem can be solved. Here are the calculations:

$$\begin{bmatrix} w, & t(c^2 + \sqrt{z}) \\ t & ,w \end{bmatrix} \begin{bmatrix} 0 & 1 \\ (c^2 + \sqrt{z})^{-1} & 0 \end{bmatrix} = M\}$$

is equal to the following spread

$$x^T = 0^T, y^T = \begin{bmatrix} w, & t(c^2 + \sqrt{z}) \\ t & ,w \end{bmatrix} \begin{bmatrix} 0 & 1 \\ (c^2 + \sqrt{z})^{-1}, & 0 \end{bmatrix} x^T.$$

The mapping $\begin{bmatrix} x^T \\ y^T \end{bmatrix} \rightarrow \begin{bmatrix} x^T \\ My^T \end{bmatrix}$ will be the right mapping. And $\begin{bmatrix} x^T \\ y^T \end{bmatrix} \rightarrow \begin{bmatrix} x^T \\ My^T \end{bmatrix}$, will place M on the left. The resulting spread is also right simultaneous.

$$x^T = 0^T, y^T = \begin{bmatrix} t, & w \\ w(c^2 + \sqrt{z})^{-1} & ,t \end{bmatrix} x^T,$$

is a right spread since the product of two matrices, each with non-zero determinants, will be non-singular, and the product is non-singular. Just as if R is singular, then ERE^{-1}, for E non-singular, must be singular. Note that this spread is right simultaneous and so produces the extra hyperbolic spread.

Now take the matrix set

$$x^T = 0^T, y^T = \begin{bmatrix} 0 & 1 \\ (c^2+\sqrt{z})^{-1}, & 0 \end{bmatrix} \begin{bmatrix} w, & t(c^2+\sqrt{z}) \\ t & , w \end{bmatrix} x^T,$$

this works out as

$$x^T = 0^T, y^T = \begin{bmatrix} t, & w \\ (c^2+\sqrt{z})^{-1}w, & (c^2+\sqrt{z})^{-1}t(c^2+\sqrt{z}) \end{bmatrix} x^T,$$

we have seen previously that if $\beta^{-1} = (c^2+\sqrt{z})^{-1}$, and $(c^2+\sqrt{z})^{-1}t(c^2+\sqrt{z}) = t^\tau$, we have a quasifibration and spread of the form

$$x^T = 0^T, y^T = \begin{bmatrix} t, & w \\ (c^2+\sqrt{z})^{-1}w, & t^\tau \end{bmatrix} x^T,$$

we have a type 3 (fake type 2) derivable net where τ is an inner automorphism, provided c^2 is not in the center. All quaternion division rings, our primary example for D, have such elements c such that c and c^2 are both not in the center of D and therefore not in the center of $D(z)(\sqrt{z})$. This spread is a right conical spread, even though it does not admit the right unwrapping principle. Also, this spread admits the right pseudo-regulus group with elements

$$(x^T, y^T) \to (x^T, \begin{bmatrix} v & 0 \\ 0 & v \end{bmatrix} y^T);$$

for all non-zero v in $D_1(\sqrt{z})$, so it is also simultaneous.

We now go back and take the transpose of the original spread to obtain the matrix set:

$$x^T = 0^T, y^T = \begin{bmatrix} w, & t \\ t(c^2+\sqrt{z}) & , w \end{bmatrix} x^T,$$

which we shall show is a quasifibration, which satisfies the right unwrapping principle. Assume there is a determinant that is 0; we take the (1, 2)-determinant, which is

$$wt^{-1}w = t(c^2+\sqrt{z}),$$

so that for $k = (t^{-1}w)$, we have

$$k^2 = c^2 + \sqrt{z},$$

a contradiction. Therefore, the transpose matrix is also a right spread that admits the left pseudo-regulus homology group

$$(x^T, y^T) \to (x^T \begin{bmatrix} v & 0 \\ 0 & v \end{bmatrix}, y^T);$$

for all non-zero v in $D_1(\sqrt{z})$.

Therefore, we obtain a quasifibration and a right spread. Note that this spread is also right simultaneous. Also, this spread is a right elation double spread.

The right map spread is

$$x^T = 0^T, y^T = \begin{bmatrix} w, & t \\ t(c^2+\sqrt{z}) & ,w \end{bmatrix} \begin{bmatrix} 0, & (c^2+\sqrt{z})^{-1} \\ 1 & ,0 \end{bmatrix} x^T$$

which is

$$x^T = 0^T, y^T = \begin{bmatrix} t, & w(c^2+\sqrt{z})^{-1} \\ w & ,t \end{bmatrix} x^T.$$

And the left map spread is:

$$x^T = 0^T, y^T = \begin{bmatrix} 0, & (c^2+\sqrt{z})^{-1} \\ 1 & ,0 \end{bmatrix} \begin{bmatrix} w, & t \\ t(c^2+\sqrt{z}) & ,w \end{bmatrix} x^T.$$

Then we obtain it now

$$x^T = 0^T, y^T = \begin{bmatrix} t^\tau, & (c^2+\sqrt{z})^{-1} w \\ w, & t \end{bmatrix} x^T.$$

for all v non-zero in $D(z)(\sqrt{z})$, so the question is whether

$$\begin{bmatrix} v^\tau, & 0 \\ 0 & ,v \end{bmatrix} \begin{bmatrix} t^\tau, & (c^2+\sqrt{z})^{-1} w \\ w & ,t \end{bmatrix}$$

is closed in the spread, which is almost immediate.

Thus, we have started with a left quasifibration that mapped under the conjugation by $\begin{bmatrix} 1 & c \\ 0 & 1 \end{bmatrix}$ to another left quasifibration that produced a right spread, which is a right simultaneous spread and a right elation double spread. There are two additional elation spreads arising from the right elation double spread. At this point, there are two hyperbolic spreads and three elation spreads. We then take the transpose matrix set, which also turns out to be a simultaneous spread and also another elation double spread. This produces another hyperbolic spread and two additional elation double spreads. In total, there are three right hyperbolic spreads, six right elation spreads and four left quasifibrations.

Quasifibrations #5 through #14.

But, because we are also interested in quasifibrations, we note that the original left quasifibration and the constructed left quasifibration also admit left pseudo-regulus-inducing homology groups. That is, consider if there is a right pseudo-regulus homology group of

$$x = 0, y = x \begin{bmatrix} u + B(t), & B(u) + t\sqrt{z} \\ t & u \end{bmatrix}.$$

Then, we would consider (notation change warning: take u to be notated by t and original $t = 1$ and ask the question if

$$\begin{bmatrix} t, & B(t) + \sqrt{z} \\ 1 & t \end{bmatrix} \begin{bmatrix} v & B(v) \\ 0 & v \end{bmatrix} = \begin{bmatrix} tv, & tB(v) + B(t)v + \sqrt{z}v \\ v, & B(v) + tv \end{bmatrix}$$

maps back into the spread. We see that $tB(v) + B(t)v = B(tv)$. To see if this matches the spread, let $w = B(v) + tv$; this transforms the quasifibration into

$$\begin{bmatrix} w + B(v) & B(w + B(v)) + \sqrt{z}v \\ v & w \end{bmatrix} =$$

$$\begin{bmatrix} w + B(v) & B(w) + B^2(v) + \sqrt{z}v \\ v & w \end{bmatrix}$$

In previous material, some of the functions did have the property that $B^2(v) = 0$. Here, we obtain

$$B(vc + cv) = (vc + cv)c + c(vc + cv) = vc^2 + c^2v.$$

When we worked with the right spreads, we would prefer that c^2 is not in the center of the division ring. For quaternion division rings, our main examples, it is possible to find examples where c^2 is not in the center or when c^2 is in the center. We choose c^2 to not be in the center, which would mean here that we do not have a simultaneous quasifibration. In the other previous material, these sorts of quasifibrations sometimes had both left and right pseudo-regulus homology groups. We do not have a right group, but we do have a left group, as we now show.

$$\begin{bmatrix} v & B(v) \\ 0 & v \end{bmatrix} \begin{bmatrix} t, & B(t) + \sqrt{z} & w \\ 1 & t & \end{bmatrix} = \begin{bmatrix} vt, & vB(t) + B(v)t + \sqrt{z}v \\ v & vt \end{bmatrix},$$

which, since $vB(t) + B(v)t = B(vt)$, will map back into the spread. So, we do have a left group, and we need a right group for a left hyperbolic quasifibration. Let $vt = w\ v = 1$, to obtain a left hyperbolic quasifibration by taking $(x, y) \to (y, x)$, to construct the left hyperbolic quasifibration:

$$x = 0, y = 0, y = x \begin{bmatrix} u & B(v) \\ 0 & u \end{bmatrix}$$

$$y = x \begin{bmatrix} w, & B(w) + \sqrt{z} \\ 1 & w \end{bmatrix}^{-1} \begin{bmatrix} v & B(v) \\ 0 & v \end{bmatrix}$$

for all v,u non-zero and for all w in $D(z)(\sqrt{z})$. Thus, we actually may construct three left quasifibrations. Again, there is another left quasifibration.

$$x^T = 0^T, y^T = \begin{bmatrix} u + tc, & t \\ t(c^2 + \sqrt{z}) & , u + tc \end{bmatrix} x^T.$$

PROBLEM 99. Find $\begin{bmatrix} w, & B(w) + \sqrt{z} \\ 1 & w \end{bmatrix}^{-1}$, and then determine the associated left hyperbolic quasifibration. Why do you not use the inverse for a corresponding right hyperbolic flock?

Thus, we actually may construct three left quasifibrations. Again, there is another left quasifibration.

$$x = 0, y = 0, y = \begin{bmatrix} u + tc, & t \\ t(c^2 + \sqrt{z}) & , u + tc \end{bmatrix}.$$

PROBLEM 100. Show that the above quasifibration admits a left pseudo-regulus homology group. Then construct the inverse mapping associated with the left quasifibration and also determine the quasiflock (see Chapter 34).

29.1. Strings of right quasifibrations and left flock spreads

Think of this section as a problem set. Use the "transpose code" given in parts IV and V. to redo the previous strings but now as right quasifibrations and then left semifield spreads and the left inversing constructions.

This chapter will essentially involve the project of changing the focus from left to right. That is, the unusual situation is that the right and left flocks seem to be connected, but not by flocks, but by quasiflocks.

By starting with the right quasifibrations, we may repeat everything to obtain another set of "left" spreads, in a manner analogous to the previous material. For example, we now obtain the following left semifield conical flock spread:

$$x = 0, y = x \begin{bmatrix} w, & (e^2 + \sqrt{z})t \\ t & w \end{bmatrix}$$

for all t, w in $D_1(\sqrt{z})$, for all e. In this case, we may apply the left unwrapping theorem. We note that the matrix set is a quasifibration by using the $(2, 1)$-row determinant.

None of the left flocks here are equivalent/isomorphic to the right flocks constructed in the previous chapter. However, there is an analogous string of left flocks and right quasiflocks.

PROBLEM 101. Using the previous left quasifibration starting spot of the previous chapter, where is the analogous right quasifibration starting spot? Once this is found, work through this material, obtaining exactly a right string of the same length – but none of the elements of the right string will be the same as the left string.

Here is what you are looking for: Here again is the previous string originating from a left quasifibration. Now we will start with a right quasifibration. Change the following string into the corresponding string. "Left quasifibration" becomes "Right quasifibration" and "right spread" becomes "left spread". Write out what the matrix sets of the new string look like. The pre-map and post-map changes will become the most delicate.

We note that a left spread can never be a right spread, but a left quasifibration can produce a right spread.

Any of the strings that we have considered may also be considered more generally using the ideas from the kaleidoscope of derivable nets and using any of the K-mappings. So, these examples are something of a very small sample of the possible quasiflock strings and flock strings, as the strings could be of infinite length.

Part 13

Switch and imposter switch

Given a left imposter pseudo-regulus net in $PG(3, D)$, where D is a skewfield, it was shown how to obtain a right imposter pseudo-net in $PG(3, D)$. Actually, this was the derivation process, which we shall explain in this part. Originally, in the finite case, T.G. Ostrom discovered the "derivation process", finding situations where the classical regulus net of the form

$$x = 0, y = x \begin{bmatrix} u & 0 \\ 0 & u \end{bmatrix}$$

for all u in $GF(q)$, the finite field, although matrix expressions were not initially used. What Ostrom noticed was that if one has a finite translation plane with spread components in $PG(3, q)$, the affine restriction of the dual of the projective extension of the translation plane could be given a coordinate system that was a right vector space over $GF(q)$, and contained a net that looked like a classical regulus net. These affine planes are called "dual translation planes", and they derive a very interesting class of finite planes called "semi-translation planes" (see, e.g., Johnson [**26**]). During that early period, a coordinate method for transferring from the dual translation plane to the semi-translation plane was developed for "switching" from one coordinate system to the other. It is that method that we consider in this part. In [**30**], it has been proved that all finite derivable nets may be represented as classical regulus nets, validating Ostrom's intuition. However, there are other forms of regulus nets, of the form $D^{\sigma - A}$, where D is a field. These are "pseudo-regulus nets" when D is a non-commutative division ring. Derivation of a classical regulus net produces another classical regulus net by the process of "switching". Since all derivable nets may be represented in this manner, either as regulus nets or pseudo-regulus nets, here is the formal definition.

DEFINITION 102. Given any derivable net, finite or infinite. Then, using the embedding/contraction method, there are two classical representations of a derivable net: The right form and the left form, respectively: points $(x, y) = (x_1, x_2, x_3, x_4)$, for all x_i in D,$i = 1, 2, 3, 4$, where the components are

$$x = 0, y = \delta x$$

for all δ in D, where D is a skewfield, (the right form) with Baer subspaces:

$$B_L(a, b) = \{(\alpha(a, b), \beta(a, b)); \alpha, \beta \text{ in } D\}$$

and the left form

$$x = 0, y = x\delta$$

for all δ in D, where D is a skewfield, with Baer subspaces:

$$B_R(a, b) = \{((a, b)\alpha, (a, b)\beta); \alpha, \beta \text{ in } D\}.$$

We define "switching" as the replacement of the set of left Baer subspaces $B_L(a, b)$ to become "left components" and of right components $x = 0, y = \delta x$ as "right Baer subspaces". The switch mapping is

$$(x_1, x_2, x_3, x_4) \to (x_1, x_3, x_2, x_4).$$

Starting with the set of Baer subspaces

$$B_L(a, b) = \{(\alpha(a, b), \beta(a, b)); \alpha, \beta \text{ in } D\},$$

we note that (a, b) is in $\{(0, 1), (1, 0), (1, \text{b})$ for b non-zero in $D\}$, using a choice of bases. Applying the switch mapping, we have the points on $B_L(0, 1)$ now have the form $x = 0$. The points on $B_L(1, 0)$ have the form $y = 0$, and the points on $B_L(1, b)$ now have the form $\{(\alpha, \beta, \alpha b, \beta b)\}$, or rather $y = xb$. The same occurs for the right components; the switch mapping with maps these components to the right Baer subspaces. So, we see that

(1) The switch mapping maps the right classical derivable net over D to the left classical net over D, and conversely.

(2) When D is a field, the switch mapping seems to recreate itself, but it really is just a derivation of the classical regulus net.

(3) The switch mapping produces derivation in the affine translation net obtained from the embedding/contraction method.

(4) We have determined the "imposter derivable nets" that share the same ambient space as the classical D-derivable nets. They are derivable nets, but if "derivation" is defined by the interchange of their component set with their Baer subplane set, this is not switching. We call this "imposter switching".

(5) Imposter switching is what you would do if you have a hyperbolic or conical spread in $PG(2, D)$. All hyperbolic or conical spreads admit a derivable net

$$x = 0, y = 0, y = x \begin{bmatrix} u^\sigma & A(u) \\ 0 & u \end{bmatrix};$$

u in D. If we would take $PG(3, \{\begin{bmatrix} u^\sigma & A(u) \\ 0 & u \end{bmatrix}; u \text{ in } D\}$, and apply the switch in this sense to the above derivable net, the vector space would be written over this skewfield, which is isomorphic but not identical to D. Applying "switch" in this case would produce the right derivable net of the form:

$$x^T = 0^T, y^T = 0^T, y^T = \begin{bmatrix} v & 0 \\ A^*(v) & v^{\sigma^{-1}} \end{bmatrix} x^T$$

for all non-zero v in D.

We shall develop the “imposter switch” in the forthcoming part.

In this part, we generalized how to present derivation over arbitrary flock spreads. It turns out that the derivation of two pseudo-regulus in hyperbolic or conical flock spreads can construct non-isomorphic spreads. This means that the ideas of “deficiency one” partial flocks of hyperbolic or conical flock spreads are constructed in much different ways than in the field/finite field cases.

CHAPTER 30

Derivation of flock spreads

All of the spreads that we construct and study in this text may be derived in a variety of various manners. For hyperbolic spreads consisting of a set of pseudo-regulus nets that share two components, any of these pseudo-regulus nets may be derived. Historically, the only net that is derived is the generic/classical left pseudo-regulus net. Consider the Thas-Walker situation over a conical spread for a quadratic cone.

We have discussed the left and right flock spreads and how to translate from one to the other. We are only considering the classical flock spreads based on the imposter generic derivable nets. However, there are other derivable nets of types 0,1,2,3 as previously considered. The ideas of "switch" apply to each of these.

Consider first any division ring $(D, +, o)$, where o represents scalar multiplication. From the 4-dimensional vector space V_4/D as $D \oplus D \oplus D \oplus D$. Consider the left vector space and write the elements of V_4 as (x_1, x_2, x_3, x_4), where x_i is in D, for $i = 1, 2, 3, 4$.

If D is a field and a pseudo-regulus is the classical regulus net, we consider that net in the form:

$$x = 0, y = xou$$

for all u in D. The Baer subspaces are of the form:

$$B(a, b) = \{(a, b)ov, (a, b)ow)$$

for all 2-vectors in V_4,and for all v, u in D. Now if D is considered a non-commutative division ring, the net changes to a pseudo-regulus net, the components are left 2-dimensional D-subspaces and the Baer subspaces $B(a, b)$ are right 2-dimensional D-subspaces. If we try to write the components using matrices, then $y = xou = x \begin{bmatrix} u & 0 \\ 0 & u \end{bmatrix}$ in the regulus case, since $xou = (x_1, x_2)\, ou = (x_1ou, x_2ou) = (x_1, x_2) \begin{bmatrix} u & 0 \\ 0 & u \end{bmatrix}$, if the notation is suppressed/understood. Now try a generic type 2 derivable net, which has a matrix of sets of the form $\begin{bmatrix} u^\sigma & 0 \\ 0 & u \end{bmatrix}$. Now consider the

DOI: 10.1201/9781003767978-30

vector space $V_4/\{(\begin{bmatrix} u^\sigma & 0 \\ 0 & u \end{bmatrix} = D^\sigma, +, o)$. Writing $xou = x\begin{bmatrix} u^\sigma & 0 \\ 0 & u \end{bmatrix}$, still allows the use of matrix multiplication to represent the change from one type of pseudo-regulus net being considered. In the most general type of left type 1 pseudo-regulus net with matrix sets of the form $\begin{bmatrix} u^\sigma & A(u) \\ 0 & u \end{bmatrix} = D^{\sigma-A}$, we use the same notation technique for writing elements in the form (x_1, x_2, x_3, x_4), x_i in $D^{\sigma-A}$ (the elements are now matrices) and use $xou = (x_1, x_2)\begin{bmatrix} u^\sigma & A(u) \\ 0 & u \end{bmatrix}$. When we do this, this is a certain problem. We may use the matrix notation without use of the "o" notation, but what do we do with the right scalar mapping: uox? This makes sense as we have a left and right $D^{\sigma-A}$ vector space. But we can't think of x as a row vector if we ignore the fact that it is a actually 2-matrix set, as in

$$(x_1, x_2) = (\begin{bmatrix} x_1^\sigma & A(x_1) \\ 0 & x_1 \end{bmatrix}, \begin{bmatrix} x_2^\sigma & A(x_2) \\ 0 & x_2 \end{bmatrix}).$$

What then works and lets us essentially ignore the "o" notation is to think of column vectors representing the right scalar multiplication as in $uox = \begin{bmatrix} u^\sigma & A(u) \\ 0 & u \end{bmatrix}\begin{bmatrix} x_1 \\ x_2 \end{bmatrix}$, where here we have further changed notation to represent this as

$$uox = \begin{bmatrix} u^\sigma & A(u) \\ 0 & u \end{bmatrix}\begin{bmatrix} x_1 \\ x_2 \end{bmatrix} = \begin{bmatrix} v & 0 \\ A^*(v) & v^{\sigma^{-1}} \end{bmatrix}\begin{bmatrix} x_1 \\ x_2 \end{bmatrix},$$

and in the general right spread notation, we also changed σ^{-1} to τ. Again, none of this becomes a problem until we have to actually derive a left spread. Here, we use the "o" notation for all possible types of pseudo-regulus nets. We illustrate the derivation of a left classical pseudo-regulus net and note that the net admits both a pseudo-regulus-inducing affine homology group with axis $y = 0$ and coaxis $x = 0$ and a pseudo-regulus-inducing an elation group with axis $x = 0$. The two different central collineation groups are used when considering left hyperbolic spreads or left conical spreads.

30.1. Imposter switching-hyperbolic flock spreads

When we have an imposter left pseudo-regulus net contained in a left flock spread, the spread is in $PG(3, D)$, so that we cannot apply switching the generic pseudo-regulus net since the derivable net is an "imposter" in this ambient space relative to D, not to the isomorphic imposter net. When we "derive" the imposter pseudo-regulus net, we shall use the same interchange mapping, but this is now "imposter switching". The form of the derived net will seem to look like a derived net. When we derived this net, we have part of the new spread as left components in $PG(3, D)$ and part of the spread as right components in $PG(3, \{\begin{bmatrix} u^{\sigma} & A(u) \\ 0 & u \end{bmatrix}; u \text{ in } D\}$. The maximum subskewfield K of so that the left and right components are K-subspaces is

$$Z(D) \cap Fix\sigma \cap \{u; A(u) = 0\}.$$

Inducing A on $Z(D) \cap Fix\sigma$ must act as an inner $\sigma - A$ derivation set, since it is acting on a field. It then follows that $\{u; A(u) = 0\} = Z(D) \cap Fix\sigma$. If σ is an inner automorphism, then $Fix\sigma$ contains $Z(D)$. If σ is an outer automorphism of the form constructed in this text, D is $E_1(\theta)$, where E_1 is a division ring and $E_1(\theta)$ is a central Galois extension of E_1, where $Fix\sigma = E_1$, and $Z(D) = Z(E_1)(\theta)$. Therefore, $Z(D) \cap Fix\sigma = Z(E_1)$, which has finite index in $Z(D)$. We are assuming in this text that D is finite-dimensional over $Z(D) \cap Fix\sigma \cap \{u; A(u) = 0\}$. Therefore, when D is a finite-dimensional division ring, we have a spread of finite-dimensions and the kernel is $K = \mathrm{Z(D)} \cap \mathrm{Fix}\sigma$.

To have an understanding of what imposter switching imposes on the structure of a derived flock spread, we provide some examples: one when D is a field and $\sigma - A$ is non-trivial, and one when D is a quaternion division ring. Ordinarily, it is not necessary to recoordinatize a derived net to realize the structure of the constructed translation plane.

THEOREM 103. *The imposter switch of the left imposter pseudo-regulus net in $PG(3, D)$, where D is a division ring, is as follows:*

$$x = 0, y = x \begin{bmatrix} u^{\sigma} & A(u) \\ 0 & u \end{bmatrix}$$

with set of Baer subspaces $B(a, b)$, where

$$B(a, b) = \{(a, b) \begin{bmatrix} w^{\sigma} & A(w) \\ 0 & w \end{bmatrix}, (a, b) \begin{bmatrix} v^{\sigma} & A(v) \\ 0 & u \end{bmatrix})\}$$

for all w, v in D.

(1) Then the imposter switch (interchanging left D-components and right $\{\begin{bmatrix} u^{\sigma} & A(u) \\ 0 & u \end{bmatrix} = D^{\sigma - A}$-components) has the following form:

$$x^T = 0^T, y^T = \left\{ \begin{bmatrix} u & 0 \\ 0 & u \end{bmatrix} \begin{bmatrix} x_1^{\sigma^{-1}} \\ x_2^{\sigma^{-1}} \end{bmatrix} + \begin{bmatrix} A(x_1^{\sigma^{-1}}) \\ A(x_2^{\sigma^{-1}}) \end{bmatrix} \right\}.$$

(2) Considering the left hyperbolic flock spread

$$x = 0, y = x \begin{bmatrix} u^\sigma & A(u) \\ 0 & u \end{bmatrix}$$

for all u in D and

$$y = x \begin{bmatrix} f(t) & , g(t) \\ 1 & t \end{bmatrix} \begin{bmatrix} v^\sigma & A(v) \\ 0 & v \end{bmatrix},$$

for all non-zero v in D. When deriving a left hyperbolic flock spread by the generic pseudo-regulus net, the form of the spread, when applying the imposter switch to the entire flock spread, produces a spread in $PG(V-1, K = Z(D) \cap (full)Fix\sigma)$ *with the following components*

$$x^T = 0^T, y^T = \{ \begin{bmatrix} u & 0 \\ 0 & u \end{bmatrix} \begin{bmatrix} x_1^{\sigma^{-1}} \\ x_2^{\sigma^{-1}} \end{bmatrix} + \begin{bmatrix} A(x_1^{\sigma^{-1}}) \\ A(x_2^{\sigma^{-1}}) \end{bmatrix} \}$$

and

$$y = x \begin{bmatrix} -f(t), & (g(t) - f(t)t)v \\ v^{-\sigma}, & v^{-\sigma}(A(v) + tv) \end{bmatrix} = M_{v,t},$$

for all u, t, and non-zero v in D. Let $v = 1$, *to obtain*

$$y = x \begin{bmatrix} -f(t), & -\Delta_t \\ 1 & t \end{bmatrix} = M_{1,t}$$

where Δ_t *is the* $(2,1)$*-determinant of* $\begin{bmatrix} f(t), & g(t) \\ 1 & t \end{bmatrix}$ *(of the original expression before the imposter switch). Therefore, we have the set of components*

$$y = xM_{v,t}$$

for all non-zero v in D as the image set under the Baer homology group, for which the elements are obtained as:

$$(x, M_{1,t}) \to (x \begin{bmatrix} 1 & 0 \\ 0 & v^\sigma \end{bmatrix}, x \left(\begin{bmatrix} 0 & 0 \\ 0 & A(v) \end{bmatrix} + M_{1,t} \begin{bmatrix} 1 & 0 \\ 0 & v \end{bmatrix} \right).$$

It is possible to rewrite this spread over $PG(V-1, K = Z(D) \cap (full)Fix\sigma)$.

The Baer subplanes of the derived generic pseudo-regulus net are:

$$B^*(1,0) = \{(\alpha(1,0), \beta(1,0)); \textit{for all } \alpha, \beta \textit{ in } D\},$$
$$B^*(0,1) = \{(\alpha(0,1), \beta(0,1)); \textit{for all } \alpha, \beta \textit{ in } D\},$$
$$B^*(1,u^\sigma) = \{(\alpha(1,u^\sigma), \beta(u,1) + \alpha(0, A(u))); \textit{for all } \alpha, \beta \textit{ in } D\},$$

for all non-zero u in D.

This is a spread that admits a Baer "homology" group of the following form:

$$\left\{ \begin{bmatrix} \begin{bmatrix} 1 & 0 \\ 0 & u^{\sigma} \end{bmatrix}, & \begin{bmatrix} 0, & 0 \\ 0, & A(u) \end{bmatrix} \\ \begin{bmatrix} 0 & 0 \\ 0 & 0 \end{bmatrix}, & \begin{bmatrix} 1 & 0 \\ 0 & u \end{bmatrix} \end{bmatrix} ; u \text{ non-zero in } D \right\}.$$

This group fixes the Baer subspace $B^*(1,0)$ pointwise, the Baer "axis", and leaves the Baer subspace $B^*(0,1)$ invariant, the Baer "coaxis" mapping

$$B^*(0,1) = \{(\alpha(0,1), \beta(0,1)); \text{for all } \alpha, \beta \text{ in } D\}, \text{ to}$$
$$\{(\alpha u^{\sigma}(0,1), \beta u(0,1)); \text{for all } \alpha, \beta \text{ in } D\} = B^*(0,1).$$

for all non-zero u in D.

The Baer group acts sharply transitive on the remaining Baer subspaces $B^*(1,u)$, for all non-zero u in D.

Let $S = \begin{bmatrix} 1 & 0 & 0 & 0 \\ 0 & 0 & 1 & 0 \\ 0 & 1 & 0 & 0 \\ 0 & 0 & 0 & 1 \end{bmatrix}$ *and realize the imposter switch as $(x,y) \to (x,y)S$. Then if the original left D-components not in the generic pseudo-regulus net are $y = xM$, then the new form of the components is of the form* $(x, xm) \begin{bmatrix} I_2 & 0_2 \\ 0_2 & \begin{bmatrix} u^{\sigma} & A(u) \\ 0 & u \end{bmatrix} \end{bmatrix} S$*. Therefore, the two components Baer axis, and co-axis, together with images for all non-zero u in D, for pseudo-regulus nets are the set of pseudo-regulus nets. The right 2-dimensional* $-\begin{bmatrix} u^{\sigma} & A(u) \\ 0 & u \end{bmatrix}$ *-subspaces are now right 2-dimensional-*

$$\left\{ \begin{bmatrix} \begin{bmatrix} 1 & 0 \\ 0 & u^{\sigma} \end{bmatrix}, & \begin{bmatrix} 0, & 0 \\ 0, & A(u) \end{bmatrix} \\ \begin{bmatrix} 0 & 0 \\ 0 & 0 \end{bmatrix}, & \begin{bmatrix} 1 & 0 \\ 0 & u \end{bmatrix} \end{bmatrix} ; u \text{ in } D \right\} - \text{subspaces};$$

as they are all fixed by the multiplicative group. Adding a zero to the above group provides a skewfield isomorphic to D, so we have a blended kernel consisting of the above skewfield $D^{\sigma - A}\,S$ and D.

PROOF. The idea of the proof is to take the set of Baer subspaces of the generic derivable net and use the imposter switch on these. This will turn

$$B(a,b)=\{(a,b)\begin{bmatrix} w^{\sigma} & A(w) \\ 0 & w \end{bmatrix},(a,b)\begin{bmatrix} v^{\sigma} & A(v) \\ 0 & u \end{bmatrix})\}$$

for all w,v in D into the right components of the derived spread. Recall we consider (a,b) in $\{(0,1)$,$(1,0)$, $(1,\mathrm{b})$, b non-zero$\}$. At a certain point, we change notation so that the imposter switch looks something like what is expected, which is something that has components $x=0$ and $y=0$ in the switched components. We will obtain $x=0$ as follows: Take $B(0,1)=(0,w,0,u)\rightarrow(0,0,w,u)$; $x=0$. However, due to the $\sigma-A$ derivative functions, we do not always obtain $y=0$; what takes its place is the following:

$$B(1,0)=(w^{\sigma},A(w),u^{\sigma},A(u))\rightarrow(w^{\sigma},u^{\sigma},A(w),A(u)).$$

Letting $w^{\sigma}=x_1,u^{\sigma}=x_2$, we obtain $y=(A(x_1^{\sigma^{-1}}),A(x_2^{\sigma^{-1}}))$. Then taking $B(1,u)$ ($u=b$ here), we obtain the following, using the column vector notation:

$$y^T=\left\{\begin{bmatrix} u & 0 \\ 0 & u \end{bmatrix}\begin{bmatrix} x_1^{\sigma^{-1}} \\ x_2^{\sigma^{-1}} \end{bmatrix}+\begin{bmatrix} A(x_1^{\sigma^{-1}}) \\ A(x_2^{\sigma^{-1}}) \end{bmatrix}\right\},$$

for all u in D. We will need to use the full expression of (x,y), so just to remind the reader, it is

$$(x_1,x_2,bx_1^{\sigma^{-1}}+A(x_1^{\sigma^{-1}}),bx_2^{\sigma^{-1}}+A(x_2^{\sigma^{-1}})).$$

The pseudo-regulus-inducing homology group now turns into a Baer homology group, which is also pseudo-regulus-inducing. The left Baer subspaces of the imposter switch image of the generic pseudo-regulus derivable net may be obtained as follows: First, use the components $x=0$ and $y=0$ and switch these to obtain the following:

$$\{(0,x_3,0,x_4);x_3,x_4\text{ in }D\}.$$

Change notation, letting $x_3=\alpha$,x_4=β and rewrite the set as:

$$B^*(0,1)=\{(\alpha(0,1),\beta(0,1));\text{for all }\alpha,\beta\text{ in }D\}.$$

Use the same technique to switch $y=0$, with analogous notation change to obtain:

$$B^*(0,1)=\{(\alpha(1,0),\beta(0,1));\text{for all }\alpha,\beta\text{ in }D\},$$

and similarly $y=x$ will switch to

$$B^*(1,1)=\{(\alpha(1,1),\beta(1,1));\text{ for all }\alpha,\beta\text{ in }D\}.$$

Use the switch map to change the homology group mapping

$$(x,y)\rightarrow(x,y\begin{bmatrix} w^{\sigma} & A(w) \\ 0 & w \end{bmatrix}),$$

for all u in D. This will change to the group

$$\left\{ \left[\begin{array}{cc} \begin{bmatrix} 1 & 0 \\ 0 & u^\sigma \end{bmatrix}, & \begin{bmatrix} 0, & 0 \\ 0, & A(u) \end{bmatrix} \\ \begin{bmatrix} 0 & 0 \\ 0 & 0 \end{bmatrix}, & \begin{bmatrix} 1 & 0 \\ 0 & u \end{bmatrix} \end{array} \right] ; u \text{ non-zero in } D \right\}.$$

Just to check that this is a group,

$$\left[\begin{array}{cc} \begin{bmatrix} 1 & 0 \\ 0 & u^\sigma \end{bmatrix}, & \begin{bmatrix} 0, & 0 \\ 0, & A(u) \end{bmatrix} \\ \begin{bmatrix} 0 & 0 \\ 0 & 0 \end{bmatrix}, & \begin{bmatrix} 1 & 0 \\ 0 & u \end{bmatrix} \end{array} \right] \left[\begin{array}{cc} \begin{bmatrix} 1 & 0 \\ 0 & v^\sigma \end{bmatrix}, & \begin{bmatrix} 0, & 0 \\ 0, & A(v) \end{bmatrix} \\ \begin{bmatrix} 0 & 0 \\ 0 & 0 \end{bmatrix}, & \begin{bmatrix} 1 & 0 \\ 0 & v \end{bmatrix} \end{array} \right]$$

is

$$\left[\begin{array}{cc} \begin{bmatrix} 1 & 0 \\ 0 & (uv)^\sigma \end{bmatrix}, & \begin{bmatrix} 0, & 0 \\ 0, & u^\sigma A(v) + A(u)v = A(uv) \end{bmatrix} \\ \begin{bmatrix} 0 & 0 \\ 0 & 0 \end{bmatrix}, & \begin{bmatrix} 1 & 0 \\ 0 & uv \end{bmatrix} \end{array} \right].$$

The remaining new Baer subspaces can then be obtained by realizing that the original homology group fixes $y = 0$ pointwise, leaves $x = 0$ invariant and maps $y = x$ to $y = x \begin{bmatrix} u^\sigma & A(u) \\ 0 & u \end{bmatrix}$, for all non-zero u in D. $y = 0$ is the "axis" and $x = 0$ is the "coaxis" of the homology group. Upon imposter switch, the image of $y = 0$ is the Baer homology group axis, and the image of $x = 0$ is the Baer homology group coaxis. Therefore, take the image of $B^*(1,1)$ under the Baer homology group to obtain the following:

$$(\alpha(1,1) \begin{bmatrix} 1 & 0 \\ 0 & u^\sigma \end{bmatrix}, \alpha(1,1) \begin{bmatrix} 0 & 0 \\ 0 & A(u) \end{bmatrix} + \beta(1,1) \begin{bmatrix} 1 & 0 \\ 0 & u \end{bmatrix}),$$

which is $B^*(1,u)$, for all non-zero u in D. We also need to change/switch the form of the remaining left D-components

$$y = x \begin{bmatrix} f(t) & ,g(t) \\ 1 & t \end{bmatrix} \begin{bmatrix} v^\sigma & A(v) \\ 0 & v \end{bmatrix},$$

for all t and non-zero v in D. Multiply out the matrix representation, then make the imposter switch. There is a notational trick, which is necessary to see how this works out. So, we obtain after the switch is: First

$$(x_1, x_2, (x_1 f(t) + x_2)v^\sigma, (x_1 f(t) + x_2)A(v) + (x_1 g(t) + x_2 t)v,$$

then the switch:

$$(x_1, (x_1 f(t) + x_2)v^\sigma, x_2, (x_1 f(t) + x_2)A(v) + (x_1 g(t) + x_2 t)v.$$

The idea is to relabel the first entries as x_1^* and x_2^*, then solve for x_2 in terms of these labels to calculate the form of the switch left D-component. First, we see that

$$x_2 = x_2^* v^{-\sigma} - x_1^* f(t).$$

Then the fourth term becomes

$$x_1^*(g(t) - f(t)t)v + x_2^*(A(v) + tv).$$

The left spread terms are then

$$y = x \begin{bmatrix} -f(t), & (g(t) - f(t)t)v \\ v^{-\sigma}, & v^{-\sigma}(A(v) + tv) \end{bmatrix}.$$

It will now follow that the Baer homology group will act on this set of components. It is helpful to see this to believe it, as the verification is intricate. Let $M_{t,v}$ denote the matrix in the above component. The group will map

$$(x, M_{t,v}) \to (x \begin{bmatrix} 1 & 0 \\ 0 & u^\sigma \end{bmatrix}, x\left(\begin{bmatrix} 0 & 0 \\ 0 & A(u) \end{bmatrix} + M_{t,v}\begin{bmatrix} 1 & 0 \\ 0 & u \end{bmatrix}\right).$$

Let $x^* = x \begin{bmatrix} 1 & 0 \\ 0 & u^\sigma \end{bmatrix}$, then the second expression becomes

$$x^* \begin{bmatrix} 1 & 0 \\ 0 & u^{-\sigma} \end{bmatrix}\left(\begin{bmatrix} 0 & 0 \\ 0 & A(u) \end{bmatrix} + M_{v,t}\begin{bmatrix} 1 & 0 \\ 0 & u \end{bmatrix}\right).$$

Then

$$\begin{bmatrix} -f(t), & (g(t) - f(t)t)v \\ v^{-\sigma}, & v^{-\sigma}(A(v) + tv) \end{bmatrix} \to$$

$$\begin{bmatrix} -f(t), & (g(t) - f(t)t)vu \\ u^{-\sigma}v^{-\sigma}, & u^{-\sigma}(v^{-\sigma}(A(v) + tv)u) + A(u)) \end{bmatrix}.$$

This last expression should be

$$\begin{bmatrix} -f(t), & (g(t) - f(t)t)vu \\ (vu)^{-\sigma}, & (vu)^{-\sigma}(A(vu) + tvu) \end{bmatrix},$$

to check that the group action is correct. Therefore, this reduces to checking if

$$u^{-\sigma}(v^{-\sigma}(A(v) + tv)u + A(u)) = (vu)^{-\sigma}(A(vu) + tvu).$$

This is valid if and only if

$$u^{-\sigma}v^{-\sigma}A(v)u + u^{-\sigma}A(u) = (vu)^{-\sigma}(A(vu).$$

And since $A(vu) = v^\sigma A(u) + A(v)u$,this follows immediately.

Finally, it is checked that the components

$$x^T = 0^T, y^T = \left\{ \begin{bmatrix} u & 0 \\ 0 & u \end{bmatrix} \begin{bmatrix} x_1^{\sigma^{-1}} \\ x_2^{\sigma^{-1}} \end{bmatrix} + \begin{bmatrix} A(x_1^{\sigma^{-1}}) \\ A(x_2^{\sigma^{-1}}) \end{bmatrix} \right\}$$

are all fixed by the Baer homology group.

So, we could dispense with the column notation, but the matrix would then not be apparent; however, to apply the Baer group, we need to write the expression as

$$x = 0, \& \ (x_1, x_2, ux_1^{\sigma^{-1}} + A(x_1^{\sigma^{-1}}), ux_2^{\sigma^{-1}} + A(x_2^{\sigma^{-1}}))$$

for all u in D. Now apply the group and note that the second component, will map to

$$(x_1, x_2 v^\sigma, ux_1^{\sigma^{-1}} + A(x_1^{\sigma^{-1}}), (ux_2^{\sigma^{-1}} + A(x_2^{\sigma^{-1}}))v + x_2 A(v)).$$

Replace the x_2- term in the expression for the u-component by $x_2 v^\sigma$, so the u-component is invariant if and only if

$$u(x_2^{\sigma^{-1}} v) + A(x_2^{\sigma^{-1}} v) = (ux_2^{\sigma^{-1}} + A(x_2^{\sigma^{-1}}))v + x_2 A(v),$$

which is true as $A(x_2^{\sigma^{-1}} v) = x_2 A(v) + A(x_2^{\sigma^{-1}})v$. This completes the proof. □

30.2. Imposter switching-conical flock spreads

In this section, we consider the derivation of a generic $D^{\sigma - A}$ pseudo-regulus net, which we call "imposter switching", and in this case, we consider the left conical flock spreads.

THEOREM 104. *Consider the left conical flock spread:*

$$x = 0, y = x \begin{bmatrix} u^\sigma + F(t), & A(u) + G(t) \\ t & u \end{bmatrix},$$

for all t, u in D. The generic net that we derive is the same as in the hyperbolic case. In this setting, we have an elation group E, where

$$E = \begin{bmatrix} I_2 & \begin{bmatrix} v^\sigma & A(v) \\ 0 & v \end{bmatrix} \\ 0_2 & I_2 \end{bmatrix}$$

for all u in D, where D is a finite-dimensional division ring.

The derived spread has a blended kernel and is also obtained in $PG(V-1, K = Z(D) \cap (full)Fix\sigma)$ by derivation/imposter switch of the generic pseudo-regulus net has the following form:

$$x^T = 0^T, y^T = \left\{ \begin{bmatrix} u & 0 \\ 0 & u \end{bmatrix} \begin{bmatrix} x_1^{\sigma^{-1}} \\ x_2^{\sigma^{-1}} \end{bmatrix} + \begin{bmatrix} A(x_1^{\sigma^{-1}}) \\ A(x_2^{\sigma^{-1}}) \end{bmatrix} \right\}$$

and by using the same technique, the left D-components not in the generic pseudo-regulus net have the following form.

$$y = x \begin{bmatrix} -(v^\sigma + F(t))t^{-1}, & G(t) + A(v) - (v^\sigma + F(t))t^{-1}v \\ t^{-1} & t^{-1}v \end{bmatrix}$$

for all t non-zero, for all u and v of D.

This spread admits the Baer elation group ES, where E is the pseudo-regulus-inducing elation group, using the notation of the hyperbolic case. This group is the Baer elation group, which is

$$\left\{ \begin{bmatrix} 1 & v^\sigma & 0 & A(v) \\ 0 & 1 & 0 & 0 \\ 0 & 0 & 1 & v \\ 0 & 0 & 0 & 1 \end{bmatrix} ; \text{for all } u \text{ in } D \right\},$$

and fixes the Baer subplane $B^*(0,1)$ pointwise, the Baer elation "axis".

The Baer elation group is sharply transitive on the remaining Baer subspaces. This group acts additively on the remaining components, which are $(x,xM)S + ES$, where M is an original matrix corresponding to a component not on the generic pseudo-regulus net, and E is the original elation group of the conical flock spread.

(3) We note that since the kernel of this spread is $Z(D) \cap (full)Fix\sigma)$, the spread is of dimension 2 if and only if $Z(D) = D$ and $\sigma = 1$ (which makes A identically zero, as we are not in a field). This makes the theorem of Jha and Johnson [**21**] *more relevant, as it is proved that "full Baer groups", such as those obtained here in a spread $PG(3,D)$, where D is a skewfield, implies that D is a field.*

PROOF. The proof is analogous to the proof for the hyperbolic spreads. Since all of this is well explained previously, and the Baer subplanes $B^*(0,1)$ and $B^*(1,u)$ for u in D are the same as in the hyperbolic proof, we need only to verify that the Baer elation group with axis $B^*(0,1)$ acts on the imposter switched spread with components

$$x^T = 0^T, y^T = \{\begin{bmatrix} u & 0 \\ 0 & u \end{bmatrix} \begin{bmatrix} x_1^{\sigma^{-1}} \\ x_2^{\sigma^{-1}} \end{bmatrix} + \begin{bmatrix} A(x_1^{\sigma^{-1}}) \\ A(x_2^{\sigma^{-1}}) \end{bmatrix}\}$$

and

$$y = x \begin{bmatrix} -(v^\sigma + F(t))t^{-1}, & G(t) + A(v) - (v^\sigma + F(t))t^{-1}v \\ t^{-1} & t^{-1}v \end{bmatrix} = N_{t,v}^u$$

for all non-zero t, and for all u and v in D. Since $x^T = 0^T$, it is clearly invariant under the Baer elation group, we consider the other components (the original Baer subspaces of the generic pseudo-regulus net), written in row form as

$$(x_1, x_2, ux_1^{\sigma^{-1}} + A(x_1^{\sigma^{-1}}), ux_2^{\sigma^{-1}} + A(x_2^{\sigma^{-1}})),$$

for all x_1, x_2, u in D (fixed in each component).

We need to verify the following:

(i) ES fixes $B^*(0,1)$ *pointwise* and is sharply transitive on the set $\{B^*(1,u); u \text{ in } D\}$.

(ii) ES fixes each component of the imposter-switched original right Baer subspaces.

(iii) ES leaves invariant the set of all $y = xN_{t,v}^u$ for all v in D, and maps $y = xN_{t,1}^u$ to $y = xN_{t,r}^u$ for all r in D.

(i) To prove this, it is clear that

$$\left\{ \begin{bmatrix} 1 & r^\sigma & 0 & A(r) \\ 0 & 1 & 0 & 0 \\ 0 & 0 & 1 & r \\ 0 & 0 & 0 & 1 \end{bmatrix}; \text{for all } r \text{ in } D \right\} = ES,$$

fixes

$$B^*(0,1) = \{(\alpha(0,1), \beta(0,1)); \text{for all } \alpha, \beta \text{ in } D\}, \text{ pointwise},$$

by matrix multiplication (row-column). Consider the image of $B^*(1,0)$ under ES, which is

$$(\alpha(1,0) \begin{bmatrix} 1 & r^\sigma \\ 0 & 1 \end{bmatrix}, (\alpha(1,0) \begin{bmatrix} 0 & A(r) \\ 0 & 0 \end{bmatrix} + \beta(1,0) \begin{bmatrix} 1 & r \\ 0 & 1 \end{bmatrix}),$$

which is

$$(\alpha(1, r^\sigma), (\alpha(0, A(r)) + \beta(1,r)) = B^*(1,r).$$

This proves (i).

Consider (ii), and the component

$$(x_1, x_2, ux_1^{\sigma^{-1}} + A(x_1^{\sigma^{-1}}), ux_2^{\sigma^{-1}} + A(x_2^{\sigma^{-1}})),$$

for u fixed and in D. Using the ES Baer elation group, multiply out and notice that the first and second components become $(x_1, x_1 r^\sigma + x_2) = (x_1^*, x_2^*)$. In the multiplication, notice that the third component is unchanged from the original. Here is the fourth component in the $*$-notation: (as $x_2^{\sigma^{-1}} = x_2^{*\sigma^{-1}} - x_1^{*\sigma^{-1}} r$)

$$(ux_1^{*\sigma^{-1}} + A(x_1^{*\sigma^{-1}}))r + u(x_2^{*\sigma^{-1}} - x_1^{*\sigma^{-1}} r) + A(x_2^{*\sigma^{-1}} - x_1^{*\sigma^{-1}} r) + x_1 A(r)).$$

This needs to be

$$ux_2^{*\sigma^{-1}} + A(x_2^{*\sigma^{-1}}),$$

which it is, recalling that A is additive and that

$$A(x_1^{*\sigma^{-1}})r + (A(-x_1^{*\sigma^{-1}}r) + x_1A(r) = 0.$$

This proves (ii).

Finally, consider (iii). Applying the group element

$$\begin{bmatrix} 1 & r^\sigma & 0 & A(r) \\ 0 & 1 & 0 & 0 \\ 0 & 0 & 1 & r \\ 0 & 0 & 0 & 1 \end{bmatrix},$$

we obtain the image of $(x, xN_{t,v})$ as

$$(x\begin{bmatrix} 1 & r^\sigma \\ 0 & 1 \end{bmatrix}, x(\begin{bmatrix} 0 & A(r) \\ 0 & 0 \end{bmatrix} + N_{t,v}\begin{bmatrix} 1 & r \\ 0 & 1 \end{bmatrix})).$$

The matrix part works to be the following:

$$\begin{bmatrix} 1 & -r^\sigma \\ 0 & 1 \end{bmatrix} U$$

where U is

$$\begin{bmatrix} -(v^\sigma + F(t))t^{-1} & G(t) + A(v+r) - (v^\sigma + F(t))t^{-1}(v+r) \\ t^{-1} & t^{-1}(v+r) \end{bmatrix},$$

using additivity of A. Using the same procedure of letting $x^* = x\begin{bmatrix} 1 & r^\sigma \\ 0 & 1 \end{bmatrix}$, we may replace x by $x^*\begin{bmatrix} 1 & r^\sigma \\ 0 & 1 \end{bmatrix}^{-1} = x^*\begin{bmatrix} 1 & -r^\sigma \\ 0 & 1 \end{bmatrix}$. Completing the matrix multiplication, we obtain:

$$\begin{bmatrix} -((v+r)^\sigma + F(t))t^{-1}, & S \\ t^{-1} & t^{-1}(v+r) \end{bmatrix},$$

$$S = G(t) + A(v+r) - ((v+r)^\sigma + F(t))t^{-1}(v+r),$$

which shows that the set of matrices with fixed t is closed under the Baer elation group. It follows also that the image of

$$y = xN_{t,1}$$

under the Baer elation group is

$$y = xN_{t,r}$$

for all r in D. This completes the proof. □

PROBLEM 105. The smallest dimension of any derived flock spread is 4, over a field, when σ has order 2 and either A is identically zero or is an inner $\sigma - A$ derivation function. For non-commutative division rings, the smallest dimension of a derived flock spread for interesting examples is 8. In the following, we provide these examples, writing over the field $Z(D) \cap (full)Fix\sigma$.

30.3. Derived – lifted skewfield spreads

The lifted skewfield spreads have the following form over a skewfield $D_1(\theta)$, where D_1 is a skewfield of finite dimension n^2 over a field $F(z_1, z_2)$, and $D_1(\theta)$ is a central Galois quadratic extension of D_1. σ fixes D_1 elementwise: The general construction is as follows, repeated here, for convenience. If D is a finite-dimensional division ring over F, for

$$D \otimes_F F(z_1, z_2),$$

where z_1, z_2 are commuting indeterminates, and σ is the mapping $z_1 \leftrightarrow z_2$ and fixing D elementwise, then define

$$D \otimes_F Fix\sigma = D_1.$$

Then, using the irreducible polynomial

$$x^2 - (z_1 + z_2)x + z_1 z_2,$$

with formal roots z_1 and z_2, then $F(z_1, z_2)$ is a Galois quadratic extension over $Fix\sigma$. The $\mathrm{Fix}_{D_1(\theta)}\sigma$ $=$Full(Fixσ) $= D_1$, and θ is a generator of $F(z_1, z_2)/Fix\sigma$. The set of lifted skewfield semifield spreads $S_{\delta,\gamma}$ is lifted from the same Desarguesian spread but using different pre-Skewfields, which can change the isomorphism of the associated semifield flock spread. All of these are conical flock spreads, and depending on the choice of representations, some are also hyperbolic flock spreads; some are simultaneous spreads. This set is represented in the following form: $S_{\delta,\gamma}$is

$$x = 0, y = x \begin{bmatrix} u^\sigma & (\theta\delta + \gamma)t^\sigma \\ t & u \end{bmatrix}$$

for all t,u in $D_1(\theta)$, δ non-zero and γ in $F(z_1, z_2)$.

We wish to represent a derived spread of a lifted conical spread over the kernel of the derived spread. However, if D is of dimension n^2 over the center F, if D is a non- center $F(z_1, z_2)$, and it follows that the dimension of $D_1(\theta)$ is of dimension $2n^2$over $Fix\sigma$. The kernel of the derived spread is

$$Z(D_1(\theta)) \cap FullFix\sigma = F(z_1, z_2) \cap D_1 = Fix\sigma.$$

Therefore, we are in $PG(4n^2 - 1, Fix\sigma)$. This means our matrices are $2n^2$by $2n^2$. So, it is impossible to actually visualize what these derived spreads look like except for very small degrees. The smallest possibility is when D is a field, in which case we have $n^2 = 1$ and the spread is a lifted Pappian spread. In this case, the projective space is $PG(3, Fix\sigma)$. This is a different extension than the normal situation for lifted Pappian spreads. What is customary is to take a field F and form an algebraic Galois quadratic extension $F(\theta)$ over F. The lifted Pappian spread is in

$PG(3, F(\theta))$ and the derived lifted plane is in $PG(7, F)$, so the matrices are 4 by 4 over $F = Fix\sigma$ in this setting.

The smallest dimension for non-commutative finite division ring is 4, for quaternion division rings. Therefore, the projective space over the kernel for the derived spread is $PG(15, Fix\sigma)$, using transcendental extensions, which makes the matrices 8 by 8 over $Fix\sigma$. In the first volume, Galois groups over quaternion division rings were considered, and it is also possible to find algebraic extensions of quaternion division ring spreads. These spread, are also lifted skewfields. In this setting, here are the two differences: If $(a, b)_F$ is a quaternion division ring of characteristic not 2 , the associated matrix representation is

$$\begin{bmatrix} u^\sigma & bu^\sigma \\ t & u \end{bmatrix}$$

for all t, u in $F(\sqrt{a})$, a Galois quadratic extension, and σ the induced involutory automorphism. As a spread

$$x = 0, y = x \begin{bmatrix} u^\sigma & bu^\sigma \\ t & u \end{bmatrix},$$

the corresponding components are in $PG(3, F(\sqrt{a}))$. If we take a quadratic algebraic extension $F(c)$ such that $F(\sqrt{a}, c)$ is a quadratic extension of $F(\sqrt{a})$, an extension of degree 4 over F where c is not $\sqrt{b}$, there is a central extension of $(a, b)_F$ equal to $(a, b)_{F(c)}$. A more general theory of this in arbitrary characteristics applies where, in characteristic 2, the extension $F(c)$ need not be Galois, and the extension of F is of the form $F(\theta)$, but still, there is an extension $F(\theta, c)$ is of dimension 4 over F.

Since the transcendental extensions work even for algebraically closed fields, we could give our illustrations using the transcendental technique of Galois quadratic extensions of finite-dimensional Pappian and of quaternion division rings. In the Pappian case, for simplicity and algebraic Galois quadratic extensions of characteristic not 2. We shall see that it would be beyond the capacity of our computer system to list the form of any of the smallest matrix systems for the derived-lifted quaternion division ring, as they would be 8 by 8 matrices, well beyond our page capacity.

30.4. Derived – lifted Pappian spreads

Let the lifted Pappian spread be

$$x = 0, y = x \begin{bmatrix} u^\sigma & \theta t^\sigma \\ t & u \end{bmatrix}$$

for all t, u in $F(\theta)$, where F is a field, and the extension is an algebraic Galois quadratic extension of characteristic not 2. These spreads are simultaneous; this representation is a conical flock spread. The hyperbolic flock spread is

$$x = 0, y = x \begin{bmatrix} u^{\sigma} & 0 \\ 0 & u \end{bmatrix}$$

and

$$y = x \begin{bmatrix} t^{\sigma} & \theta \\ 1 & t \end{bmatrix} \begin{bmatrix} v^{\sigma} & 0 \\ 0 & v \end{bmatrix}$$

for all u, t and for all non-zero elements v of $F(\theta)$.

The derived conical spreads are as follows:

$$x = 0, y = (x_1^{\sigma^{-1}} u, x_1^{\sigma^{-1}} u),$$

for all u in $F(\theta)$, and

$$y = x \begin{bmatrix} -v^{\sigma}t^{-1}, & \theta t^{\sigma} - v^{\sigma}t^{-1}v \\ t^{-1} & t^{-1}v \end{bmatrix}$$

for all v and non-zero t in $F(\theta)$. The kernel of this spread is $Fix\sigma = F$. Therefore, to write the spread over the kernel, we let *an* element $k{=}k_0 + k_1\theta$, where k_0, k_1 in F and, since the characteristic is not 2, we may choose a basis so that $\theta^{\sigma} = -\theta$, and $\theta^2 = d$, where d is a non-square in F.

We write the elements in the basis for 2-vectors $(k_0, k_1) = k_0 + k_1\theta$.

The set of elements (x_1, x_2, x_3, x_4), where $x_i = x_{0i} + x_{1i}$, for $i = 1, 2, 3, 4$. Therefore, we are now in an 8-dimensional vector space over F.

To determine the matrices for the derived net, we must find

$(x_{0i} - x_{1i}\theta)(u_0 + u_1\theta) = (x_{0i}u_0 - x_{1i}) + (x_{0i}(-u_1 d + x_{1i}(-u_0)\theta.$

Letting

$$W = \begin{bmatrix} u_0 & u_1 \\ -u_1 d, & -u_0 \end{bmatrix}$$

The derived net is

$$x = 0, y = x \begin{bmatrix} W & ,0_2 \\ 0_2 & W \end{bmatrix},$$

for all u_0, u_1in F. This derived net is expressed as a 4-dimensional regulus net.

To express the components

$$y = x \begin{bmatrix} -v^{\sigma}t^{-1}, & \theta t^{\sigma} - v^{\sigma}t^{-1}v \\ t^{-1} & t^{-1}v \end{bmatrix},$$

we first express $t = t_0 + t_1\theta$, then we need to find $t^{-1} = s_0 + s_1\theta$. The way to find s_0,s_1 in terms of t_0, t_1, is just to multiply the two elements together and set them equal to 1. Then solve the simultaneous equations to show that

$$s_0 = \frac{t_0}{(t_0^2 - t_1^2 d)},\ s_1 = \frac{-t_1}{(t_0^2 - t_1^2 d)}.$$

The components are, remembering that $x = (x_{01}, x_{11}, x_{02}, x_{12})$:

$$y = xR,$$

where R is:

$$y = x\left[r_{ij}\right]_{4x4},$$

$$r_{11}, r_{12} = \frac{-(v_0t_0 + v_1t_1d)}{(t_0^2 - t_1^2 d)}, \frac{v_0t_1 + v_1t_0}{(t_0^2 - t_1^2 d)},$$

$$r_{13} = \frac{-(v_0t_0 + v_1t_1d) + (v_0t_1 + v_1t_0)v_1d}{(t_0^2 - t_1^2 d)},$$

$$r_{14} = -(v_0t_0 + v_1t_1d)v_1 + (v_0t_1 + v_1t_0)v_0$$

$$r_{21}, r_{22} = \frac{(v_0t_1 + v_1t_0)d}{(t_0^2 - t_1^2 d)}, \frac{-(v_0t_0 + v_1t_1d)}{(t_0^2 - t_1^2 d)},$$

$$r_{23} = \frac{-(v_0t_0 + v_1t_1d)}{(t_0^2 - t_1^2 d)}, \frac{-(v_0t_0 + v_1t_1d)v_1 + (v_0t_1 + v_1t_0)v_0d}{(t_0^2 - t_1^2 d)},$$

$$r_{24} = \frac{-(v_0t_0 + v_1t_1d)v_0 + (v_0t_1 + v_1t_0)v_1d}{(t_0^2 - t_1^2 d)},$$

$$r_{31}, r_{32} = \frac{t_0}{(t_0^2 - t_1^2 d)}, \frac{-t_1}{(t_0^2 - t_1^2 d)},$$

$$r_{33}, r_{34} = \frac{(t_0v_0 - t_1v_1d)}{(t_0^2 - t_1^2 d)}, \frac{(t_0v_1 - t_1v_0)}{(t_0^2 - t_1^2 d)},$$

$$r_{41}, r_{42} = \frac{-t_1d}{(t_0^2 - t_1^2 d)}, \frac{t_0}{(t_0^2 - t_1^2 d)},$$

$$r_{43}, r_{44} = \frac{(t_0v_1 - t_1v_0)d}{(t_0^2 - t_1^2 d)}, \frac{(t_0v_0 - t_1v_1d)}{(t_0^2 - t_1^2 d)},$$

for each t_0, t_1 in F and for all v_0, v_1 in F.

The derived lifted hyperbolic spreads:

PROBLEM 106. Try the derived lifted left hyperbolic spreads, using the previous material. This sort of computation is not theoretically necessary but rather is esoterically beautiful! You will notice the nice symmetry in the matrices, just as one can see the symmetry in the matrices obtained for the derived lifted conical spreads.

30.5. Derivation of right flock spreads

In this case, there are no field cases, only non-commutative division rings. The methods on the right are analogous but different and distinct. First, consider the right hyperbolic flock spreads.

Right hyperbolic flock spreads:

THEOREM 107. *Assume that*

$$x^T = 0^T,\ y^T = \begin{bmatrix} u & 0 \\ A^*(u), & u^\tau \end{bmatrix} x^T, y^T = \begin{bmatrix} v & 0 \\ A^*(v) & v^\tau \end{bmatrix} \begin{bmatrix} r(t) & 1 \\ s(t) & t \end{bmatrix} x^T,$$

is a right hyperbolic spread in $PG(3, D)_{right}$, where D is a finite-dimensional division ring, for all u, t in D and for all non-zero v in D. And, where r s are functions on D, and τ is an automorphism of D, and we have a τ-$A^*(u)$ derivative function.

Here is the form for the imposter switch for any right hyperbolic flock spread.

$$x = 0, y = \left\{ (x_1^\tau, x_2^\tau) \begin{bmatrix} u & 0 \\ 0 & u \end{bmatrix} + (A^*(x_1), A^*(x_2)) \right\},$$

for all u in D and

$$y^T = \begin{bmatrix} vr(t), & v^{-1} \\ v^\tau s(t)(1 - r(t)), & (A^*(v) + v^\tau s(t))v^{-1} \end{bmatrix} x^T,$$

for all t, for all non-zero v in D.

The spread is in $PG(V-1, Z(D) \cap (full)(Fix\tau))$.

PROOF. It is possible to use the codes from left to right to prove this result. It would be recommended for the reader to try the proof without the use of the codes, as there are a few twists and turns. $\square$

We now turn to the

Right conical flock spreads:

THEOREM 108. *Let D be a finite-dimensional division ring, and let the following conical flock spread be represented as follows:*

$$x^T = 0^T, y^T = \begin{bmatrix} u + R(t) & t \\ S(t) + A^*(u), & u^\tau \end{bmatrix} x^T,$$

for u, t in D,R,S functions on D. The flock

$$\Sigma_t^{right}:$$

$$-(R(t)^\tau + A^*(t))x_1 - t^\tau x_2 + (S(t) - A^*(R(t))x_3 - x_4 = 0.$$

The imposter switch is as follows:

$$x = 0, y = \left\{ (x_1^\tau, x_2^\tau) \begin{bmatrix} u & 0 \\ 0 & u \end{bmatrix} + (A^*(x_1), A^*(x_2)) \right\},$$

for all u in D, and

$$y^T = \begin{bmatrix} -t^{-1}(R(t)+v), & t^{-1} \\ -v^\tau t^{-1}(R(t)+v) + S(t) + A^*(v), & v^\tau t^{-1} \end{bmatrix} x^T,$$

for all v and for all non-zero t in D.

The spread is in $PG(V-1, Z(D) \cap (full)(Fix\tau))$.

PROOF. The proof is left as a problem for the reader. □

30.5.1. The coordinates question. Given any hyperbolic or conical flock spread. The generalized theory of spreads and flocks does not depend on a geometric theory, as is the case with finite hyperbolic flocks and flocks of quadratic cones.

For example, consider the quadratic cone case and how the geometry works prior to the theory of coordinates. Take a finite projective 3-dimensional projective space and choose any point v_o, called the vertex of the cone. Choose any plane that does not intersect v_o and choose any non-singular conic in the plane. This conic, when represented in the associated 4-dimensional affine geometry, is a regulus net.

Now try the same construction with a division ring D. Take a 3-dimensional projective space and choose any point v_o, called the vertex. Choose any plane that contains – what must it contain? Some set of points that corresponds to the projective version of a derivable net. Maybe, but does it have any particular structure? The problem is, we don't know. The fact is that we are restricted to a theory using coordinates.

So, we take a coordinate approach and construct from a matrix spread set of hyperbolic or quadratic type certain flocks of a quadratic cone. Now, take any of the corresponding derivable nets that are defined by the central collineation groups. Recoordinatize this derivable net so that it becomes isomorphic to the original net in such a way that the central collineation group is preserved. This recoordinatized spread is isomorphic to the original but has different functions defined in the matrix set. We obtain another flock. Why is it isomorphic to the flock of the original?

This is the coordinates question:

Given any hyperbolic spread. Recoordinatize by choosing one of the derivable nets involved in the construction of the hyperbolic spread and a method that rearranges the derivable nets so that we obtain two different hyperbolic flocks. The two spreads are isomorphic; are the hyperbolic flocks isomorphic?

Given any conical spread, repeat the above construction. Are the two conical flocks isomorphic?

Except that when one has a conical flock spread that is a semifield spread, then there are coordinate changes that are identical, which shows that the two flocks are isomorphic.

In the following part, we will show that the hyperbolic coordinates question may be answered in the negative.

Part 14

Baer groups over skewfields

This part is more fundamental than much of the previous material and deals with Baer subspaces in a fundamental manner. In the first volume, the foundations of derivable nets over skewfields are established. In the last part of this text, the idea of trying to discuss flocks of hyperbolic quadrics and flocks of quadratic cones but over arbitary skewfields is brought out. There were a variety of problems to overcome. For example, a derivable net has Baer subplanes. A Baer subplane of a projective plane is a subplane with two properties:

(1) Point-Baer property: Every line of the projective plane is incident with at least one point of the subplane.

(2) Line-Baer property: Every point of the projective plane is incident with at least one line of the subplane.

In the finite case, a counting argument will show that (1) is equivalent to (2). However, for infinite projective planes, the situation is much different, as shown by Barlotti [**2**].

For finite translation planes of dimension 2,

$$(spread\ is\ in\ PG(3, GF(q)),$$

the affine restriction of the dual of the projective extension always admits a derivable net (which is actually a regulus net). The derivation of these affine planes produces what is called a "semi-translation plane", which was discovered/invented by T.G. Ostrom sixty-five years ago. There are infinite translation planes of dimension 2; are the analogous affine planes derivable? The problem of point-Baer and line-Baer arises. The author [**29**] has shown that there are dual translation planes that contain a derivable net that is not derivable in the plane, in the sense that the Baer subplanes of the net are point-Baer subplanes of the plane but not line-Baer subplanes.

So, we need to be concerned with the nature of derivable nets over infinite fields or division rings. For the flock spreads that we will be considering, the spread is a union of derivable nets that share with one component, conical flocks, or two components, hyperbolic flocks. Since the Baer subspaces are defined using the Baer subspaces of a "primary derivable net", we need to show that in a flock spread, the Baer subspaces satisfy the point-Baer and line Baer properties. Each Baer subspace is a two-dimensional right $\{\begin{bmatrix} u^{\sigma} & A(u) \\ 0 & u \end{bmatrix}; u \text{ in } D\}=D_{\sigma-A}$-subspace.

Then we discuss "deficiency one" flocks; this issue of Baer subspaces of the net and of the plane comes up again. Basically, a deficiency in one partial flock has one pseudo-regulus net missing. However, the way that flock spreads is defined as a group acting on the partial spread that fixes a 2-dimensional right and left D-subspace pointwise and acts as if it were a central collineation group; an elation group when the pseudo-regulus nets share one component and a homology group when the pseudo-regulus nets share two components; both of these assumptions ensure that the fixed point space has exactly two possibilities, that of a line or a Baer subspace. But, to realize that the fixed point space is actually a "Baer" subspace as defined above requires some material on point-Baer subspaces acting on a translation plane over a skewfield.

Here we provide an introduction to what occurs when subplanes of an affine translation plane are point-Baer and line-Baer. And what sorts of Baer collineations can act on the sorts of flocks? The work here is adapted from the more general work in Jha-Johnson [**20**],[**21**],[**22**],[**24**] Johnson-Ostrom [**41**].

CHAPTER 31

Point-Baer subplanes of planes

PROPOSITION 109. *(Proposition 43 of* [**20**]*, p. 265). Let π be a translation plane of unspecified kernel. Let π_0 and π_1 be subplanes sharing the zero vector of π such that $\pi = \pi_0 \oplus \pi_1$. Assume that both subplanes share their parallel classes.*

Then both suplanes are point-Baer subplanes.

PROOF. We need to show that given any point P, there is a line of either subplane sharing P. If P is one of the subplanes, then if δ is an infinite point of that subplane, this is also a line of the second subplane. Therefore, assume that P is not a point of either subplane. Then consider $\pi_0 + P$, the coset of the additive group with respect to the group of π_0. Then, since $P = P_0 + P_1$, for P_i in π_i,$i = 0, 1$, and these two points are distinct affine points. Let PP_1 be the associated line and $\pi_0 + P = \pi_0 + P_1$. Since we are in a translation plane and the translation group fixes each point on the line at infinity, we see that $\pi_0 + P_1$ is a subplane that shares all infinite points with π_0 and π_1. Therefore, PP_1 is a line that intersects a point of infinity of π_1 so that this line projectively shares two points of π_1. Therefore, π_1 is a point-Baer subplane, and by analogy, π_0 is a point-Baer subplane. This completes the proof. □

PROPOSITION 110. *(Proposition 42, p. 264* [**20**]*)*

Let π be a translation plane, and let π_0 be a point-Baer subspace. Let M be any component that is disjoint from π_0, except for the zero vector. Then π_0 is a Baer subplane of π if and only if $\pi = M \oplus \pi_0$.

PROOF. We note that the set of all components forms an exact cover of the plane. The remaining lines are translates $M + P$ of a component M, where P is a point not in M. We need to show that $M + P$ and π_0 share a point. Since $M \oplus \pi_0 = \pi$. Then $M + P$ must meet π_0, so that π_0 is a Baer subplane. □

COROLLARY 111. *Let π be a left translation plane with components in $PG(3, D)$, where D is a skewfield, whose spread consists of a set of pseudo-regulus derivable nets that are:*

DOI: 10.1201/9781003767978-31

(1) Share exactly two components, L and M, and admit a pseudo-regulus-inducing homology group, or

(2) Share exactly one component L and admit a pseudo-regulus-inducing an elation group.

(3) Assume that D is finite-dimensional over $K = Z(D) \cap Fix\sigma \cap \{u; A(u) = 0\}$.

Then all Baer subplanes of the derivable nets are Baer subplanes of the plane.

PROOF. By the embedding theory and contraction,

we have a 4-dimensional left and right D-module V_4, and the components are 2-dimensional left D-subspaces. The derivable net containing $x = 0$, $y = 0$, $y = x$ may be assumed to be the generic imposter pseudo-regulus net. In this case, the Baer subspaces of that net are right 2-dimensional $D^{\sigma - A}$-subspaces, where $D^{\sigma - A}$ is isomorphic to D. It follows from proposition 222 that the subplanes in the generic pseudo-regulus derivable net are point-Baer subspaces. All of the other net Baer subspaces are images of the Baer subspaces of the generic pseudo-regulus derivable by basis change. This makes all net Baer subspaces to be point-Baer subspaces. Take a component M disjoint from a point-Baer subspace. Consider the spread written over $\mathrm{K} = \mathrm{Z}(D) \cap Fix\sigma \cap \{u; A(u) = 0\}$. Since the Baer subspaces of the classical generic net are right 2-dimensional $D^{\sigma - A}$, which is isomorphic to D, we see that the spread components and Baer subspaces of the classical generic pseudo-regulus net are written as finite-dimensional $2s$-vector subspaces over the same field K. The associated 4-dimensional vector space V_4 over D is then $4s$-dimensional over K. It now follows from proposition 223, and that by basis change, we may realize each of the derivable nets as generic; we have that all Baer subspaces over the nets are Baer subspaces of the plane. This completes the theorem. □

CHAPTER 32

Baer collineations in translation planes

When we consider deficiency one flocks over skewfields D, we note that any derived flock spread admits Baer collineations. However, we shall see that for all of those spreads, the Co spreads are in $PG(3,D)$ exactly when D is a field. So, derived flock spreads over fields admit Baer collineations. Are there any translation planes over skewfields that are not fields that admit Baer collineations? In the following theorem, we list a few such spreads.

THEOREM 112. *(1) Any quaternion division ring $(a,b)_F$ written in the form:*

$$x=0, y=x\begin{bmatrix} u^{\sigma} & bt^{\sigma} \\ t & u \end{bmatrix}$$

for all t,u in $F(\theta)$, a Galois quadratic extension of the field F, and σ the induced automorphism, where a,b are in F.

(2) Let D be any division ring of finite dimension over a field F. Then there is a central Galois-quadratic extension of a division ring extension of D, D_1, so that the set of "lifted skewfield spreads" is given by

$$x=0, y=x\begin{bmatrix} u^{\sigma} & (\theta\delta+\gamma)t^{\sigma} \\ t & u \end{bmatrix},$$

for all non-zero δ, and for all γ in $Z(D_1(\theta)$, in $PG(3,D_1(\theta))$.

(3) Let D be any division ring of finite dimension n over a field F. Then there is an extension D_1 of D that admits a central Galois cyclic extension of degree n, $D_1(\theta)$. Then consider the following spread in $PG(3,D_1(\theta))$:

$$x=0, y=x\begin{bmatrix} u^{\sigma} & \theta t^{\sigma^{-1}} \\ t & u \end{bmatrix},$$

for all t,u in $D_1(\theta)$. In both of the above sets of spreads over division rings of finite dimension, there is a Baer group defined by the following elements:

$$(x,y)\rightarrow(x\begin{bmatrix} 1 & 0 \\ 0 & c \end{bmatrix}, y\begin{bmatrix} 1 & 0 \\ 0 & c \end{bmatrix}),$$

DOI: 10.1201/9781003767978-32

for all c in F, in part (1), and for all c in $Z(D_1(\theta))$, *in parts (2) or* (3), *such that* $c^{\sigma^{-1}+1}=1$.

PROOF. For the sample matrix $\begin{bmatrix} u^\sigma & \rho t^{\sigma^{-1}} \\ t & u \end{bmatrix}$, for ρ a central element, just work out

$$\begin{bmatrix} 1 & 0 \\ 0 & c^{-1} \end{bmatrix}\begin{bmatrix} u^\sigma & \rho t^{\sigma^{-1}} \\ t & u \end{bmatrix}\begin{bmatrix} 1 & 0 \\ 0 & c \end{bmatrix}=\begin{bmatrix} u^\sigma & \rho t^{\sigma^{-1}}c \\ c^{-1}t & u \end{bmatrix},$$

which indicates we have a collineation.

The fixed point set $\{(x_1,0,x_2,0)$; for all x_1,x_2 the respective skewfield. All of these spreads are simultaneous flock spreads. This set is one of the Baer subplanes $B(1,0)$ in the derivable net

$$x=0, y=x\begin{bmatrix} u^\sigma & 0 \\ 0 & u \end{bmatrix},$$

in each case. This completes the proof. □

Even though there are infinitely many spreads over skewfields that admit Baer collineations, if the Baer group is sufficiently large, the skewfield is forced to be a field.

In Jha and Johnson [**21**], Baer groups acting on the spread in $PG(3,D)$, for D a division ring, and showed that when the so-called "full Baer groups", such as those in derived flock spreads, D is forced to be a field. This result also used the technique of direct products of subplanes considered by Johnson and Ostrom [**41**]. Due to space requirements, we are not able to cover these works, but they are of relatively close interest to this text and are recommended reading.

CHAPTER 33

Derived spreads and Baer groups

In part 12, we have shown that given any division ring D of finite dimension n over the center F, we may extend the multiple replacement theorem, which extends André and Ostrom. This theorem constructs an infinite variety of new translation planes obtained from either multiple derivation or multiple replacement of sets of derivable nets. In this case, the automorphisms are always of order 2, and are outer automorphisms. In the general setting over left Desarguesian spreads, there is a set of mutually disjoint pseudo-regulus nets, with components in $PG(3, D)_{left}$. The derived nets have components in $PG(3, D)_{right}$. So, when the division ring is non-commutative, and of vector dimension n^2 over the center $Z(D)$, then for all replacements subsets of pseudo-regulus nets, the constructed spread has a "blended kernel", and a standard kernel in PG. The finite André's planes play a vital role in the classification of all finite hyperbolic flocks, for which all of the flocks are "nearfield" flocks. In the infinite field case, there are interesting studies of when a flock could be of the André's variety and when that spread is a nearfield spread. However, in the non-commutative skewfield case, there are no classification results and essentially nothing is known.

There are important applications to the field of coding theory using division rings of finite dimension. Are there similar applications using the spreads constructed from division ring planes using multiple derivation or multiple replacement? For regulus net replacements, the dimension of the translation plane does not change, but for pseudo-regulus replacement, there is a left/right kernel that is called a "blended kernel". It would seem that such exceedingly complex structures would be a natural and important area for additional research and applications.

In this part, we consider the continuation of the theory of Baer groups and Baer theory hyperbolic and conical flocks/spreads. Consider a translation plane that admits a pseudo-regulus-inducing homology group or elation group. All of the translation planes of this work have that assumption, and all of these produce hyperbolic or conical flocks. In the non-commutative skewfield case, all of the examples of such spreads that we have given are "simultaneous spreads", which,

DOI: 10.1201/9781003767978-33

of course, means that the spreads produce both hyperbolic and conical spreads. Furthermore, all of the simultaneous spreads are semifield spreads. In particular, given any division ring D of finite dimension, there is a central Galois quadratic division ring extension. This means there is always a "lifted skewfield" extension, which is actually is a semifield spread. The semifield spread is never a division ring spread but is always a simultaneous spread. In this case, and in all cases of spreads corresponding to either hyperbolic or conical flocks, any derivation of a pseudo-regulus or regulus net will construct a translation plane that admits a corresponding group that fixes a Baer subplane pointwise.

The variation of the hyperbolic and conical flocks based on a finite regulus net has been generalized to the variety of a pseudo-regulus net containing three components $x = 0, y = 0, y = x$. The partial spread and subsequent spread have left space components in $PG(3, D)$, where D is a division ring of finite dimension over the center. Thus, in this setting, we still have a "ruled quadric" in the hyperbolic case. However, the components of the pseudo-regulus net are left 2-dimensional D-subspaces, and the Baer subspaces are right 2-dimensional D-subspaces. Thus, the ruled surface has a "left ruling" corresponding to the components of the left pseudo-regulus net and a "right ruling" corresponding to the set of Baer subspaces. In this variety, we consider points of the flock as the Baer subspaces of the corresponding spread, so lying on the right ruling. And, of course, the planes of the flock are the planes considered in $PG(3, D)$, corresponding to the subspaces of the associated left 4-dimensional vector space.

In a series of articles by Jha and Johnson ([**20**],[**22**],[**23**],[**24**]), there is a general study of point-Baer, line-Baer and Baer subspaces in affine planes. This set of articles forms a sort of nucleus for the three texts on the Geometry of Derivation. This was the first study of the Thas-Walker theorem over arbitrary skewfields. This means that in the general setting of $\sigma - A$ flocks, $\sigma = 1$ and A is identically zero. The idea of "deficiency one" is also studied, but only when D is a field. One reason for this restriction is that "full Baer groups" exist in spreads in $PG(3, D)$; where D is a division ring, it is proved by Jha and Johnson that D must be a field [**21**]. In that study, the ideal of a "deficiency one conical or hyperbolic flock" is studied. This idea is also discussed in the first book . We will extend those studies here for non-commutative skewfields D, and for fields where σ is not 1 or A is not zero. In this case, it is also the case that when there are Baer groups called "full Baer groups" acting on spreads, the spreads cannot be in $PG(3, D)$.

For our purposes, a "full Baer group" corresponds under derivation to a pseudo-regulus-inducing central collineation group.

When we defined what changes should be made in the skewfield case so that a hyperbolic or conical flock can construct a corresponding spread, the type of derivable net and the associated functions would be required. In this setting, the points are the Baer subspaces of two different types of translation planes: hyperbolic translation planes and conical translation planes. In the first case, there is a set of pseudo-regulus nets that share two components, and in the second, there is a set of pseudo-regulus nets that share one component. Suppose we have a subset of a flock so that all but one of the required pseudo-regulus nets exist in an associated net, just as in the flock/spread case. The associated part of a flock would then be said to be of "deficiency one". When this occurs, there is a unique spread extending this partial flock. There are a few partial flocks of deficiency one in the hyperbolic case and all in the finite case, but there are no deficiency one conical flocks in finite or infinite cases (see [**34**] for a discussion of this topic).

DEFINITION 113. Let π be a translation plane with left spread components in $PG(3, D)_{left}$ where D is a division ring of finite dimension n over the center F, that corresponds to a hyperbolic or conical flock. The translation plane π is called a hyperbolic or conical flock spread, accordingly as the corresponding flock is hyperbolic or conical. Choose any derivable pseudo-regulus net of the spread. The constructed translation plane admits a Baer group that is said to be either a hyperbolic Baer group or a conical Baer group, again corresponding to whether the spread is a hyperbolic or conical flock spread. The derived spread does not have the spread in $PG(3, D)_{left}$, as the derived pseudo-regulus net has components that are right 2-dimensional D-subspaces, whereas the remaining components are left 2-dimensional D-subspaces. The spread is said to have a "blended kernel". The actual kernel mappings would then be associated with $Z(D) \cap Fix\sigma \cap \{u; A(u) = 0\}$, $Z(D)$ is the center of D, when the flock spread is of the type $\sigma - A$ flock. Then the spread would have components in

$$PG(4n-1, F = Z(D) \cap Fix\sigma \cap \{u; A(u) = 0\}).$$

For infinite-dimensional skewfields D, the theory of derivable nets is also valid. If there are flock spreads in $PG(3, D)$, let $Z(D)$. Then the derived spreads have kernel $Z(D) \cap Fix\sigma \cap \{u; A(u) = 0\}$. If $D \oplus D \oplus D \oplus D$ is the associated vector spread V and there are left flock spreads, then the derived spreads would have components in PG$(V-1, Z(D) \cap Fix\sigma \cap \{u; A(u) = 0\})$.

Therefore, the only time there is a derived net in a $PG(3, D)$,is when we are in the original hyperbolic and conical setting of finite or infinite fields of Thas-Walker. Therefore, to consider a generalization of "deficiency one" in the more general situations, we need to deal with very general situations. We begin with the finite case.

CHAPTER 34

Deficiency one flocks of order p^4

When we consider hyperbolic spreads, we mean that there is a derivable net containing the components

$$x = 0, y = 0, y = x,$$

then the expression of the derivable is always defined by a skewfield; the net is coordinatized by a skewfield. And, when we consider hyperbolic flocks, there is a pseudo-regulus-inducing homology group that is defined using the multiplicative group of the skewfield. We use the right action of the group on left spreads and the left action of the group on right spreads. And, in this way, for example, in the finite case of order q^2 translation planes with spread in $PG(3, q)$, the orbits define a set of $q + 1$ regulus nets, sharing two components $x = 0, y = 0$.

In this chapter, we present two finite spreads that are derivable and admit a cyclic homology group of order $q - 1$ (corresponding to the multiplicative group of a finite field $GF(q)$), and there is a set of $q + 1$ derivable nets sharing two components $x = 0, y = 0$. What is the difference? If you read the two previous sentences again, there is one thing missing. The question is, can you have a homology group of order $q - 1$ with axis $y = 0$ and coaxis $x = 0$, and have $q + 1$ derivable nets sharing two components, where the homology group is regulus-inducing on some of the derivable nets – say all but one – but not regulus-inducing? If you could, and if the subset of q derivable was regulus-induced by a classical regulus form, but the last net was not? What would occur?

If you derived the last net, the homology group would turn into a Baer group, and the derived plane would be a deficiency-one hyperbolic flock! Not possible. Unless the translation plane is not of dimension 2, that is, the spread is not in $PG(3, GF(q))$.

The idea is that if you have a derivable net of the form

$$x = 0, y = x \begin{bmatrix} u^\sigma & 0 \\ 0 & u \end{bmatrix},$$

this is a type 2 regulus net if σ is non-trivial. And, the partial spread is in $PG(3, GF(q))$. But it is not the classical regulus net which would require σ to be 1. So, if you derived a net like this, the form would look

DOI: 10.1201/9781003767978-34

like:

$$x = 0, y = x^{\sigma^{-1}} \begin{bmatrix} u & 0 \\ 0 & u \end{bmatrix}$$

but as we have seen, this spread is not in $PG(3, GF(q))$ but in $PG(4s-1, GF(h))$ for $GF(h)$ a subfield of index s in $GF(q)$.

Consider a spread of the form:

$$x = 0, y = x \begin{bmatrix} u^\sigma & bt^{-1} \\ t & u \end{bmatrix},$$

for all u, t in $GF(q^2)$, and σ an automorphism of $GF(q^2)$, and b a constant. This is a spread in $PG(3, GF(q^2))$, and there is a derivable net obtained by taking $t = 0$. The Baer subspaces are

$$B(a,b) = \{((a,b) \begin{bmatrix} u^\sigma & 0 \\ 0 & u \end{bmatrix}, (a,b) \begin{bmatrix} v^\sigma & 0 \\ 0 & v \end{bmatrix})\}$$

for all u, v in $GF(q^2)$. We claim that there is a Baer homology group of the following form:

$$\{ \begin{bmatrix} \begin{bmatrix} 1 & 0 \\ 0 & w \end{bmatrix} & 0_2 \\ 0_2 & \begin{bmatrix} 1 & 0 \\ 0 & w \end{bmatrix} \end{bmatrix} \};$$

for all non-zero w in $GF(q^2)$, which fixes $B(1,0)$ pointwise (the Baer axis), leaves $B(0,1)$ invariant (the coaxis), and is transitive on the remaining Baer subspaces.

To see that the indicated group is a collineation group, the action of the group, fixes $x = 0$ and $y = x \begin{bmatrix} u^\sigma & 0 \\ 0 & u \end{bmatrix}$ componentwise and maps

$$y = x \begin{bmatrix} u^\sigma & bt^{-1} \\ t & u \end{bmatrix}$$

onto

$$y = x \begin{bmatrix} 1 & 0 \\ 0 & , w^{-1} \end{bmatrix} \begin{bmatrix} u^\sigma & bt^{-1} \\ t & u \end{bmatrix} \begin{bmatrix} 1, & 0 \\ 0, & w \end{bmatrix} = \begin{bmatrix} u^\sigma & bt^{-1}w \\ w^{-1}t & u \end{bmatrix}$$

and since $(w^{-1}t)^{-1} = t^{-1}w$, we see that we do have a collineation group of the spread.

Now consider the derived net obtained by switching. First, we see that the Baer group in the switch is of the following form:

$$\left\{ \begin{bmatrix} \begin{bmatrix} 1 & 0 \\ 0 & 1 \end{bmatrix} & 0_2 \\ 0_2 & \begin{bmatrix} w & 0 \\ 0 & w \end{bmatrix} \end{bmatrix} \right\}$$

for all non-zero elements of $GF(q^2)$; which is a homology group whose orbits define derivable nets. We notice that in the derived spread, and if the derived spread is a hyperbolic spread in the Thas-Walker sense, the derived plane would be a hyperbolic spread that creates a hyperbolic flock. However, the derived spread is first of all not in $PG(3, GF(q))$, and secondly, the homology group has q orbit classical regulus nets, but the derived net is not a classical regulus net. It needs one more regulus net!

THEOREM 114. *Let π be a translation plane in $PG(3, q^2)$ of the following form:*

$$x = 0, y = x \begin{bmatrix} u^\sigma & bt^{-1} \\ t & u \end{bmatrix}$$

for all t, u in $GF(q^2)$, where σ is a non-trivial automorphism. Then the plane admits a Baer homology group and corresponds to a non-extendable hyperbolic flock.

There are exactly two examples of this sort of spread due to Johnson [**32**]and Johnson-Pomareda [**40**] both of order p^4, where $p = 2$ and 3. There should be additional examples. Unfortunately, there are none of order p^4 exist as proved by Johnson and Cordero [**36**], for p a prime.

PROBLEM 115. Find other examples of spreads of this form or prove that none can exist. See the argument of Johnson-Cordero. There are four other partial hyperbolic flocks two of order 25 and two of order 49, obtained by computer program by Gordon F. Royle, [**46**] and see the discussion of these see spreads in [**32**], p. 814.

EXAMPLE 116. Complete the problem above, derive the planes using the imposter switch. Now the imposter switch will need to be employed over the entire spread.

34.0.1. The Associated σ-conical Flocks. Since all spreads of the form

$$x = 0, y = x \begin{bmatrix} u^\sigma & bt^{-1} \\ t & u \end{bmatrix},$$

for all t, u in $GF(q^2)$, are also conical flock spreads, there is an associated conical flock of the following form:

$$\Sigma_t : -x_1 0 - x_2 t^{\sigma^{-1}} + x_3 b t^{-1} - x_4 = 0,$$

for all t in $GF(q^2)$. This spread is nonlinear, where

$$\cap_t \Sigma_t = \{(1, 0, 0, 0)\}.$$

CHAPTER 35

t_o-Interchange-hyperbolic spreads

In the listing theorem for finite spreads of dimension 2, the result of Hiramine, Matsumoto and Oyama creates derivable extensions of any spread in $PG(3,q)$, by taking a Galois quadratic extension and using the representation of the finite spread to create new classes of derivable spreads in $PG(3,q^2)$. The new spreads depend upon the representation of the original spread and are said to be "lifted spreads". If we change the representation, another lifted spread is created, which may not be isomorphic to the first lifted spread. This "many from one" theory has been mentioned in the text in several places; two interrelated geometries may have the property that isomorphism of one geometry may not respect an isomorphism in the second related geometry. In this chapter, a "recoordinatization" of a hyperbolic or conical spread that respects the type i of the classical pseudo-regulus in either spread is most likely of this many-to-one category. Here is the idea. First, take a non-Desarguesian $\sigma - A$ hyperbolic spread of the form.

Let D be any division ring, and let π denote a left spread in $PG(3,D)$, that is, the union of a set of pseudo-regulus nets that share two components. The left spread has the following form:

$$x = 0, y = 0, y = x \begin{bmatrix} u^\sigma & A(u) \\ 0 & u \end{bmatrix},$$

$$y = x \begin{bmatrix} f(t) & g(t) \\ 1 & t \end{bmatrix} \begin{bmatrix} v^\sigma & A(v) \\ 0 & v \end{bmatrix},$$

for all non-zero u and v and for all t in D. Note that the pseudo-regulus nets sharing the two components $x = 0$ and $y = 0$ are

$$R_t : x = 0, y = 0, y = x \begin{bmatrix} f(t), & g(t) \\ 1 & t \end{bmatrix} \begin{bmatrix} v^\sigma & A(v) \\ 0 & v \end{bmatrix},$$

for all non-zero elements v in D. We shall denote the generic pseudo-regulus net

$$x = 0, y = 0, y = x \begin{bmatrix} u^\sigma & A(u) \\ 0 & u \end{bmatrix}$$

for all non-zero u in D by *Gnet*.

DOI: 10.1201/9781003767978-35

(1) Then using elementary properties of a spread, $f(t)$ is bijective. Furthermore, $g(t) - A(t) = r(t)$ is non-zero for all t and $g(t) - A(f(t^{\sigma^{-t}}) = s(t)$ is also non-zero for all t. Note that σ is an automorphism that could be inner, outer or trivial. And A is trivial, a derivative function or a σ-derivative function, which is inner or outer.

(2) There is a corresponding left hyperbolic flock with the following planes:

$$\Sigma_t^L : a_1(g(t) - A(f(t)^{\sigma^{-1}}) - a_2 f(t)^{\sigma^{-1}} + a_3 t - a_4 = 0,$$

and

$$\varrho^L : x_2 = x_3.$$

Now we notice that we choose any non-zero element t_o in D, the coordinate change mapping

$$(x, y) \to (x \begin{bmatrix} f(t_o), & g(t_o) \\ 1 & t_o \end{bmatrix}, y)$$

will map R_{t_o} to the Gnet and preserve the pseudo-regulus-inducing homology group. That is, this is another left hyperbolic spread. Here is what it looks like:

$$x = 0, y = 0, y = x \begin{bmatrix} u^\sigma & A(u) \\ 0 & u \end{bmatrix},$$

$$y = x \begin{bmatrix} f(t_o) & g(t_o) \\ 1 & t_o \end{bmatrix}^{-1} \begin{bmatrix} f(t) & g(t) \\ 1 & t \end{bmatrix} \begin{bmatrix} v^\sigma & A(v) \\ 0 & v \end{bmatrix}, \text{for all } t \text{ not } t_o\,,$$

and for all v, u non-zero in D, and

$$y = x \begin{bmatrix} f(t_o) & g(t_o) \\ 1 & t_o \end{bmatrix}^{-1} \begin{bmatrix} w^\sigma & A(w) \\ 0 & w \end{bmatrix},$$

for all non-zero w in D. Since we have the same generic pseudo-regulus net, and we define the generalized hyperbolic quadric using this net, and, by the structure of the three main nets, we see we also have a pseudo-regulus-inducing homology group. In order to see what the associated flock looks like, we need to write the nets so that they all of the nets except the generic pseudo-regulus nets; we need to express the other two nets with a "1" in the $(2, 1)$-entry of all of the matrices and generate the new functions required for the hyperbolic flock.

Here is the procedure.

$$\begin{bmatrix} f(t_o) & g(t_o) \\ 1 & t_o \end{bmatrix}^{-1} = \begin{bmatrix} t_o\Delta_{t_o}^{-1}, & 1 - t_o\Delta_{t_o}^{-1} f(t_0) \\ -\Delta_{t_o}^{-1}, & \Delta_{t_o}^{-1} f(t_o) \end{bmatrix}.$$

To create a "1" in the $(2,1)$-entry, there exists a non-zero w^{σ} such that $-\Delta_{t_o}^{-1}=w^{-\sigma}$. Multiplication on the right by $\begin{bmatrix} w^{\sigma} & A(w) \\ 0 & w \end{bmatrix}$, will transform the matrix into the correct form. That is, we now know what w is, which means that

$$\begin{bmatrix} w^{\sigma} & A(w) \\ 0 & w \end{bmatrix} = \begin{bmatrix} -\Delta_{t_o} & -A(\Delta_{t_{o_1}}^{\sigma^{-1}}) \\ 0 & -\Delta_{t_o}^{\sigma^{-1}} \end{bmatrix}$$

then transforms the set into the following:

$$y = x \begin{bmatrix} -t_o, & -\Delta_{t_o}^{\sigma^{-1}} + t_o\Delta_{t_o}^{-1}(f(t_o)\Delta_{t_o}^{\sigma^{-1}} - A(\Delta_{t_o}^{\sigma^{-1}})) \\ 1 & , -\Delta_{t_o}^{-1}(f(t_o)\Delta_{t_o}^{\sigma^{-1}} - A(\Delta_{t_o}^{\sigma^{-1}})) \end{bmatrix} \begin{bmatrix} r^{\sigma} & A(r) \\ 0 & r \end{bmatrix},$$

for all non-zero r in D. Although this might feel that w is not used, it is and re-expresses the original term. And, since the determinant Δ_{t_o} is not zero, a non-zero w^{σ} is possible. The main part to recognize is that when $t = t_0$, this now becomes the primary pseudo-regulus net; the following calculation is when t is not t_0.

$$\begin{bmatrix} f(t_o) & g(t_o) \\ 1 & t_o \end{bmatrix}^{-1} \begin{bmatrix} f(t) & g(t) \\ 1 & t \end{bmatrix} \begin{bmatrix} v^{\sigma} & A(v) \\ 0 & v \end{bmatrix} =$$

$$= \begin{bmatrix} t_o\Delta_{t_o}^{-1}(f(t) - f(t_o)) + 1, & t + t_o\Delta_{t_o}^{-1}(g(t) - f(t_o)t) \\ -\Delta_{t_o}^{-1}(f(t) - f(t_o)), & \Delta_{t_o}^{-1}(g(t) - f(t_o)t) \end{bmatrix} \begin{bmatrix} v^{\sigma} & A(v) \\ 0 & v \end{bmatrix}.$$

Let

$$\begin{bmatrix} w^{\sigma} & A(w) \\ 0 & w \end{bmatrix} =$$

$$= \begin{bmatrix} (-\Delta_{t_o}^{-1}(f(t) - f(t_o))^{-1}, & A((-\Delta_{t_o}^{-1}(f(t) - f(t_o))^{-\sigma^{-1}})) \\ 0 & , (-\Delta_{t_o}^{-1}(f(t) - f(t_o))^{-\sigma^{-1}}) \end{bmatrix}$$

which is valid since $-\Delta_{t_o}^{-1}(f(t) - f(t_o))$ is non-zero, as t is not equal to t_o and f is bijective. Multiplying on the right by this matrix produces

$$\begin{bmatrix} F(s), & G(s) \\ 1 & s \end{bmatrix}$$

such that

$$s = -\Delta_{t_o}^{-1}(f(t) - f(t_o))A((-\Delta_{t_o}^{-1}(f(t) - f(t_o))^{-\sigma^{-1}})) +$$
$$+\Delta_{t_o}^{-1}(g(t) - f(t_o)t)(-\Delta_{t_o}^{-1}(f(t) - f(t_o))^{-\sigma^{-1}}),$$

$$F(s) = -t_o - (\Delta_{t_o}^{-1}(f(t) - f(t_o))^{-1}),$$

and

$$G(s) = (t_o\Delta_{t_o}^{-1}(f(t) - f(t_o) + 1)A((-\Delta_{t_o}^{-1}(f(t) - f(t_o))^{-\sigma^{-1}})) +$$

$$+(t+t_o\Delta_{t_o}^{-1}(g(t)-f(t_o)t))(-\Delta_{t_o}^{-1}(f(t)-f(t_o))^{-\sigma^{-1}}.$$

We see that this becomes a hyperbolic spread as s is bijective, $F(s)$ is bijective, and $G(s)$ is never 0 for all s. Again, s is defined in this way when t is not t_0. Then t is t_0, the original part of the spread with the coordinate change is

$$x=0, y=0, y=x\begin{bmatrix} f(t_o) & g(t_o) \\ 1 & t_o \end{bmatrix}^{-1}\begin{bmatrix} u^\sigma & A(u) \\ 0 & u \end{bmatrix}.$$

Going through the same process to find a "1" in the (2,1)-entry, we have the following:

$$x=0, y=0,$$

$$y=x\begin{bmatrix} -t_0, & t_0\Delta_{t_0}^{-1}(f(t_0)\Delta_{t_0}^{\sigma^{-1}}-A(\Delta_{t_0}^{\sigma^{-1}}))-\Delta_{t_0}^{\sigma^{-1}} \\ 1, & -\Delta_{t_0}^{-1}(f(t_0)\Delta_{t_0}^{\sigma^{-1}}-A(\Delta_{t_0}^{\sigma^{-1}})) \end{bmatrix}\begin{bmatrix} u^\sigma & A(u) \\ 0 & u \end{bmatrix}$$

The corresponding hyperbolic flock is:

Then this part, we let the final part of the coordinate change, when

$$s^*=-\Delta_{t_0}^{-1}(f(t_0)\Delta_{t_0}^{\sigma^{-1}}-A(\Delta_{t_0}^{\sigma^{-1}})),$$

$$F^*(s^*)=-t_0, G^*(s^*)=t_0\Delta_{t_0}^{-1}(f(t_0)\Delta_{t_0}^{\sigma^{-1}}-A(\Delta_{t_0}^{\sigma^{-1}}))-\Delta_{t_0}^{\sigma^{-1}}.$$

$$\Sigma_s^{t_o,L}: a_1(G(s)-A(F(s)^{\sigma^{-1}})-a_2F(s)^{\sigma^{-1}}+a_3s-a_4=0, s \text{ not } s^* \text{and}$$

$$\Sigma_{s^*}^{t_0,L}: b_1(G^*(s^*)-A(F^*(s^*))-b_2F^*(s^*)+b_3s^*-b_4=0.$$

and

$$\varrho^{t_o,L}: a_2=a_3.$$

DEFINITION 117. The left hyperbolic flock $\Sigma_s^{t_o,L}, \Sigma_{s^*}^{t_0,L}, \varrho^{t_o,L}: x_2=x_3$. for each t_o not equal to zero shall be called the

"t_o-*recoordinatization*"

of the hyperbolic flock $\Sigma_t^L, \varrho^L: x_2=x_3$. In particular, if Σ_t^L corresponds to a semifield spread, the t_o-recoordinatization is still a semifield spread. In the finite case of a $\sigma-A$ spread, where $D=GF(q)$, there are q additional spreads. In any infinite case of $\sigma-A$ spreads, there are infinitely many additional spreads, which are highly different, as they depend on the complete variety of elements in D.

THEOREM 118. *(1) Let π denote any generalized hyperbolic spread in $PG(3, D)$, where π is non-Desarguesian, where D is a division ring. Assume then that any t_o-interchange spread defines a generalized hyperbolic flock.*

(2) If π is a simultaneous plane, then the t_0-interchange spread is a semifield plane that is a hyperbolic flock spread, but it loses the conical flock spread representation.

PROOF. If π is a semifield spread, then there is an elation group E with axis $x = 0$. If F_0 is $\begin{bmatrix} f(t_0) & ,g(t_0) \\ 1 & t_0 \end{bmatrix}$ then if G represents the entire matrix spread, then the original spread is

$$x = 0, y = 0, y = xG,$$

and the t_0-interchange may be written as

$$x = 0, y = 0, y = xF_0^{-1}G.$$

Since this last set is additive, we have an additive group, and there we have an isomorphic semifield spread. If the original spread is simultaneous, then since the pseudo-regulus-inducing homology group acts on the right side, we still have a hyperbolic spread, but we have lost the form required for this to be a conical spread. This completes the proof. □

As examples of the t_0-interchange, we consider any lifted skewfield flock spread. This class of spreads is valid for all division rings that are of finite dimension over a field F. We choose $t_0 = 1$ to illustrate the procedure. One proof of this result is just to take the general theorem and insert the current situation. However, a more instructive proof is to repeat the theorem in this special case given here.

THEOREM 119. *Let D be any finite-dimensional division ring over a field F. Then there is a central Galois quadratic extension D_2 of an extension D_1 of D and an involutory automorphism σ that fixes D_1 pointwise. If $F(\theta)$ is the corresponding Galois quadratic extension of F, so that $D_2 = D_1(\theta)$.*

The set of lifted skewfield simultaneous flock spreads $S_{\delta,\gamma}$ for δ non-zero and γ in $F$are represented as follows:

$$x = 0, y = x \begin{bmatrix} u^\sigma & (\theta\delta + \gamma)t^\sigma \\ t & u \end{bmatrix}$$

for all t, u in $D_1(\theta)$. Previously, we have shown that for characteristic not 2, we may choose a representative generator θ so that $\theta^\sigma = -\theta$. It *then is immediately clear that if* $\theta^* = (\theta\delta + \gamma)$, *then* $\theta^{*\sigma} = -\theta\delta + \gamma$, *so that* $\theta^{*\sigma}$ *is never equal to* θ^*.

(1) The hyperbolic spread may be written in the following form:

$$x = 0, y = 0, y = x \begin{bmatrix} u^\sigma & 0 \\ 0 & u \end{bmatrix}$$

$$y = x \begin{bmatrix} t^\sigma & \theta^* \\ 1 & t \end{bmatrix} \begin{bmatrix} v^\sigma & 0 \\ 0 & v \end{bmatrix}$$

for all non-zero u, v *and for all* t *in* $D_1(\theta)$. *The associated hyperbolic flock is:*

$$\Pi_t : x_1\theta^* - x_2(t^\sigma)^{\sigma^{-1}} + x_3 t - x_4 = 0$$

and

$$\varrho : x_2 = x_3.$$

Then

$$\cap_t \Pi_t \cap \varrho = (x_1, x_2, x_2, x_1\theta^*); x_1, x_2 \text{ in } D_2.$$

that is, this is a linear hyperbolic flock.

Take the $t_0 = 1$*-interchange, mapping*

$$(x, y) \to (x \begin{bmatrix} 1 & \theta^* \\ 1 & 1 \end{bmatrix}, y).$$

The hyperbolic flock spread now has the following form:

$$x = 0, y = 0, y = x \begin{bmatrix} 1 & \theta^* \\ 1 & 1 \end{bmatrix}^{-1} \begin{bmatrix} u^\sigma & 0 \\ 0 & u \end{bmatrix},$$

$$, y = x \begin{bmatrix} 1 & \theta^* \\ 1 & 1 \end{bmatrix}^{-1} \begin{bmatrix} t^\sigma & \theta^* \\ 1 & t \end{bmatrix} \begin{bmatrix} v^\sigma & 0 \\ 0 & v \end{bmatrix},$$

for all u, v *non-zero and for all* t *in* $D_1(\theta^*)$. *When* $t = 1$, *we obtain:*

$$y = x \begin{bmatrix} v^\sigma & 0 \\ 0 & v \end{bmatrix}.$$

When t *is not* 1, *we obtain, noting that the elements in the matrix*

$$\begin{bmatrix} (1-\theta^*)^{-1}, & -\theta^*(1-\theta^*)^{-1} \\ -(1-\theta^*)^{-1} & (1-\theta^*)^{-1} \end{bmatrix} = \begin{bmatrix} 1 & \theta^* \\ 1 & 1 \end{bmatrix}^{-1}$$

are all central,

$$y = x \begin{bmatrix} (t^\sigma - \theta^*)(1-\theta^*)^{-1}, & -(t-1)(1-\theta^*)^{-1}\theta^* \\ -(t^\sigma - 1)(1-\theta^*)^{-1}, & (t-\theta^*)(1-\theta^*)^{-1} \end{bmatrix} \begin{bmatrix} v^\sigma & 0 \\ 0 & v \end{bmatrix}.$$

This is equivalent to

$$y = xR \begin{bmatrix} v^\sigma & 0 \\ 0 & v \end{bmatrix}$$

where R is

$$\begin{bmatrix} -(t^{\sigma}-\theta^{*})(t^{\sigma}-1)^{-1}, & (1-\theta^{*})^{-1}\theta^{*}(1-\theta^{*\sigma}) \\ 1, & -(t-\theta^{*})(t-1)^{-1}(1-\theta^{*})^{-1}(1-\theta^{*\sigma}) \end{bmatrix},$$

(as a set). The set

$$y = x\begin{bmatrix} 1 & \theta^{*} \\ 1 & 1 \end{bmatrix}^{-1}\begin{bmatrix} u^{\sigma} & 0 \\ 0 & u \end{bmatrix}$$

is equivalent to

$$y = x\begin{bmatrix} -1, & (1-\theta^{*})^{-1}\theta^{*}(1-\theta^{*\sigma}) \\ 1, & -(1-\theta^{*})^{-1}(1-\theta^{*\sigma}) \end{bmatrix}\begin{bmatrix} u^{\sigma} & 0 \\ 0 & u \end{bmatrix},$$

(as a set).

The hyperbolic flock arising from the 1-interchange is as follows:

$$\begin{gathered} \Pi_s^{t_0=1}\&\ t \text{ not } 1: \text{for } s=-((t-\theta^{*})(t-1)^{-1}(1-\theta^{*})^{-1}(1-\theta^{*\sigma})): \\ a_1(\theta^{*}(1-\theta^{*})^{-1}(1-\theta^{\sigma}))-a_2(-(t-\theta^{*\sigma})(t-1)^{-1})+ \\ +a_3(-(t-\theta^{*})(t-1)^{-1}(1-\theta^{*})^{-1}(1-\theta^{*\sigma}))-a_4=0. \end{gathered}$$

$$\Pi_s^{t_0=1}\&\ t=1: \text{for} s=-(1-\theta^{*})^{-1}(1-\theta^{*\sigma}):$$

$$a_1(\theta^{*}(1-\theta^{*})^{-1}(1-\theta^{*\sigma}))-a_2(-1)+a_3(-(1-\theta^{*})^{-1}(1-\theta^{*\sigma}))-a_4=0.$$

$$\varrho: x_2 = x_3.$$

Then

$$\cap_s(\Pi_s^{t_0=1}\&\ t \text{ not } 1), (\Pi_s^{t_0=1}\&\ t=1), \varrho$$

is

$$(1,0,0,(\theta^{*}(1-\theta^{*})^{-1}(1-\theta^{*\sigma})).$$

As this is a nonlinear hyperbolic flock, the two flocks cannot be isomorphic. This shows that the hyperbolic coordinates question may be answered in the negative.

Corollary 120. *The previous theorem is more general than stated when considering hyperbolic flock spreads over fields. Since the Hiramine-Matsumoto-Oyama lifting theorem is valid for finite or infinite fields that admit a Galois quadratic extension, any lifted Desarguesian spread has the same form as in the above theorem. There are a number of spreads that may be lifted from a given Desarguesian or Pappian spread that admits a Galois quadric extension, as noted by the $\theta\delta+\gamma$ set of possibilities. Therefore, there are finite and infinite examples over fields where changing coordinates changes the flock obtained from linear to nonlinear.*

Here are a few problems arising from the previous results.

PROBLEM 121. (1) Does it matter if θ^* is changed? Is the lifted skewfield spread then also not isomorphic to the original choice? In the finite case, look at Cardinal et al. [**4**].

(2) Change t_0-interchange and apply the same result. For example, take t_0 to be any element in F. What can be said about the resulting flocks? Are these always nonlinear? If so, are these sets of nonlinear flocks isomorphic?

Now, some other problems.

Recall from part III:

THEOREM 122. *Let S be a cyclic division ring of degree 3 with center $(\mathbb{Q}\varsigma_3)$, $S = ((\mathbb{Q}\varsigma_3)(\tau)/(\mathbb{Q}\varsigma_3), \rho, g)$, and let $S(\theta)$ be a central Galois extension of dimension 3. Let $(\mathbb{Q}\varsigma_3)(\theta)$ be a cyclic Galois field extension of $(\mathbb{Q}\varsigma_3)$, such that $\mathbb{Q}\varsigma_3(\tau)$ is not isomorphic to $(\mathbb{Q}\varsigma_3)(\theta)$, with associated automorphism group $\langle\sigma\rangle$. Then θ may be taken as $\sqrt[3]{\delta}$ for $\delta \in \mathbb{Q}\varsigma_3$ and δ is non-square, and τ may be taken as $\sqrt[3]{\gamma}$, where γ is a non-square. Then the center of $S(\theta)$ is $(\mathbb{Q}\varsigma_3)(\theta)$. Note that the $S(\theta)$ is $((\mathbb{Q}\varsigma_3)(\tau,\theta)/(\mathbb{Q}\varsigma_3)(\theta), \rho, g)$.*

PROBLEM 123. Take any simultaneous spread as follows:

$$x = 0, y = x \begin{bmatrix} u^{\sigma} & \theta t^{\sigma^{-1}} \\ t & u \end{bmatrix}$$

for all t, u in $S(\theta)$, so σ^{-1} is not σ, where the extension is central of dimension $n>2$. Since this spread is simultaneous, there is a corresponding hyperbolic spread.

(1) Find the representation, then determine the associated hyperbolic flock. Show that it is linear.

(2) Take $t_0 = 1$ and show that the proof of the previous theorem basically shows that the t_0-interchange indicates that the constructed hyperbolic spread is nonlinear.

There are many other possibilities for problems.

PROBLEM 124. (Hard) (1) Take any constructed nonlinear hyperbolic flock obtained from left-inversing process found in the previous parts. Try any t_0-interchange. What does the new hyperbolic flock look like? Is it isomorphic to the original?

(2) Take any other simultaneous spread using a $\sigma - A$ hyperbolic spread, where A is an inner σ-derivation function. These also produce

linear flocks. Show that the resolution may not create a lifted skewfield spread.

$$x = 0, y = x \begin{bmatrix} u^{\sigma} + A_c(t), & A_c(u) + t^{\sigma}\theta \\ t & u \end{bmatrix},$$

where $A_c(u) = (u^{\sigma} - u)c$, where $c^{\sigma} = -c$, and c is in the center of the division ring D, and $\sigma^2 = 1$.

Take any t_0-interchange and show that there is a resulting non-linear hyperbolic flock.

- **So, the t_0-interchange of hyperbolic flock spreads does provide another construction technique for new non-isomorphic hyperbolic flocks.**

Problem set on the analogous right hyperbolic flock spreads:
There are a set of problems regarding the analogous right hyperbolic flock spreads.

PROBLEM 125. (1) Determine the t_0-interchange for right hyperbolic flock spreads.

(2) Use the set of right lifted skewfield spreads: Let D be a non-commutative finite-dimensional division ring of dimension n^2 over the center F and form the construction $D_1(\theta)$, which is a central Galois -2-dimensional extension of $D_1 = D \oplus_F Fix\tau$.

$$x^T = 0^T, y^T = \begin{bmatrix} v & t \\ \theta\delta + \gamma, & v^{\tau} \end{bmatrix}$$

for all δ non-zero and γ in $F(\theta)$. Show that this is, in fact, a spread using the right unwrapping principle and the fact that the matrix has a non-zero (1, 2)-determinant.

(3) Prove the previous theorem on right hyperbolic spreads regarding the t_0-replacement implies right hyperbolic flock that is not isomorphic to the original right hyperbolic flock.

CHAPTER 36

t_o-Interchange-conical spreads

In the previous section, the t_0-interchange of hyperbolic flock spreads is considered. Furthermore, examples are given of semifield hyperbolic flock spreads. Many of the spreads are simultaneous spreads. However, for the analogous construction for conical spreads does not lead to new conical flocks when the conical flock spreads are semifields. However, the question is completely open as to what occurs when the conical flock spread is not a semifield. The idea of σ-cones, spreads and flocks was introduced in the article [**43**] by Kantor and Pentilla. We consider one of the σ-flock spreads determined in this article.

Consider the Kantor-Pentilla conical flock spread:

$$x = 0, y = x \begin{bmatrix} u^{\sigma^{-1}} + t^5, & t^{14} \\ t & u \end{bmatrix}; t, u \text{ in } GF(2^e = q),$$

where 3 does not divide e, and the automorphism defined by $x \to x^2$ is denoted by σ. The expression used here to represent the spread is changed here to fit the transform method.

The associated spread is

$$\Sigma_t : x_1 t^{14} + x_2 t^{10} + x_3 t^2 + x_4 = 0.$$

Take any t_o not equal to 0 in $GF(q)$, the mapping

$$(x, y) \to (x, -x \begin{bmatrix} t_o^5 & t_o^{14} \\ 0 & t_o \end{bmatrix} + y),$$

which creates the new conical isomorphic flock spread:

$$x = 0, y = x \begin{bmatrix} u^{\sigma^{-1}} + t^5 - t_o^5, & t^{14} - t_0^{14} \\ t - t_0 & u \end{bmatrix};$$

for all t and u in $GF(q)$. The associated conical flock (using the method here developed) (letting $s = t - t_0$ then using the t-notation) is

$$x = 0, y = x \begin{bmatrix} u^{\sigma^{-1}} + s^5 + st_o^4 + s^4 t_0, & \Sigma_{i=1}^{7} s^{2i} t_0^{14-2i} \\ s & u \end{bmatrix};$$

where

$$\Sigma_{i=1}^{7} s^{2i} t_0^{14-2i} =$$
$$= s^{14} + s^{12} t_0^2 + s^{10} t_0^4 + s^8 t_0^6 + s^6 t_0^8 + s^4 t_0^{10} + s^2 t_0^{12}.$$

DOI: 10.1201/9781003767978-36

Therefore, the associated spread is:

$$\Sigma_s^{t_0} : a_1(s^{10} + s^2 t_o^8 + s^8 t_0^2) + a_2 s^2 + a_3 \Sigma_{i=1}^7 s^{2i} t_0^{14-2i} + a_4 = 0.$$

The original mapping does not map the first flock to the second.

PROBLEM 126. Are the flocks isomorphic? This is an open problem.

For all of these conical flocks, the cones are the same:

$$x_2 x_3^\sigma = x_1^{\sigma+1}$$

with vertex $(0, 0, 0, 1)$.

PROBLEM 127. (1) It is noted that t^{14} is bijective in $GF(q)$, so the transpose spread is:

$$x = 0, y = x \begin{bmatrix} u^{\sigma^{-1}} + t^5, & t \\ t^{14} & u \end{bmatrix}; t, u \text{ in } GF(2^e = q),$$

and the associated flock with the t_0-interchange (note the interchange here has a different function form in the transpose) is:

$$\Sigma_s^{t_0,transpose} : b_1(s^{10} + s^2 t_o^8 + s^8 t_0^2) + b_2 \Sigma_{i=1}^7 s^{2i} t_0^{14-2i} + b_3 s^2 + b_4 = 0.$$

Are any of these flocks isomorphic?

(2) (Difficult) Apply any of the $K = \begin{bmatrix} 1 & c \\ 0 & 1 \end{bmatrix}$, mappings to the any of the conical flock spreads, then construct the conical flocks. Are any of these isomorphic?

THEOREM 128. *Let D be a division ring, and consider a general left conical spread:*

$$x = 0, y = x \begin{bmatrix} u^\sigma + F(t), & A(u) + G(t) \\ t & u \end{bmatrix}$$

for all t, u in D and functions F and G on D, A a σ-derivative function, and σ an automorphism. Choose any non-zero t_0 in D and consider the isomorphism

$$(x, y) \to (x, x \begin{bmatrix} I_2 & \begin{bmatrix} -F(t_0), & -G(t_0) \\ -t_0 & 0 \end{bmatrix} \\ 0_2 & I_2 \end{bmatrix} + y).$$

The new form is

$$x = 0, y = x \begin{bmatrix} u^\sigma + F(t) - F(t_0), & A(u) + G(t) - G(t_0) \\ t - t_0 & u \end{bmatrix}$$

for all t in D. The original flock is given by:

$$\Sigma_t^L : -x_1(F(t)^{\sigma^{-1}} + A(t^{\sigma^{-1}})) - x_2 t^{\sigma^{-1}} + x_3(G(t) - A(F(t)^{\sigma^{-1}}) - x_4 = 0.$$

The t_0 interchange is

$$\Sigma_s^{t_0,L} : -x_1(F(s+t_0) - F(t_0))^{\sigma^{-1}} + A(s^{\sigma^{-1}}) - x_2 s^{\sigma^{-1}} +$$
$$+x_3(G(s+t_0) - G(t_0) - A((F(s+t_0) - F(t_0))^{\sigma^{-1}}) - x_4 = 0.$$

PROOF. The proof is left for the reader to complete. □

- Note that if the original left conical flock spread is a semifield spread, all t_0-interchange flocks are isomorphic to the original, as they are identical.

PROBLEM 129. Determine the t_0-interchanges for right conical spreads.

CHAPTER 37

Left inversing "Minus One"

Consider now a left conical flock spread. Also, suppose that this admits a left-acting pseudo-regulus-inducing homology group. As we have mentioned, the mapping $(x, y) \to (y, x)$ will map the conical flock spread to a left hyperbolic flock spread and in many instances creates a new nonlinear left hyperbolic flock. In this chapter, we realize that there is another construction possible that does not inverse one of the pseudo-regulus nets; that is, we do inverse all of the pseudo-regulus nets "minus one".

Consider a left conical flock spread with spread:

$$x = 0, y = x \begin{bmatrix} u^{\sigma} + F(t), & G(t) + A(u) \\ t & u \end{bmatrix},$$

for all t,u in a division ring D. Now assume that there is a homology group, which is pseudo-regulus-inducing with coaxis $x = 0$ and axis $y = 0$ and of the form:

$$\left\{ \begin{bmatrix} \begin{bmatrix} u^{\sigma} & A(u) \\ 0 & u \end{bmatrix} & 0_2 \\ I_2 & I_2 \end{bmatrix} ; u \text{ non-zero in } D \right\}.$$

The spread will now look as follows:

$$x = 0, y = 0, y = x \begin{bmatrix} u^{\sigma} & A(u) \\ 0 & u \end{bmatrix},$$

$$y = x \begin{bmatrix} v^{\sigma} & A(v) \\ 0 & v \end{bmatrix} \begin{bmatrix} u^{\sigma} + F(t), & G(t) + A(u) \\ t & u \end{bmatrix},$$

for all t in D. This is valid if and only if we obtain the spread form:

$$= \begin{bmatrix} (vu)^{\sigma} + v^{\sigma}F(t) + A(v)t, & v^{\sigma}G(t) + v^{\sigma}A(u) + A(v)u \\ vt & vu \end{bmatrix},$$

which is true if and only if

$$v^{\sigma}F(1) + A(v) = F(v)$$

and

$$v^{\sigma}G(t) = G(vt),$$

DOI: 10.1201/9781003767978-37

for all v non-zero and for all u,t in D. Therefore, it follows, since $F(0) = 0$ and $G(0) = 0$, that

$$F(r) = r^{\sigma}F(1) + A(r),$$

for all r in D, and

$$G(s) = s^{\sigma}G(1)$$

for all s in D.

Therefore, we have the following theorem.

THEOREM 130. *A left $\sigma - A$ conical flock spread admits a left-acting pseudo-regulus homology group of type $\sigma - A$ if and only if the conical flock spread has the following form:*

$$x = 0, y = 0, y = x\begin{bmatrix} u^{\sigma} + F(t), & G(t) + A(u) \\ t & u \end{bmatrix},$$

such that $F(t) = t^{\sigma}F(1) + A(t)$, and $G(t) = t^{\sigma}G(1)$. We note that since A and t^{σ} is additive, we have a set of semifield conical flock spreads.

Now we continue with the ideas of this chapter: Since letting $F(1) = f$ and $G(1) = g$, we have the matrix spread set as:

$$\begin{bmatrix} u^{\sigma} + t^{\sigma}f + A(t), & t^{\sigma}g + A(u) \\ t & u \end{bmatrix}.$$

Notice now that we can rewrite the spread as

$$x = 0, y = 0, y = x\begin{bmatrix} u^{\sigma} & A(u) \\ 0 & u \end{bmatrix},$$

$$y = x\begin{bmatrix} v^{\sigma} & A(v) \\ 0 & v \end{bmatrix}\begin{bmatrix} t^{\sigma} + f, g & +A(t) \\ 1 & t \end{bmatrix},$$

for all u, v non-zero and for all t in D.

At this point, we may take the mapping $(x, y) \rightarrow (y, x)$ to create the related left hyperbolic flock spread called the "left inverting" flock spread. But, before we do, we choose any t_0 in D and take the mapping

$$(x, y) \rightarrow (x, y\begin{bmatrix} t_0^{\sigma} + f, g & +A(t_0) \\ 1 & t_0 \end{bmatrix}^{-1}),$$

noting also that this creates an isomorphic left spread. The isomorphic left spread has the following form:

$$x = 0, y = 0, y = x\begin{bmatrix} w^{\sigma} & A(w) \\ 0 & w \end{bmatrix},$$

$$, y = x\begin{bmatrix} u^{\sigma} & A(u) \\ 0 & u \end{bmatrix}\begin{bmatrix} t_0^{\sigma} + f, g & +A(t_0) \\ 1 & t_0 \end{bmatrix}^{-1},$$

$$y = x \begin{bmatrix} v^{\sigma} & A(v) \\ 0 & v \end{bmatrix} \begin{bmatrix} t^{\sigma} + f, g & +A(t) \\ 1 & t \end{bmatrix} \begin{bmatrix} t_0^{\sigma} + f, g & +A(t_0) \\ 1 & t_0 \end{bmatrix}^{-1},$$

for all non-zero w, u, v and for all t not equal to t_0. We note from the form that this "minus one" left spread admits a left-acting pseudo-regulus-inducing homology group also and therefore may also be used to construct a new left hyperbolic flock spread.

Here is the form of that spread:

$$y = 0, y = x \begin{bmatrix} r^{\sigma} & A(r) \\ 0 & r \end{bmatrix},$$

$$y = x \begin{bmatrix} t_0^{\sigma} + f, g & +A(t_0) \\ 1 & t_0 \end{bmatrix} \begin{bmatrix} s^{\sigma} & A(s) \\ 0 & s \end{bmatrix},$$

and

$$y = x \begin{bmatrix} t_0^{\sigma} + f, g & +A(t_0) \\ 1 & t_0 \end{bmatrix} \begin{bmatrix} t^{\sigma} + f, g & +A(t) \\ 1 & t \end{bmatrix}^{-1} \begin{bmatrix} m^{\sigma} & A(m) \\ 0 & m \end{bmatrix},$$

for all non-zero r, s, m and for all t not equal to t_0.

DEFINITION 131. The constructed spread using t_0 is called the "minus one" left spread from the left conical flock that admits a left inversing pseudo-regulus homology group. We note that this is isomorphic to but is not a left conical flock spread. However, it admits the above homology group. There, the minus one left spread using the inversing mapping creates a hyperbolic flock spread called the "t_0-minus one left inversing hyperbolic flock spread". We also have the left inversing hyperbolic flock spread of "the parent".

PROBLEM 132. Although the above looks like a new concept, what we have done is use the t_0-hyperbolic interchange of the left inversing hyperbolic flock spread. This is left as a problem.

When $F(t) = t^{\sigma}F(1)+A(t)$, and $G(t) = t^{\sigma}G(1)$, there are an infinite number of examples. Let $F(1) = 0, A(t) = (t^{\sigma} - t)e$, for any *non-zeroe* in $Z(D)$, and $G(1) = \theta\delta + \gamma$ for all δ non-zero and γ in $Z(D)$, and σ has order 2.

PROBLEM 133. From part XIII, look at theorems 176 and 183. Choose any example and calculate a minus one left inversive left hyperbolic flock. Is this new hyperbolic flock nonlinear?

CHAPTER 38

Deficiency one

In this chapter, the idea of partial hyperbolic flock spreads and hyperbolic flocks are discussed and partial conical flock spreads and conical flock spreads are studied in the general cases. While this could be analyzed in a more general setting, we consider those partial flocks missing exactly one derivable net.

DEFINITION 134. We shall provide the main results for deficiency one first for hyperbolic and then for conical cases. We state the results for left flocks and spreads. From a hyperbolic spread, there are a set of derivable pseudo-regulus nets sharing two components. Any of these may be derived, which constructs another spread, but not in $PG(3, D)_{left}$. One left derivable net is derived to a right derivable net consisting of Baer subspaces of the left derivable net as right 2-dimensional components, and the rest of the left spread components remain as left 2-dimensional components. However, this spread no longer has components represented as $x = 0$, $y = 0$ and $y = x$. We may change it so that the spread looks like the classical right pseudo-regulus net of a right hyperbolic/conical spread. We shall go over how this works but is complex in the following two manners: First, there is the representation of Baer subspaces. It is mentioned in the first volume that the vectors (a, b) that represent a 2-dimensional subspace/the Baer subspace need to be written using the basis of whatever type of pseudo-regulus that is under consideration. So, in order to see how Baer groups arise from central collineation groups, we need to recall how the basis needs to be considered. Then, there is the use of matrices to represent left and right spreads. The use of these is the standard use, but when forming derivations, we need to resort back to the Baer group setting of bases. We shall begin with the original constructions of hyperbolic flocks and of flocks of quadratic cones.

We may always derive the prime pseudo-regulus using a combination of t_0-interchange coupled with imposter switching. While the t_0-interchange produces an isomorphic flock spread, the coordinate change will alter the accompanying flock, which can be non-isomorphic to the

DOI: 10.1201/9781003767978-38

original flock. It might seem clear that the two derived spreads are isomorphic. However, if the system of derivable nets is not the original Thas-Walker sets over fields, then assume that two of these translation planes are isomorphic by an isomorphism g. Then, due to blended kernels, it follows that g must map one of the two derivable nets (the derived nets) to the other and then map the remaining subspaces of the original spreads to each other. This means that g is a collineation of the original spread that maps one of the derivable nets that share $x = 0$ and $y = 0$ to the other.

Therefore, for hyperbolic flock spreads, we note the following:

THEOREM 135. *(1) Let D be a division ring and let π be a hyperbolic flock spread. Assume that either D is not a field or we do not have a Thas-Walker system. Let G be the collineation group that leaves invariant the set of pseudo-regulus nets sharing two components $x = 0, y = 0$. Let $OrbG$ denote the set of orbits of these pseudo-regulus nets.*

Then there are $Card(OrbG)$ mutually non-isomorphic derived translation planes obtained by the replacement of a derivable net from each orbit of G.

PROBLEM 136. In terms of derived spreads, the question is what happens to the formulation of a placeholder to determine putative deficiency in one type of spreads?

Due to the fact that we can use t_0-replacement on either conical flock spreads or hyperbolic flock spreads, a given flock spread cannot be used to discuss deficiency in one type of spreads. There is a new theory that could be developed in the more general flock spreads; formulate that theory.

DEFINITION 137. A partial spread in $PG(3, D)$, where D is a skewfield, is said to be a $D^{\sigma-A}$-deficiency one partial spread if and only if the form corresponding to a flock is either hyperbolic or conical that does not contain $\varrho : x_2 = x_3$ for a hyperbolic flock or Σ_0 for the conical conical flock. That is, we are missing one plane, which would correspond in both cases to the generic pseudo-regulus net.

If we assume that a partial spread in $PG(3, D)$, admits a pseudo-regulus-inducing central collineation group that covers all "lines" of a generalized ruled surface but one in the hyperbolic case, or all lines but one of a generalized cone (since the generalized cone is embedded in the generalized ruled surface), then the above partial spread may be so defined.

So, we have the following situation: Since the groups exist, the fixed point sets and cosets are in the hyperbolic case $x = 0$ and $y = 0$, and

$x = 0$ in the conical case. Either $x = 0$ and $y = 0$ are components of $PG(3, D)$, or they are subplanes. Since either one of the ruled lines is missing or on every line of the cone through the vertex is missing exactly, adding the points to the existing partial spread produces a spread containing the subplanes indicated as Baer subplanes, since the preimage sets then are mutually non-isomorphic $D^{\sigma-A}$, right 2-dimensional subspaces. The components of the partial spread in $PG(3, D)$, the kernel of the spread are $Z(D) \cap Fix\sigma \cap \{u : A(u) = 0\}$.

THEOREM 138. *Let there be a partial spread in $PG(3, D)$, for D a skewfield of finite dimensions that admits a central pseudo-regulus-inducing central collineation group, and then consists of a set of pseudo-regulus derivable nets either sharing one component or two components.*

If the corresponding partial flock is missing exactly one plane, then there is a projective space containing the fixed point subspaces with coaxis in the hyperbolic case are Baer subspaces, whose kernel is $Z(D) \cap Fix\sigma \cap \{u : A(u) = 0\}$.

There are three possibilities:

(1) The adjoined partial spread to the left partial hyperbolic spread is a pseudo-regulus such that the derived spread is a flock spread of a hyperbolic or conical flock.

(2) If the adjoined partial spread does not have a derivable net derived in the manner of (1), we obtain a "deficiency one hyperbolic" flock or a "deficiency one conical flock".

(3) When $Z(D) \cap Fix\sigma \cap \{u : A(u) = 0\} = K$, $D = K$, and we have a Thas-Walker hyperbolic flock type or a flock of a quadratic cone.

PROOF. We may recoordinatize using t_0-*interchange*, so that the missing pseudo-regulus net in the partial flock of deficiency one has the representation previously given for flocks but is missing the generic pseudo-regulus net. There are two cases: the deficiency one hyperbolic partial spread and the deficiency one conical partial spread. In the representation given, we consider the components $x = 0, y = 0$ and note that these do not appear in the partial hyperbolic flock of deficiency one, as the points of the flock are Baer subspaces/subplanes. Since there is exactly one missing right point in each of the left planes in the projective space, we see that these points are 1-dimensional -$\{\begin{bmatrix} u^\sigma & A(u) \\ 0 & u \end{bmatrix}; u \text{ in } D\}$, subspaces, the preimage sets of are a set of mutually disjoint 2-dimensional-$\{\begin{bmatrix} u^\sigma & A(u) \\ 0 & u \end{bmatrix}; \text{u in } D\}$, subspaces, noting that we have generalized ruled quadric with the set of Baer subspaces of the generic pseudo-regulus net taking the place of the lines in a standard ruled quadric. Therefore, we have a spread π^* with two types of components: The set $\Gamma_{u^\sigma - A}$, the partial spread whose components are right 2-dimensional $\{\begin{bmatrix} u^\sigma & A(u) \\ 0 & u \end{bmatrix}; u \text{ in } D\}$-subspaces, and the partial spread Γ_D, of components that are left 2-dimensional D-subspaces. Our assumption of D being a finite-dimensional division ring over a field F is required to ensure that the direct sum of any two components is the entire associated vector space V. So, we see that X and Y also direct sum to the vector space V and are subspaces. In this more general setting, it is not immediate that X and Y are subplanes, much less Baer subplanes. So, let S_Y denote the set of slopes of the components that non-trivially intersect Y. In this case, the pseudo-regulus homology group H has elements with the form:

$$(x, y) \to (x, y\begin{bmatrix} u^\sigma & A(u) \\ 0 & u \end{bmatrix}),$$

for all non-zero u in D. We define the lines of a putative subplane defined by $Y = (y = 0)$ as follows: The lines of Y are the lines of π^* that are an "infinite point"; element of S_Y and an element of Y. Choose any two of these lines, and we need to show that they intersect at a point. Since all lines are then fixed by H, including the line at infinity, it follows that the natural intersections are also fixed by H, and the intersections are either on S_Y or are affine. But the only affine points are those of Y, which shows all lines of the putative plane corresponding to Y uniquely intersect. Since any pair of elements of Y are points of Y, and there is the line of π^* union the line at infinity joining this pair, we see that Y is a subplane. To show that X is a subplane, we see that the group H induces a 1-dimensional $\{\begin{bmatrix} u^\sigma & A(u) \\ 0 & u \end{bmatrix}; u \text{ in } D\} = L$-subspaces, which is defined by the points $y\{\begin{bmatrix} u^\sigma & A(u) \\ 0 & u \end{bmatrix}; u \text{ in } D\}$-subspaces. Since X is a right subspace of dimension 2 over L whenever it intersects a 2-dimensional L-subspace M of S_Y, it must intersect in a 1-dimensional L-subspace. Assume that there is a right component N that does not intersect Y. Then $Y \oplus N$ by dimensionality. Then every line $N + y$, for y in Y unique intersects Y. This shows that Y is a line-Baer subplane. Let P be a point of the plane that is not in Y. Let g be an element of H, and consider the line PP^g, since Y is line-Baer PP^g intersects Y is a point Q. Considering $\{P, P^g, Q\}$ on PP^g, since P maps to P^g and Q is fixed by g, we see PP^g is fixed by g, which implies that PP^g is a line of Y, so P is incident with a line of Y, showing that Y is point-Baer, and therefore is Baer. Now consider the subspace $X = (x = 0)$. Consider the action of H on X, it is evident that there is a partition of X by 1-dimensional $D^{\sigma-A}$-subspaces. Since X itself is a 2-dimensional right $D^{\sigma-A}$-subspace, and the components that X intersects non-trivially are also right 2-dimensional $D^{\sigma-A}$-subspaces, we see that there X becomes a subplane. Since H leaves invariant any 1-dimension subspace of X, it follows that X is a subplane of π^* that shares the slopes (shares the infinite line) of Y. Since $X \oplus M = \pi^*$ for all components M disjoint from X (again, by dimensionality), it follows that X is a line-Baer subplane. It then is a Baer subplane. Also, here is the argument: We need to show that X is a point-Baer subplane; that every point P of π^* must lie in a line of X. Recalling the definition of point and line-Baer on the associated projective plane, clearly we may assume that P is an affine point. Recalling that $X \oplus Y = \pi^*, P = x_0 + y_0$,where x_0 is in X and y_0 is in Y. If P is in Y, then $P0$, where 0 is the zero vector, intersects the infinite line at δ, which is also a point of X, as is 0. This is also a line of X, so that P is incident with a line of X. If P is not in Y,then PP^g, for g in H, is a line and X is line Baer, so PP^g contains a point Q of X. And, a point T of Y. It follows that g leaves PP^g invariant but does not fix Q. So, g maps Q on PP^g to another of X,unless Q is an infinite point δ. Then PP^g is invariant, and there is a line of Y, which has an affine point T. The lines Tδ are lines of X, since the translation group will map T to 0, so the common lines of Y and X incident with 0, map to common lines on T. P is on a line of X. Therefore, X is a Baer subplane. This completes the proof of the deficiency of one hyperbolic flock part. Now assume that we have a pseudo-regulus-inducing elation group with axis X. Here there is exactly one missing right point of each left plane; it follows there is exactly one missing point from each of the cone lines incident with the vertex, and therefore, we have the preimage 2-dimension right L-spaces together with the left 2-dimensional left D-subspaces forming a spread. Since X is also a right 2-dimensional L-space, and there is a collineation group that acts as an elation group on the deficiency one partial spread, the same argument as previously shows that X is a Baer subplane of partial spread consisting of 2-dimensional right L-subspaces. This completes both parts of the theorem. We see that the kernel of spread is the common subfield $Z(D) \cap Fix\sigma \cap \{u : A(u) = 0\} = K$. And, this is D if and only if D is a field $\sigma = 1$ and A is identically zero. That is, this is the partial hyperbolic and conical flock situation over fields. This completes the proof. □

COROLLARY 139. *The t_0-interchanges provide isomorphic flock spreads but not always isomorphic flocks. Therefore, the previous theorem may be considered for all such interchanges and obtaining possible different deficiency one partial hyperbolic or conical flocks.*

38.1. Deficiency one, on the right

Consider this section as a problem set that duplicates the theory of deficiency one on the left – but for right flock spreads. Show the analogous results that can be obtained on the right.

38.1.1. Games with quasiflocks over fields. The structure of a quasifibration in $PG(3, D)$, for D a skewfield is identical to that of a spread, with the exception that the covering condition does not hold. In parts of the first text on this subject, there were several questions left open concerning a quasifibration that actually turned out to be a spread. And, we have determined a condition for additive quasifibration to actually be a spread. However, there are infinitely many quasifibrations that are not spreads and are of interest due to this fact. Consider the quasifibrations that are unions of pseudo-regulus nets obtained from pseudo-regulus-inducing homology groups or elation groups; the hyperbolic or conical spread type, when the quasifibrations are not spreads. When this occurs for quasifibrations that are not spreads, we call the associated hyperbolic or conical partial flocks, quasi-hyperbolic or quasi-conical flocks.

Given a quasifibration within $PG(3, K)$ for K a field that admits a Galois quadratic extension, there is a lifted quasifibration. If the initial quasifibration is proper (not a spread), is the lifted quasifibration also not a spread? In [**34**], chapter 9 is devoted to this question. Once a quasifibration is lifted, the lifted quasifibration is derivable by a regulus, and the form of a quasifibration that admits a regulus-inducing elation group. Therefore, there will be a quasi-conical flock.

EXERCISE 140. (1) Choose any lifted quasifibration (see [**34**], chapter 9), and determine the associated quasi-conical flock. Show that these quasi-flocks are linear; the planes contain a common line.

(2) There is a discussion of the Cherowitzo, Johnson, Vega quasifibration on page 189 of [**34**]. Note the following:

$$x = 0, y = x \begin{bmatrix} u, & -t^{3\alpha^{-1}} \\ t & u^{\alpha} \end{bmatrix}.$$

for t, u in an ordered field K that does not contain all cube roots, is a quasifibration if and only if

$$x = 0, y = x \begin{bmatrix} u, & -t^{\alpha/3} \\ t & u^{\alpha} \end{bmatrix},$$

is a quasifibration. The proof in [**34**], p. 189, shows that both of these quasifibrations are proper quasifibrations.

Determine the associated quasi-conical flock from both quasifibrations.

Show that both quasi-conical flocks are nonlinear; the planes of the flock do not contain a line. Show that the quasi-flocks are not simultaneous (do not support a corresponding quasi-hyperbolic flock).

There are T-copies from finite semifields that are also simultaneous quasi-hyperbolic and quasi-conical flocks.

PROBLEM 141. Consider the Hughes-Kleinfield finite semifield spread

$$(*)_1\colon x = 0, y = x \begin{bmatrix} u & gt^{\varrho} \\ t & u^{\sigma} \end{bmatrix},$$

for all $t, u \epsilon GF(q), q$ odd, g a non-square and σ, ϱ are automorphisms of $GF(q)$. Extend to $GF(q)(z)$, the rational function field of $GF(q)$. Show that this is a simultaneous quasifibration leading to quasi-hyperbolic flocks and quasi-conical flocks if and only if $\varrho = \sigma$. It is also true here that there is another corresponding quasifibration:

$$(*)_2\colon x = 0, y = x \begin{bmatrix} u^{\sigma} & gt^{\varrho} \\ t & u \end{bmatrix}.$$

EXERCISE 142. Now determine both the quasi-conical and the quasi-hyperbolic flocks. When are they isomorphic?

Note, we have mentioned the simultaneous spreads when $(**)$: $x = 0, y = x \begin{bmatrix} u^{\sigma^{-1}} & gt^{\sigma} \\ t & u \end{bmatrix}$, previously, which exists over a variety of skewfields. Now the difference between the two is hidden. In the first example $(*)_1$, in $GF(q)(z)$, the rational function field GF(q), ϱ is a power and an endomorphism but not an automorphism. Whereas, in the second example $(**)$, $\varrho = \sigma$ is an automorphism of $GF(q)(\mathrm{z})$. The first example is called a "T-extension" and the second example could be called a "Twisted T-extension". The twisted version is when the corresponding automorphism of $GF(q)$ is assumed to fix z. When this happens, all of these structures before spreads and the one obtains flocks instead of quasi-flocks. Thus, it is often very difficult to determine when a quasifibration actually is a spread.

PROBLEM 143. For an interesting project, take any chain of lifted quasifibrations over fields. Read the material in [**34**] on lifting quasifibrations. Take any of these quasifibrations at any state in the chain. Determine the associated quasi-conical flock. Are any of these quasifibrations simultaneous?

Now read the material in [**34**] on "lifting skewfields" , which shows how to lift finite division rings by forming the tensor product of central Galois quadratic extensions? Can you weaken this idea to "lifting 2-dimensional quasifibrations over skewfields" using central Galois quadratic extensions. What goes wrong? First of all, you can't necessarily use a tensor product construct. Can you get around this? Some of this project will lead to new unresolved/open material.

We have shown that for any finite-dimensional division ring D, using an inner $\sigma - A$ derivation set and any element c in $Fix\sigma \cap Z(D)$, such that c^2not -1, there is a semifield spread in $PG(3, D_1(\theta))$, which is not a simultaneous spread. In this setting, σ has order 2. What can be said for situations where σ has order k for $k > 2$? Note that the setup for the proof is basically the same. If one stays with the possibilities that satisfy the fundamental unwrapping principle for quasifibrations, then one just needs to show non-zero determinants.

CHAPTER 39

Hyperbolic skew-Desarguesian flocks

In this chapter, we consider the hyperbolic flocks that may be obtained from the left skew-Desarguesian spreads. These hyperbolic flocks are linear, but the recoordinatized t_0-interchange hyperbolic flocks are nonlinear.

Here is the hyperbolic flock spread:

$$x = 0, y = 0, y = x \begin{bmatrix} u & 0 \\ 0 & u \end{bmatrix},$$

$$y = x \begin{bmatrix} t + \delta, & \gamma \\ 1 & t \end{bmatrix} \begin{bmatrix} v & 0 \\ 0 & v \end{bmatrix},$$

for u, v non-zero and for all t in a division ring D. The associated left hyperbolic flock is as follows:

$$\Pi_t :$$

$$x_1 \gamma - x_2(t + \delta) + x_3 t - x_4 = 0,$$

and

$$\varrho : x_2 = x_3.$$

This is a linear semifield non-Desarguesian flock provided at least one of δ and γ is not in $Z(D)$, the center of D.

Now take the t_0-interchange, which creates the following hyperbolic flock spread:

$$x = 0, y = 0, y = x \begin{bmatrix} u & 0 \\ 0 & u \end{bmatrix},$$

$$y = x \begin{bmatrix} t_0 + \delta, & \gamma \\ 1 & t_0 \end{bmatrix}^{-1} \begin{bmatrix} v & 0 \\ 0 & v \end{bmatrix},$$

$$y = x \begin{bmatrix} t_0 + \delta, & \gamma \\ 1 & t_0 \end{bmatrix}^{-1} \begin{bmatrix} t + \delta, & \gamma \\ 1 & t \end{bmatrix} \begin{bmatrix} w & 0 \\ 0 & w \end{bmatrix}$$

for all non-zero u, v, w, for all t not equal to t_0. To construct the associated hyperbolic flock, we need to write the elements in the last two expressions with matrix elements equal to 1 (2,1)-entries in the

DOI: 10.1201/9781003767978-39

most left matrices. This is done using the matrices involved by selecting v and w appropriately. At this point, the hyperbolic flock may be constructed. Here is what all of that will look like, recalling that

$$\Delta_{t_0} = (t_0^2 + \delta t_0 - \gamma),$$

which is the $(2,1)$-left determinant.

$$x = 0, y = 0, y = x \begin{bmatrix} u & 0 \\ 0 & u \end{bmatrix},$$

$$y = x \begin{bmatrix} t_0 \Delta_{t_0}^{-1}, & 1 - t_0 \Delta_{t_0}^{-1}(t_0 + \delta) \\ -\Delta_{t_0}^{-1} & \Delta_{t_0}^{-1}(t_0 + \delta) \end{bmatrix} \begin{bmatrix} v & 0 \\ 0 & v \end{bmatrix},$$

$$y = x \begin{bmatrix} t_0 \Delta_{t_0}^{-1}, & 1 - t_0 \Delta_{t_0}^{-1}(t_0 + \delta) \\ -\Delta_{t_0}^{-1} & \Delta_{t_0}^{-1}(t_0 + \delta) \end{bmatrix} \begin{bmatrix} t + \delta, & \gamma \\ 1 & t \end{bmatrix} \begin{bmatrix} w & 0 \\ 0 & w \end{bmatrix}.$$

Just to remind the reader of a notational problem. When we multiply out this expression, we may choose the $(2,1)$-matrix entries on the most left matrices to be 1. Doing this by choosing v as $-\Delta_{t_0}$, will then affect all of the matrix entries. We then still use the same expressions for v, since this is for all v. We will then obtain for

$$y = x \begin{bmatrix} t_0 \Delta_{t_0}^{-1}, & 1 - t_0 \Delta_{t_0}^{-1}(t_0 + \delta) \\ -\Delta_{t_0}^{-1} & \Delta_{t_0}^{-1}(t_0 + \delta) \end{bmatrix} \begin{bmatrix} v & 0 \\ 0 & v \end{bmatrix},$$

the expression

$$y = x \begin{bmatrix} -t_0, & -\Delta_{t_0} + t_0 \Delta_{t_0}^{-1}(t_0 + \delta)\Delta_{t_0} \\ 1 & -\Delta_{t_0}^{-1}(t_0 + \delta)\Delta_{t_0} \end{bmatrix} \begin{bmatrix} v & 0 \\ 0 & v \end{bmatrix},$$

for all non-zero v in D.

Now consider

$$y = x \begin{bmatrix} t_0 \Delta_{t_0}^{-1}, & 1 - t_0 \Delta_{t_0}^{-1}(t_0 + \delta) \\ -\Delta_{t_0}^{-1} & \Delta_{t_0}^{-1}(t_0 + \delta) \end{bmatrix} \begin{bmatrix} t + \delta, & \gamma \\ 1 & t \end{bmatrix} \begin{bmatrix} w & 0 \\ 0 & w \end{bmatrix}.$$

This becomes

$$y = x \begin{bmatrix} t_0 \Delta_{t_0}^{-1}(t - t_0) + 1, & t + t_0 \Delta_{t_0}^{-1}(\gamma - (t_0 + \delta)t) \\ -\Delta_{t_0}^{-1}(t - t_0), & -\Delta_{t_0}^{-1}(\gamma - (t_0 + \delta)t) \end{bmatrix} \begin{bmatrix} w & 0 \\ 0 & w \end{bmatrix},$$

for t not t_0. Now take

$$w = -(t - t_0)^{-1} \Delta_{t_0}.$$

This will then create the following expression:

$$y = xM \begin{bmatrix} w & 0 \\ 0 & w \end{bmatrix},$$

where M is

$$\begin{bmatrix} -t_0-(t-t_0)^{-1}\Delta_{t_0}, & -(t+t_0\Delta_{t_0}^{-1}(\gamma-(t_0+\delta)t))(t-t_0)^{-1}\Delta_{t_0}) \\ 1, & \Delta_{t_0}^{-1}(\gamma-(t_0+\delta)t)(t-t_0)^{-1}\Delta_{t_0} \end{bmatrix},$$

for all non-zero w in D. If we write

$$s=\Delta_{t_0}^{-1}(\gamma-(t_0+\delta)t)(t-t_0)^{-1}\Delta_{t_0}$$

and the other terms as

$$f(s)=-t_0-(t-t_0)^{-1}\Delta_{t_0},$$

and

$$g(s)=-(t+t_0\Delta_{t_0}^{-1}(\gamma-(t_0+\delta)t))(t-t_0)^{-1}\Delta_{t_0}.$$

And when $t=t_0$, we let

$$r=-\Delta_{t_0}^{-1}(t_0+\delta)\Delta_{t_0}$$

and

$$f^*(r)=-t_0$$

and

$$g^*(r)=-\Delta_{t_0}+t_0\Delta_{t_0}^{-1}(t_0+\delta)\Delta_{t_0},$$

we may write the associated hyperbolic flock as

$$\Pi_s : t \text{ not } t_0,$$

$$x_1(-(t+t_0\Delta_{t_0}^{-1}(\gamma-(t_0+\delta)t))(t-t_0)^{-1}\Delta_{t_0})-x_2(-t_0-(t-t_0)^{-1}\Delta_{t_0})+$$
$$+x_3(\Delta_{t_0}^{-1}(\gamma-(t_0+\delta)t)(t-t_0)^{-1}\Delta_{t_0})-x_4=0,$$

then

$$\Pi_r : t=t_0,$$

$$x_1(-\Delta_{t_0}+t_0\Delta_{t_0}^{-1}(t_0+\delta)\Delta_{t_0})-x_2(-t_0)+$$
$$+x_3(-\Delta_{t_0}^{-1}(t_0+\delta)\Delta_{t_0})-x_4=0.$$

and

$$\varrho : x_2=x_3.$$

To show that we have a nonlinear hyperbolic flock, abstract the problem, first by taking $\varrho\cap\Pi_r$, which will produce

$$x_4=x_1A+x_2B,$$

Since $x_2=x_3$, this shows that these two left planes share a line L. Again, since $x_2=x_3$ in the general intersection, we now have $x_4=x_1C_t+x_2D_t$, where t is not equal to t_0. Therefore, we would obtain

$$x_1(C_t-A)=x_2(B-D_t).$$

For any t not equal to t_0,if either of the coefficients of x_1 or x_2 is 0 but not both, we cannot have a linear flock. Similarly, we cannot solve for

x_1 in terms of x_2 or vice versa. Therefore, we may assume that for all t not equal to t_0, we have both

$$C_t = A$$

and

$$D_t = B.$$

So,

$$x_1(-\Delta_{t_0} + t_0\Delta_{t_0}^{-1}(t_0+\delta)\Delta_{t_0}) - x_2(-t_0)+$$
$$+x_2(-\Delta_{t_0}^{-1}(t_0+\delta)\Delta_{t_0}) = x_4.$$

Therefore,

$$A = -\Delta_{t_0} + t_0\Delta_{t_0}^{-1}(t_0+\delta)\Delta_{t_0}$$

and

$$B = t_0 - \Delta_{t_0}^{-1}(t_0+\delta)\Delta_{t_0}.$$

Similarly,

$$x_1(-(t+t_0\Delta_{t_0}^{-1}(\gamma-(t_0+\delta)t))(t-t_0)^{-1}\Delta_{t_0}) - x_2(-t_0-(t-t_0)^{-1}\Delta_{t_0})+$$
$$+x_2(\Delta_{t_0}^{-1}(\gamma-(t_0+\delta)t)(t-t_0)^{-1}\Delta_{t_0}) = x_4.$$

Then

$$C_t = -(t+t_0\Delta_{t_0}^{-1}(\gamma-(t_0+\delta)t))(t-t_0)^{-1}\Delta_{t_0}$$

and

$$D_t = t_0 + (1+\Delta_{t_0}^{-1}(\gamma-(t_0+\delta)t))(t-t_0)^{-1}\Delta_{t_0}.$$

And, we obtain:

$$C_t = -(t+t_0\Delta_{t_0}^{-1}(\gamma-(t_0+\delta)t))(t-t_0)^{-1}\Delta_{t_0} = A =$$
$$= -\Delta_{t_0} + t_0\Delta_{t_0}^{-1}(t_0+\delta)\Delta_{t_0}$$

for all t not equal to t_0, and

$$D_t = t_0 + (1+\Delta_{t_0}^{-1}(\gamma-(t_0+\delta)t))(t-t_0)^{-1}\Delta_{t_0} = B =$$
$$= t_0 - \Delta_{t_0}^{-1}(t_0+\delta)\Delta_{t_0},$$

also for all t not equal to t_0.

LEMMA 144. $D_t = t_0 - \Delta_{t_0}^{-1}(t_0+\delta)\Delta_{t_0}$. *For all t not equal to t_0.*

PROOF. Cancel the left and right t_0 and then cancel the Δ_{t_0} terms. Then right multiply the equation by $(t-t_0)$, to obtain

$$(1+\Delta_{t_0}^{-1}(\gamma-(t_0+\delta)t)) = -\Delta_{t_0}^{-1}(t_0+\delta)(t-t_0).$$

Note that the terms involving t drop out, and then we obtain:

$$1+\Delta_{t_0}^{-1}\gamma = \Delta_{t_0}^{-1}(t_0+\delta)t_0.$$

And, this is valid if and only if

$$1 = \Delta_{t_0}^{-1}((t_0 + \delta)t_0 - \gamma),$$

which is then clean, completing the proof of the lemma. □

LEMMA 145. *$C_t = A$, if and only if $t = t_0 + 1$.*

PROOF. To prove this, consider

$$C_t = -(t + t_0 \Delta_{t_0}^{-1}(\gamma - (t_0 + \delta)t))(t - t_0)^{-1} \Delta_{t_0} = A =$$
$$= -\Delta_{t_0} + t_0 \Delta_{t_0}^{-1}(t_0 + \delta)\Delta_{t_0}.$$

Cancel out the Δ_{t_0} terms and then multiply on the right by $(t - t_0)$. Noting that the elements involving $(t_0 + \delta)t$, cancel on both sides of the equation, leaving the following:

$$(t + t_0 \Delta_{t_0}^{-1}\gamma) = 1 + t_0 \Delta_{t_0}^{-1}(t_0 + \delta)t_0.$$

From here it follows that this forces $t = t_0 + 1$, which completes the proof of the lemma. □

This completes the proof of the following theorem.

THEOREM 146. *Let π be a left skew-Desarguesian flock spread in $PG(3, D)$,where D is a non-commutative division ring. Note that the corresponding hyperbolic flock is linear. Choose any non-zero t_0 in D and form the t_0-interchange. Then this constructs an isomorphic flock spread but leads to a nonlinear hyperbolic flock.*

Hence, from a linear hyperbolic skew-Desarguesian flock spread, there are infinitely many isomorphic hyperbolic flock spreads that correspond to nonlinear hyperbolic flocks.

Part 15

Three line problem

In this work, we have emphasized the path of pseudo-regulus nets contained in spreads over skewfields using pseudo-regulus-inducing central collineation groups. There is a more general way to study generalized hyperbolic and conical flocks. This work was initiated by Jha and Johnson [**20**] and [**21**]. In this part, we shall fully develop flock theory using this approach. Regulus nets and pseudo-regulus nets, while bearing a natural similarity in representation, are very different in terms of intersection properties, and this is important to bring out. For example, two regulus nets that share three components are identical. But, there are infinitely many pseudo-regulus nets that can share three components. Given a set of mutually distinct regulus nets, each pair of such nets can share at most two components. There are beautiful results on t-nests of regulus, which we discussed in the first volume, that are simply not approachable using sets of pseudo-regulus nets. Then flocks of quadratic cones and flocks of hyperbolic quadrics may be and were initially discussed projectively without mention of the elation or homology regulus-inducing groups, since these groups do not exist in the projective space where the flocks are defined. With appropriate assumptions, one can make sets of pseudo-regulus nets behave like regulus nets, but there may be many other classes of pseudo-regulus nets that do not so conform. With sets of regulus nets, all of which share a single component, the elation regulus-inducing groups arise in a natural manner. But this idea is also deceiving, since all finite derivable nets are, in fact, regulus nets but may have an automorphism of type 2 form. There does not seem to be an obvious way to obtain the elation group that is a σ-regulus-inducing group without essentially assuring this initially.

Therefore, we here try to instill a bit of rigor into the theory of generalized flocks by stepping back a step from the central collineation assumptions. We gratefully acknowledge Pure and Applied Mathematics (lecture note series) and the Bulletin of the Belgian Mathematical Society for permission to revise the works [**20**] and [**21**].

CHAPTER 40

Do three components define a pseudo-regulus?

One property of derivable nets that are regulus nets is discussed in the part on bilinearity in [**34**] (also circle geometry p. 236) is emphasized by noticing that there is a unique circle on three distinct points. Thinking of the Miquellian inversive plane as the line at infinity of a quadratic extension of a field $K(\gamma)$-Pappian translation plane, there is then a connection between inversive planes and sets of regulus nets. This works since three components define a unique regulus net. We call this a "three for regulus" property.

In [**34**], there is a study of pseudo-regulus results of types 0,1,2, and 3. A pseudo-regulus net, which is not a regulus net, is never the less closest to being a regulus net when it is represented in type 3 form. Is there a "three for pseudo-regulus" type understanding?

Recall the initial setup:

Choose any skewfield K that is non-commutative. Consider the pseudo-regulus net arising from $PG(3, K)_N$, where N is a fixed line. The affine contraction may be given two forms: a right and a left. There is a 4-dimensional vector space V, which is a left and right space over K. The classical right pseudo-regulus net has the following components form:

$$x = 0, y = \delta x,$$

for all δ in K, where each component is a right 2-dimensional K-space. The left classical right pseudo-regulus net has a component form as,

$$x = 0, y = x\delta,$$

for all δ in K. Let $Z(K)$ denote the center of K. Since $Z(K)$ always contains $0, 1$, this implies that all left and right classical pseudo-reguli share three components. But this is not quite the point, as we are interested in discussing how left imposter pseudo-reguli can sit in the affine space wherein sits the classical right pseudo-regulus; in particular, what is the nature of left imposter pseudo-reguli of type 3? Although there are five types of type 3 derivable nets, using inner automorphisms, we refer here to the more-or-less standard type where the shared Baer

DOI: 10.1201/9781003767978-40

subspaces are referred to as (0,1),(1,0),(1,1). The left classical pseudo-regulus is of type 3 since it shares the three "generic" Baer subspaces. It has been shown in the first volume that the group $P\Gamma L(4,K)_N$ is triply transitive on the set of left and right Baer subspaces of the classical right pseudo-regulus and leaves invariant the classical right pseudo-regulus. What this means is that there is a set of left pseudo-regulus nets of type 3, which are related to the classical left pseudo-regulus as conjugate structures: The type 3 left pseudo-regulus nets are

$$x = 0, y = xM;$$

M in γ, a skewfield isomorphic to K if and only if

$$D_R^{-1}MD = \begin{bmatrix} u & 0 \\ 0 & u \end{bmatrix},$$

for all u in K, where D_R^{-1} is the right inverse of D and D is the mapping that sets a designated set of three left and right Baer subspaces shared with the designated classical right pseudo-regulus net. The cardinality of such type 3 pseudo-regulus nets then depends exactly on the cardinality of $Z(K)$. In particular, if $Z(K)$ is the field of rational numbers, there are infinitely many pseudo-regulus nets that share three components; there is only one situation/where three is pseudo-regulus, and that is when $Z(K)$ is $GF(2)$.

We now take this up in what might be considered a more projective setting.

CHAPTER 41

Three component – three point construction

Since the original consideration of regulus nets is from the projective space point of view, in this chapter, we study how the pseudo-regulus nets might be viewed also from the projective space. We have avoided this viewpoint due to the inherent complexity when discussing type 1 or 2 pseudo-regulus nets, especially when dealing with spreads containing such derivable nets and then how to consider the associated flocks. In fact, it has been more natural to consider the affine spreads that give rise to the hyperbolic and conical flocks, rather than from the flock perspective. Since in the regulus/flock case, the reguli originate as conics, this would, of course, be the natural progression, but the pseudo-regulus/flock case, there is no conics, but shades of such, as generalizations to certain varieties; the other direction made more sense.

In this, we consider again Desarguesian planes from quadratic Galois extensions, as in the Multiple Replacement Theorem – Redux. But here we construct the skew-Desarguesian planes, that are not always Desarguesian, by trying to find quadratic extensions that are not central; which amount to extensions not involving tensor products. The non-Desarguesian skew-Desarguesian planes are semifield planes. The examples that are constructed use any quaternion division ring but could, in theory, use infinite-dimensional skewfields and infinite-dimensional translation planes of type 3. The notation in the following differs from that in [**21**], but produces the same results.

DEFINITION 147. Let V_4 be a left vector space over a skewfield K. The 2-dimensional left K-subspaces are said to be "left lines" of $PG(3,K)_{left}$, and the 1-dimensional left K-subspaces are "left points" of $PG(3,K)_{left}$. Let S be a set of mutually disjoint 2-dimensional left K-subspaces, a partial spread. A "right transversal" to S is a right 2-dimensional K-subspace T such that given any pair D and E of left subspaces of S, T is the direct sum over $Z(K)$ of $(T \cap D) \oplus (T \cap E)$. A "$Z(K)$-transversal" to S is a right transversal that is also a left

DOI: 10.1201/9781003767978-41

2-dimensional K-subspaces. When T is a right transversal or a $Z(K)$-transversal contained in the set of intersections of left subspaces of S, we shall say that T is a "Baer subspace" of S.

Here is what we are trying to get ultimately: Let S contain three components $\{L, M, N\}$ (mutually disjoint 2-dimensional left K-subspaces). Choose a basis for L,M and N that has the form $x = 0, y = 0, y = x$ (see "carrier nets" in [**34**]). Then S has the following general form:

$$x = 0, y = 0, y = x, y = xM$$

where M is in a set γ in $GL(2, K)_{left}$ (see the main notation section of [**34**]). We know that γ (the union of the zero matrix) is a skewfield if and only if S is a left pseudo-regulus. A $Z(K)$-transversal T to S, which is contained in the intersections, is a Baer subspace. We are trying to study type 3 pseudo-regulus nets in a synthetic manner, without assuming the embedding theory. We know that the center of K contains $\{0, 1\}$.

Choose three $Z(K)$-transversals T_i,$i = 1, 2, 3$ and choose the intersections on $x = 0, y = 0, y = x$ to have forms

$$((0, 1)M, (0, 1)R)),$$

for all M and R in $\gamma \cup \{0\} = T_1$,

$$((1, 0)M, (1, 0)R),$$

for all M and R in $\gamma \cup \{0\} = T_2$, and

$$((1, 1)M, (1, 1R),$$

for all M and R in $\gamma\cup\{0\} = T_3$. Then the set of matrices M of S will have diagonal form: It would then follow, as we shall detail in the following, that $M = \begin{bmatrix} u & 0 \\ 0 & u \end{bmatrix}$ for all u in K. It will have the appearance of a classical left pseudo-regulus net. The only problem would be to show that there is a set of Baer subspaces comprising the derived net. These Baer subspaces would be simply

$$B(a, b) = \{(a, b)M, (a, b)N; M, N \text{ in } \gamma\cup\{0\}.$$

With these ideas in mind, we now provide the projective definition of a pseudo-regulus, free from the embedding theory, but not forgetting it either.

DEFINITION 148. *The three line – three point configuration.* Let R be the set of mutually disjoint left lines of $PG(3, K)_{left}$. Choose three left lines of R, $\{L, M, N\}$ and three $Z(K)$-points $\{U, V, W\}$ on L (that is, left points that are also right points). Assume that there are three mutually disjoint $Z(K)$-transversals to R such that any such transversal intersects U, V or W, and the transversals are contained in the set of intersections. We call this $R_{\{L,M,N\}}^{\{U,V,W\}}$, the "three line-three point configuration".

THEOREM 149. *The three line – three point configuration* $R_{\{L,M,N\}}^{\{U,V,W\}}$ *in* $PG(3,K)$ *is a left pseudo-regulus net of type* 3. *The proof shows that every three right and left points on L will produce a left pseudo-regulus net of type 3.*

PROOF. Since R is a set of mutually disjoint left 2-dimensional K-subspaces, $M \oplus L = V_4$. Therefore, a left basis for V_4 may be chosen (x, y), where x and y are left 2-vectors. Let $x = (x_1, x_2)$ and $y = (y_1, y_2)$ so that the fundamental left basis is

$$\{(1,0,0,0),(0,1,0,0),(0,0,1,0),(0,0,0,1)\}.$$

Therefore, we left L by $y = 0$ and M by $x = 0$. Let

$$K_L^* : \{(x_1, x_2, y_1, y_2) \to (ux_1, ux_2, ux_3, ux_4); u \text{ in } K^*.$$

Since all remaining elements of R may now be represented as linear mappings over K_L^*, they must have the form, for some elements e, f, g, h of K, as

$$y = x \begin{bmatrix} e & f \\ g & h \end{bmatrix},$$

such that when $g = 0$, then eh is not zero, and when g is not zero, then $(eg^{-1}h - f)g$ is not zero. Take the right inverse of this matrix, assuming the left 2-dimensional space is N, and apply this to all elements of R. That is, apply $(x,y) \to (x,y)N^{-1}$(the right inverse of N). This leaves $x = 0, y = 0$ with the same representation, and now we have N represented as $y = x$. Consider now the left and right 1-dimensional K-subspaces on $x = 0$ (that is, L), $\{U, V, R\}$. Since L is a 2-dimensional left K subspace that may be considered also a 2-dimenisonal right K space, we may represent $U = (0,0,1,0)$, $V = (0,0,0,1)$ and $W = (0,0,1,1)$ on L. Let T_U denote the assumed $Z(K)$-transversal that contains U. Assume that T_U intersects $y = 0$ in the right 1-space generated by $(a, b, 0, 0)$ and intersects $y = x$ in the right 1-space generated by (r, n, r, n). There exist elements of K, α and β such that $x = 0$ and $y = 0$, 1-right space intersections given by the following:

$$(0,0,1,0)\beta + (a,b,0,0)\alpha = (r,n,r,n).$$

This implies that $n = 0$, and $b = 0$. Therefore, we may choose $a = 1$, as this choice will generate the same 1-dimensional right K-space. Therefore,

$$T_U = ((1,0)\alpha, (1,0)\beta; \text{ for all } \alpha, \beta \text{ in } K.$$

We now use the same notation with the argument for T_V, following the previous argument to conclude, in this case, that $a = 0$ and, without loss of generality, that $b = 1$. Therefore,

$$T_V = ((0,1)\alpha, (0,1)\beta; \text{ for all } \alpha, \beta \text{ in } K.$$

Then again, with the same notation and argument for T_W, we would conclude that $a = b$ and, without loss of generality, that we may assume that $a = b = 1$. Therefore, we have:

$$T_W = ((1,1)\alpha, (1,1)\beta; \text{ for all } \alpha, \beta \text{ in } K.$$

We take a generic element of R

$$y = x \begin{bmatrix} e & f \\ g & h \end{bmatrix},$$

other than $x = 0, y = 0, y = x$, and note that the three $Z(K)$-transversals that intersect this element of R in a 1-dimensional right K-subspace.

The intersection with T_U will force $f = 0$ since $((1,0)\alpha, (1,0)\beta)$ is equivalent to $((1,0),(1,0)\beta\alpha^{-1})$ for some α, β both non-zero in K, and for a 1-dimensional right K-space, then

$$(1,0)\beta\alpha^{-1} = (1,0) \begin{bmatrix} e & f \\ g & h \end{bmatrix} = (e, f),$$

we see that $f = 0$. Similarly, using T_V, the analogous argument will show that

$$(0,1)\delta\gamma^{-1} = (0,1) \begin{bmatrix} e & 0 \\ g & h \end{bmatrix} = (g, h),$$

so that $g = 0$. Then, using T_W, the analogous argument will show that

$$(1,1)\epsilon\eta^{-1} = (1,1) \begin{bmatrix} e & 0 \\ 0 & h \end{bmatrix} = (e, h),$$

which shows that $e = h$. Therefore, the three line – three point configuration $R^{\{U,V,W\}}_{\{L,M,N\}}$ may be represented in the form:

$$x = 0, y = x \begin{bmatrix} u & 0 \\ 0 & u \end{bmatrix}$$

for all u in K. This certainly has the correct form to be a left pseudo-regulus net, but this needs to be proved. But, consider the following set of putative Baer subspaces

$$B(a,b) = \{(a,b) \begin{bmatrix} u & 0 \\ 0 & u \end{bmatrix}, (a,b) \begin{bmatrix} v & 0 \\ 0 & v \end{bmatrix}$$

for each non-zero 2-vector (a, b) and for all u, v in K. So, the actual check is just to see that these 2-dimensional right vector spaces $B(a, b)$ have the correct intersection properties to show that we have a left pseudo-regulus. $\square$

Part 16

The flocks and spreads

In this part, we provide examples of possible types of flocks by giving the corresponding flock spreads. We first indicate the types of examples that have been obtained and then consider the particular type of skewfield. We begin with some examples of inner flocks.

Easy Examples of Inner Flocks.

The kaleidoscopic mapping will provide infinitely many inner flocks. Here, we use the left non-commutative types of flock spreads with e non-central, $f = 0$ and $h = 1$.

Consider #12. $N_{hy}^{L}(1)^{1-A}$: $x\begin{bmatrix} f(t), & g(t) \\ 1 & t \end{bmatrix}\begin{bmatrix} v, & A(v) \\ 0 & v \end{bmatrix}$; A an outer derivation function. Then the full hyperbolic flock spread is:

$$x = 0, y = x\begin{bmatrix} e^{-1} & 0 \\ 0 & 1 \end{bmatrix}\begin{bmatrix} u, & A(u) \\ 0 & u \end{bmatrix}\begin{bmatrix} e, & 0 \\ 0 & 1 \end{bmatrix} = \begin{bmatrix} u^{\tau_{e^{-1}}} & e^{-1}A(u) \\ 0 & u \end{bmatrix}$$

and

$$y = xJ\begin{bmatrix} e, & 0 \\ 0 & 1 \end{bmatrix},$$

where J is

$$\begin{bmatrix} e^{-1} & 0 \\ 0 & 1 \end{bmatrix}\begin{bmatrix} f(t), & g(t) \\ 1 & t \end{bmatrix}\begin{bmatrix} e, & 0 \\ 0 & 1 \end{bmatrix}\begin{bmatrix} e^{-1}, & 0 \\ 0 & 1 \end{bmatrix}\begin{bmatrix} v, & A(v) \\ 0 & v \end{bmatrix}$$

works out to the following:

$$y = x\begin{bmatrix} e^{-1}f(t), & e^{-1}f(t)A(e) + g(t)^{\tau_{e^{-1}}} \\ 1 & te \end{bmatrix}\begin{bmatrix} v^{\tau_{e^{-1}}}, & e^{-1}A(v) \\ 0 & v \end{bmatrix}.$$

An interesting point is that we have a new $\tau_{e^{-1}} - e^{-1}A$ outer derivative function.

#5. $N_{hy}^{L}(3)$: $x\begin{bmatrix} f(t), & g(t) \\ 1 & t \end{bmatrix}\begin{bmatrix} v, & 0 \\ 0 & v \end{bmatrix}$. Apply the same mapping to obtain:

$$x = 0, y = x\begin{bmatrix} e^{-1} & 0 \\ 0 & 1 \end{bmatrix}\begin{bmatrix} u, & 0 \\ 0 & u \end{bmatrix}\begin{bmatrix} e, & 0 \\ 0 & 1 \end{bmatrix} = \begin{bmatrix} u^{\tau_{e^{-1}}} & 0 \\ 0 & u \end{bmatrix}$$

and

$$y = xH\begin{bmatrix} e, & 0 \\ 0 & 1 \end{bmatrix},$$

where H is

$$\begin{bmatrix} e^{-1} & 0 \\ 0 & 1 \end{bmatrix}\begin{bmatrix} f(t), & g(t) \\ 1 & t \end{bmatrix}\begin{bmatrix} e, & 0 \\ 0 & 1 \end{bmatrix}\begin{bmatrix} e^{-1}, & 0 \\ 0 & 1 \end{bmatrix}\begin{bmatrix} v, & 0 \\ 0 & v \end{bmatrix},$$

which provides:

$$y = x\begin{bmatrix} e^{-1}f(t), & g(t)^{\tau_{e^{-1}}} \\ 1 & te \end{bmatrix}\begin{bmatrix} v^{\tau_{e^{-1}}}, & 0 \\ 0 & v \end{bmatrix}.$$

Now 15. $N_{co}^{L}(1)^{1-A}$: $x\begin{bmatrix} u+F(t), & G(t)+A(u) \\ t & u \end{bmatrix}$, A an outer derivation function. This also provides an example of a $\tau_{e^{-1}} - e^{-1}A$ **outer derivative function.**

Try this as a problem!

CHAPTER 42

Anti-isomorphic flocks

When σ is outer or 1: There are three examples out of the four possibilities. The classes of anti-isomorphic left and right conical flocks are given by their pairs of classes of left conical flock spreads and their transpose spreads. These come in two varieties: when the isomorphism is σ or 1 and then when σ is an inner automorphism. This is also the case where one may also find the Desarguesian quaternion. left and right anti-isomorphic flocks. Therefore, there are eight cases.

1. $$N(2)^{F,\sigma-A}: \begin{bmatrix} u^{\sigma}+F(t), & G(t)+(u^{\sigma}c-cu)) \\ t & u \end{bmatrix},$$

σ *non-trivial.*

A "a fake type 1" derivable net.

$$\& \ N_{co}^{L,trans,F,\sigma^{-1}-A^{*}} x \begin{bmatrix} u^{\sigma^{-1}}-F(t)^{\sigma^{-1}}, & A^{*}(v)-A^{*}(F(t))+G(t) \\ t & u \end{bmatrix},$$

where, $A^{*}(u)=A(u^{\sigma^{-1}})$, indicating the transpose. When σ has order 2 and $A(w)=(w^{\sigma}-w)c$, where $c^{\sigma}=-c, F(t)=A(t), G(t)=t^{\sigma}\theta$, there is an example where the transpose is isomorphic to the original. There is a central Galois quadratic extension $D_1(\theta), D_1$ a finite-dimensional division ring. Furthermore, in this case the transpose admits the left-acting pseudo-regulus-inducing homology group.

2. $$N_{co}^{L}(2)^{\sigma}: \begin{bmatrix} u^{\sigma}+F(t), & G(t) \\ t & u \end{bmatrix}, \ \sigma \text{ non-trivial},$$

$$\& \ N_{co}^{L}(2)^{\sigma,trans}: x \begin{bmatrix} u^{\sigma^{-1}}-F(t)^{\sigma^{-1}}, & G(t) \\ t & u \end{bmatrix}, \ \sigma \text{ non-trivial}.$$

The transpose spread of the previous. In this case, if $F(t)=0$ and $G(t)=\theta t^{\sigma^{-1}}$, there is a solution. In this case, the transpose exists and admits the left-acting pseudo-regulus-inducing homology group. Here, there is a central Galois cyclic extension of degree n $D_1(\theta)$, for $n=3$ or 4.

DOI: 10.1201/9781003767978-42

3. $N_{co}^{L}(1)^{1-A}\colon x\begin{bmatrix} u+F(t), & G(t)+A(u) \\ t & u \end{bmatrix}$, A an outer derivation function

& $N_{co}^{L}(1)^{trans.,1-A}\colon x\begin{bmatrix} w-F(t), & G(t)+A(w)-A(F(t)) \\ t & w \end{bmatrix}$,

A an outer derivation function. For characteristic 2, $F(t) = A(t), G(t) = t\sqrt{z}$,

4. $N^{L}(2)_{co}^{F,1-A}\colon x\begin{bmatrix} u+F(t), & G(t)+A(u) \\ t & u \end{bmatrix}$, A an inner derivation function; a "fake type 1"

& $N^{L}(2)_{co}^{trans,F,1-A}\colon x\begin{bmatrix} w-F(t), & G(t)+A(w)-A(F(t)) \\ t & w \end{bmatrix}$,

A an inner derivation function; a "fake type 1". There is a central non-Galois extension of a quaternion division ring $(a,b)_F$ of characteristic 2 by $F(z)(\sqrt{z})$, where $F(z)$ is a rational function field of the field F, turning the quaternion division ring into $(a,b)_{F(z)(\sqrt{z})}$. $A(w) = dw/d\sqrt{z}$, where $F(t) = A(t)$, and $G(t) = t\sqrt{z}$. The transpose is identical to the original flock spread and admits the left-acting pseudo-regulus-inducing homology group.

5. $N_{co}^{L}(1)^{\sigma-A}\colon x\begin{bmatrix} u^{\sigma}+F(t), & G(t)+A(u) \\ t & u \end{bmatrix}$;

A is a non-trivial

outer σ-derivation function for σ, a non-trivial outer automorphism

& $N_{co}^{L}(1)^{trans.,\sigma-A}\colon x\begin{bmatrix} w^{\sigma^{-1}}-F(t)^{\sigma^{-1}}, & G(t)+A^{*}(w)-A^{*}(F(t)) \\ t & w \end{bmatrix}$;

$A^{*}(u) = A(u^{\sigma^{-1}})$ is a non-trivial outer σ-derivation function for σ, a non-trivial outer automorphism.

When σ is inner. There are no examples.

1. ${}^{Inner}N_{co}^{L}(2)^{F,\sigma-A}:\begin{bmatrix} u^{\sigma}+F(t), & G(t)+A(u) \\ t & u \end{bmatrix}$, *$\sigma$ non-trivial.* A "a double fake type 1" derivable net for σ-A, where both σ and A are inner &${}^{Inner}N_{co}^{L}(2)^{F,\sigma-A}\colon x\begin{bmatrix} u^{\sigma^{-1}}-F(t)^{\sigma^{-1}}, & L \\ t & u \end{bmatrix}$;

$$L = A^{*}(v) - A^{*}(F(t)) + G(t),$$

the transpose.

2. ${}^{Inner}N_{co}^{L}(2)^{\sigma}\colon \begin{bmatrix} u^{\sigma}+F(t), & G(t) \\ t & u \end{bmatrix}$, σ non-trivial. A fake type 2

$\&{}^{Inner}N_{co}^{L}(2)^{trans.,\sigma}x \begin{bmatrix} w^{\sigma^{-1}}-F(t)^{\sigma^{-1}}, & G(t) \\ t & w \end{bmatrix}$, *$\sigma$ non-trivial.*

The transpose spread of the previous.

3. ${}^{Inner}N_{co}^{L}(1)^{\sigma-A}\colon x \begin{bmatrix} u^{\sigma}+F(t), & G(t)+A(u) \\ t & u \end{bmatrix}$; A is a non-trivial outer σ-derivation function for σ, a non-trivial inner automorphism. A fake type 1

$\&{}^{Inner}N_{co}^{L}(2)^{F,\sigma^{-1}-A^{*}}\colon x \begin{bmatrix} u^{\sigma^{-1}}-F(t)^{\sigma^{-1}}, & A^{*}(v)-A^{*}(F(t))+G(t) \\ t & u \end{bmatrix}$,

the transpose, $A^{*}(u)=A(u^{\sigma^{-1}})$, A is an outer $\sigma-A$ derivation function, for σ a non-trivial inner automorphism.

CHAPTER 43

Constructions-generalized lifted

THEOREM 150. *If D is a central cyclic extension of a division ring E, so $E(\theta) = D$, and σ is the induced automorphism, assume that E contains a primitive nth root of unity and that n does not divide the characteristic of K, and n is a unit.*

(1) Then for any integer n,

$$x = 0, y = x \begin{bmatrix} u^{\sigma}, & \theta t^{\sigma^{-1}} \\ t & u \end{bmatrix},$$

for all t, u in D, is a semifield flock spread.

For examples, when the characteristic is 0, it is possible to choose various generators of different central extensions that are non-isomorphic, for which the associated flock spreads are then non-isomorphic. The associated flocks are also isomorphic in this setting. Therefore, even as these examples provide the same type in the general classification scheme, there are infinitely many non-isomorhic examples.

We have seen in the transpose and inverse and semifield parts that the following provides a simultaneous spread:

$$x = 0, y = x \begin{bmatrix} u^{\sigma} & \theta t^{\sigma^{-1}} \\ t & u \end{bmatrix}$$

for all u, tin D,D a division ring, where θ is in the center of D and

$$k^{\sigma} k \text{ not} \theta^{\sigma}.$$

for all k in D, σ and outer automorphism of order >2 .

Furthermore, we have also noted that the transpose is a quasifibration and also provides a left spread using the left unwrapping procedure. Furthermore, the inversing of the spread is also a hyperbolic spread. Therefore, this produces four left flock-type left spreads: two hyperbolic and two conical. In addition, we see that these examples produce four right-flock types spreads, also two hyperbolic and two conical. To see

DOI: 10.1201/9781003767978-43

this explicitly, we note that the transpose spread is

$$x = 0, y = x\begin{bmatrix} u^{\sigma} & t \\ \theta t^{\sigma^{-1}} & u \end{bmatrix}$$

for all t, u in D. Write $\theta t^{\sigma^{-1}} = s$, so that $t = \theta^{-\sigma} s^{\sigma}$. It has been checked that the skewfield determinant set is also non-zero for (t, u) not $(0, 0)$. Thus, we now have the set (replacing s again by the symbol t)

$$x = 0, y = x\begin{bmatrix} u^{\sigma} & \theta^{-\sigma} t^{\sigma} \\ t & u \end{bmatrix},$$

for all u, t in D, which is also a conical spread, but it is not a simultaneous spread. But now there is a left mapping by $\begin{bmatrix} v^{\sigma} & 0 \\ 0 & v \end{bmatrix}$ for all non-zero v in D that preserves the spread.

$$\begin{bmatrix} v^{\sigma} & 0 \\ 0 & v \end{bmatrix}\begin{bmatrix} u^{\sigma} & \theta^{-\sigma} t^{\sigma} \\ t & u \end{bmatrix} = \begin{bmatrix} (vu)^{\sigma} & \theta^{-\sigma}(vt)^{\sigma} \\ vt & vu \end{bmatrix},$$

since $\theta^{-\sigma}$ is also central (why? prove this). This is a pseudo-regulus-inducing homology group with axis $x = 0$ and co-axis $y = 0$, instead of the normal pseudo-regulus-inducing homology group with axis $y = 0$ and co-axis $x = 0$. Since the mapping $(x, y) \to (y, x)$ transforms the existing spread to another spread that admits the standard pseudo-regulus homology group, this means we have another hyperbolic spread/flock. This produces the associated hyperbolic spread:

$$x = 0, y = x\begin{bmatrix} u^{\sigma} & \theta^{-\sigma} t^{\sigma} \\ t & u \end{bmatrix}^{-1}.$$

This last spread is

$$x = 0, y = x\begin{bmatrix} v^{\sigma} & 0 \\ 0 & v \end{bmatrix},$$

$$y = x\begin{bmatrix} t^{\sigma} & \theta^{-\sigma} \\ 1 & t \end{bmatrix}^{-1}\begin{bmatrix} w^{\sigma} & 0 \\ 0 & w \end{bmatrix},$$

for all v, t,w non-zero in D. The left skewfield determinant is $t^{\sigma} t - \theta^{-\sigma}$(why? check this)$=\Delta_t$. And,

$$\begin{bmatrix} t^{\sigma} & \theta^{-\sigma} \\ 1 & t \end{bmatrix}^{-1} = \begin{bmatrix} t\Delta_t^{-1} & 1 - t\Delta_t^{-1} t^{\sigma} \\ -\Delta_t^{-1} & \Delta_t^{-1} t^{\sigma} \end{bmatrix}$$

(check this, also).

We therefore have the following left hyperbolic flocks: We note that

$$x = 0, y = x\begin{bmatrix} u^{\sigma} & \theta t^{\sigma^{-1}} \\ t & u \end{bmatrix}$$

is

$$x=0,y=x\begin{bmatrix} v^{\sigma} & 0 \\ 0 & v \end{bmatrix}, y=x\begin{bmatrix} t^{\sigma} & \theta \\ 1 & t \end{bmatrix}\begin{bmatrix} w^{\sigma} & 0 \\ 0 & w \end{bmatrix}$$

which is equivalent to the left hyperbolic flock:

$$\Gamma_t^{1,left}\colon x_1\theta^{\sigma}-x_2t^{\sigma}+x_3t^{\sigma}-x_4=0 \text{ and } \rho^{left}\colon x_2=x_3,$$

This is a left linear flock with a common line $(x_1,x_2,x_2,x_1\theta^{\sigma})$, for all x_1,x_2 in D.

Now considering

$$x=0,y=x\begin{bmatrix} v^{\sigma} & 0 \\ 0 & v \end{bmatrix}, y=x\begin{bmatrix} t^{\sigma}, & \theta^{-\sigma} \\ 1 & t \end{bmatrix}^{-1}\begin{bmatrix} w^{\sigma} & 0 \\ 0 & w \end{bmatrix},$$

to transform to the form with a 1 in the $(2,1)$-entry,

$$x=0,y=x\begin{bmatrix} v^{\sigma} & 0 \\ 0 & v \end{bmatrix}, y=x\begin{bmatrix} - & - \\ 1 & - \end{bmatrix}\begin{bmatrix} r^{\sigma} & 0 \\ 0 & r \end{bmatrix},$$

we know that for every t, there is an element w so that $-\Delta_t^{-1}=w^{-\sigma}$, which then transforms the spread into the form:

$$x=0,y=x\begin{bmatrix} v^{\sigma} & 0 \\ 0 & v \end{bmatrix}, y=x\begin{bmatrix} t & ,(-1+t\Delta_t^{-1}t^{\sigma})\Delta_t^{\sigma^{-1}} \\ 1 & -\Delta_t^{-1}t^{\sigma}\Delta_t^{\sigma^{-1}} \end{bmatrix}\begin{bmatrix} r^{\sigma} & 0 \\ 0 & r \end{bmatrix},$$

for all v,t in D and for all non-zero r in D. Let

$$k=-\Delta_t^{-1}t^{\sigma}\Delta_t^{\sigma^{-1}}, f(k)=t, g(k)=(-1+t\Delta_t^{-1}t^{\sigma})\Delta_t^{\sigma^{-1}}.$$

This then is equivalent to the following left hyperbolic flock

$$\Sigma_t^{2,left}\colon x_1g(k)^{\sigma}-x_2f(k)+x_3k^{\sigma}-x_4=0 \text{ and } \Pi\colon x_2=x_3 .$$

$$\Gamma_t^{2,left}\colon x_1((-1+t\Delta_t^{-1}t^{\sigma})\Delta_t^{\sigma^{-1}})^{\sigma}-x_2t+(-\Delta_t^{-1}t^{\sigma}\Delta_t^{\sigma^{-1})^{\sigma}}x_3-x_4=0$$
$$\text{and } \rho^{left}\colon x_2=x_3.$$

A nonlinear flock.

Then there are the two left conical flocks.

$$x=0,y=x\begin{bmatrix} u^{\sigma} & \theta t^{\sigma^{-1}} \\ t & u \end{bmatrix}$$

is equivalent to

$$\Sigma_t^{3,left}\colon x_1(\theta t^{\sigma^{-1}})^{\sigma}+x_2(0)-x_3+x_4t=0,$$

a left linear conical flock.

And, there is the spread

$$x=0,y=x\begin{bmatrix} u^{\sigma} & \theta^{-\sigma}t^{\sigma} \\ t & u \end{bmatrix}$$

and the corresponding left conical flock:

$$\Sigma_t^{4,left}\colon x_1(\theta^{-\sigma}t^{\sigma})^{\sigma}+x_2(0))-x_3+x_4t=0.$$

Since σ^2 is not 1, this is a nonlinear conical flock.

Since D could be finite, infinite field or a non-commutative skewfield, we have examples of two types of left hyperbolic and two types of left conical flocks for finite, infinite field or non-commutative skewfield (the reader may check that all three of the possible types of skewfields are represented by spread examples listed in the text).

These types are from the list of 109 types: each of which may be considered a left or right flock, and are as follows:

$$(3)\ F_{hy}(2)^{\sigma},(4)F_{hy}(2)^{\sigma,inverse},(8)\ F_{co}(2)^{\sigma},(9)\ F_{co}(2)^{trans,\sigma},$$
$$(11)F_{hy\&co}(2)^{F,\sigma-A}.$$

Therefore, since

$$(5)F(3)_{hy}and(10)F(3)_{co},$$

are the original Thas-Walker types, we have examples for all but (1), (2), (6), (7) in the finite case, which we shall obtain in the next few sections.

In the infinite field cases and the left flock spreads, we similarly have examples of

$$(3)\ IF_{hy}(2)^{\sigma},(4)IF_{hy}(2)^{\sigma,inverse},(8)\ IF_{co}(2)^{\sigma},(9)\ IF_{co}(2)^{trans,\sigma},$$
$$(11)IF_{hy\&co}(2)^{F,\sigma-A}.$$
$$(3)\ N_{hy}^{L}(2)^{\sigma},(4)N_{hy}^{L}(2)^{\sigma,inverse},(8)\ N_{co}^{L}(2)^{\sigma},(8)\ N_{co}^{L}(2)^{trans,\sigma},$$
$$(11)N_{hy\&co}^{L}(2)^{F,\sigma-A}.$$

Furthermore, from this procedure, there should be five additional right flocks. We would be looking for the right τ-hyperbolic and τ-conical flocks. We consider first the right hyperbolic flocks, where the associated right hyperbolic spreads are of the form

$$x^T=0^T,y^T=\begin{bmatrix} u & 0\\ 0 & u^{\tau}\end{bmatrix}x^T,$$

$$\begin{aligned} y^T=\begin{bmatrix} v & 0\\ 0 & v^{\tau}\end{bmatrix}\begin{bmatrix} r(t) & 1\\ s(t) & t^{\tau}\end{bmatrix}\begin{bmatrix} x_1\\ x_2\end{bmatrix}\\ =\begin{bmatrix} x_3\\ x_4\end{bmatrix}.\end{aligned}$$

And the corresponding hyperbolic flock has the form:

$$\Gamma_t^{right}\colon (s(t))p_1-t^{\tau}p_2+r(t)^{\tau}p_3-p_4, and\ \varrho: x_2=x_3.$$

We note that we changed the notation slightly so as not to confuse the transforms with the affine points. But, just for comparison of flocks, p_i may be represented as $x_i, i = 1, 2, 3, 4$.

Now the right conical spreads and flocks are

$$x^T = 0^T, y^T = \begin{bmatrix} u + R(t) & t \\ S(t) + A^*(u), & u^\tau \end{bmatrix} x^T,$$

and the flock

$$\Sigma_t^{right}: (S(t) - A^*(R(t))x_1 - (R(t)^\tau + A^*(t))x_2 + x_3 - t^\tau x_4 = 0,$$

Now it is claimed that in the situation considered in this example, we may take $r(t) = f(t), s(t) = g(t)$ and $\tau = \sigma^{-1}$ for the hyperbolic spread, and $R(t) = F(t), S(t) = G(t), \tau = \sigma^{-1}$, and we would obtain a corresponding set of four right flocks.

The reason is that both the left matrix spread and the transposed matrix sets have non-singular skewfield determinants and therefore quasifibrations. Then, using the left unwrapping procedure and realizing the term θ is in the center of the division ring D, we must have a situation where the transposed matrix gives rise to a right conical spread. But then the transposed matrix of the transposed matrix must also give rise to a right conical spread. So, the non-intuitive situation is that the left transposed spread of T, σ-type 2, produces a $\tau = \sigma^{-1}$ right spread. Then the transpose of this right spread, a right transposed T, σ^{-1} right spread produces a σ-type 2 left spread. The inversing procedure also works, it would seem, "backwards". The reason for this is that the transposed left simultaneous spread, while not "left" simultaneous, is, in fact, "right simultaneous". There would be a right hyperbolic spread and a right conical spread from the transpose of a left simultaneous spread.

So, a sort of unpacking is required to visualize what is going on the right side. Start with a left simultaneous spread. Take the transposed left matrix set that now turns out to be a spread (just using the skewfield determinant and the unwrapping procedure). Forget about left and right for a moment, and just concentrate on the matrix set. Recall that upon transposing, we have a left pseudo-regulus mapping; this is why the left inversing procedure works. But now go back to thinking of a right spread, this is what produces a right hyperbolic spread. And the transposed matrix set satisfies the criterion to have pseudo-regulus-inducing elation group, just by the form of the matrix set. So, the transposed left matrix set of a left simultaneous spread produces a right spread, which is a right simultaneous spread. Just as a note, it is not a left simultaneous spread – but is a right simultaneous spread.

So, starting with the right simultaneous spread, we have a right hyperbolic flock and a right conical flock. Now transpose the matrix set, you just back to the matrix set of the original left simultaneous spread. But, not as a right spread, it admits a "right" pseudo-regulus homology group. This is just the original right pseudo-regulus homology group, but now, thinking on the right side, we have an "inversing construction" that produces another left pseudo-regulus homology group, which is equivalent to the existence of the right hyperbolic flock.

So, we have constructed a set of four right spreads from one left simultaneous spread. One of these right spreads is simultaneous and constructs a right hyperbolic spread and a right conical spread. The transpose matrix set of this right hyperbolic spread has a pseudo-regulus homology group "on the opposite side" and therefore inversing and creates another right hyperbolic spread, which is the original matrix set of the left simultaneous spread and is therefore a right conical spread.

Probably, we need to go through the right side from the associated matrices to illustrate the process again.

This would start from the transposed matrix set of the original left simultaneous spread, but using this to construct a right simultaneous spread. Note that the notation is slightly changed so that (u, t) is the top row. This would be

$$x^T = 0^T, y^T = \begin{bmatrix} u & t \\ \theta t^{\sigma^{-1}} & u^{\sigma^{-1}} \end{bmatrix} x^T$$

for all t, u in D, a division ring, θ in the center, and σ an outer automorphism of order >2. Assume that $k^\sigma k$ is never equal to θ^σ. Using the skewfield determinant and the unwrapping procedure, this clearly produces a σ^{-1}-right conical flock. So to see that we have a right simultaneous spread, we decide if there is a left-acting pseudo-regulus-inducing homology group. Since

$$\begin{bmatrix} v & 0 \\ 0 & v^{\sigma^{-1}} \end{bmatrix} \begin{bmatrix} u & t \\ \theta t^{\sigma^{-1}} & u^{\sigma^{-1}} \end{bmatrix} = \begin{bmatrix} vu & vt \\ \theta (vt)^{\sigma^{-1}} & (vu)^{\sigma^{-1}} \end{bmatrix},$$

as θ is central, this is now clear. Therefore, we have the following two right flocks. For the right hyperbolic flock, we write the spread as

$$x^T = 0^T, y^T = \begin{bmatrix} u & 0 \\ 0 & u^\tau \end{bmatrix} x^T, y^T = \begin{bmatrix} v & 0 \\ 0 & v^\tau \end{bmatrix} \begin{bmatrix} t & 1 \\ \theta & t^{\sigma^{-1}} \end{bmatrix} \begin{bmatrix} x_1 \\ x_2 \end{bmatrix}$$

$$= \begin{bmatrix} x_3 \\ x_4 \end{bmatrix}.$$

The corresponding right hyperbolic flock is

$$\Gamma_t^{2,right}:$$

$$\Gamma_t^{1,right}\colon \theta p_1 - t^{\sigma^{-1}} p_2 + t^{\sigma^{-1}} p_3 - p_4, \mathit{and}\ \varrho\colon x_2 = x_3,$$

a linear right hyperbolic flock. Therefore, we obtain the four right flock spreads.

For the inverse-type hyperbolic flock, we take the transpose of

$$x^T = 0^T, y^T = \begin{bmatrix} u & t \\ \theta t^{\sigma^{-1}} & u^{\sigma^{-1}} \end{bmatrix} x^T$$

(and make the notational change to $u^\sigma$for the

$(1,1)$-entry (we are letting $u = v^\sigma$, then reverting to the $u -$ notation)

and

we obtain

$$x^T = 0^T, y^T = \begin{bmatrix} u^\sigma & \theta t^{\sigma^{-1}} \\ t & u \end{bmatrix} x^T.$$

Since we know that we obtain a right-acting pseudo-regulus homology group, we take the inverse and construct another right hyperbolic spread/flock, with spread

$$x^T = 0^T, y^T = \begin{bmatrix} w & 0 \\ 0 & w^{\sigma^{-1}} \end{bmatrix} x^T, y^T = \begin{bmatrix} v & 0 \\ 0 & v^{\sigma^{-1}} \end{bmatrix} \begin{bmatrix} t & \theta \\ 1 & t^{\sigma^{-1}} \end{bmatrix}^{-1} x^T,$$

for all w, k in D and for all v non-zero in D. Let $\Delta_t^* = tt^{\sigma^{-1}} - \theta$. Then

$$\begin{bmatrix} t & \theta \\ 1, & t^{\sigma^{-1}} \end{bmatrix}^{-1} = \begin{bmatrix} t^{\sigma^{-1}} \Delta_t^{*^{-1}} & 1 - t^{\sigma^{-1}} \Delta_t^{*^{-1}} t \\ -\Delta_t^{*^{-1}} & \Delta_t^{*^{-1}} t \end{bmatrix}.$$

To find the proper functions, for each $1 - t^{\sigma^{-1}} \Delta_t^{*^{-1}} t$, since it is non-zero, for all values of t, there is a v such that $v^{-1} = 1 - t^{\sigma^{-1}} \Delta_t^{*^{-1}} t$. To verify this statement, just set to 0, so note that t is not zero, and then

$$t^{-\sigma^{-1}} t^{-1} = \Delta_t^{*^{-1}}$$

if and only if $tt^{\sigma^{-1}} = \Delta_t^* = tt^{\sigma^{-1}} - \theta$, a contradiction. We therefore obtain the required form as

$$x^T = 0^T, y^T = \begin{bmatrix} w & 0 \\ 0 & w^{\sigma^{-1}} \end{bmatrix} x^T,$$

$$y^T = \begin{bmatrix} s & 0 \\ 0 & s^{\sigma^{-1}} \end{bmatrix} J x^T,$$

where J is

$$\begin{bmatrix} (1-t^{\sigma^{-1}}\Delta_t^{*^{-1}}t)^{-1}t^{\sigma^{-1}}\Delta_t^{*^{-1}}, & 1 \\ -(1-t^{\sigma^{-1}}\Delta_t^{*^{-1}}t)^{-\sigma}\Delta_t^{*^{-1}}, & (1-t^{\sigma^{-1}}\Delta_t^{*^{-1}}t)^{-\sigma}\Delta_t^{*^{-1}}t \end{bmatrix}.$$

Thus, we obtain the second right hyperbolic flock

$$\Gamma_k^{2,right}\colon (s(k))p_1 - k^{\tau}p_2 + r(k)^{\tau}p_3 - p_4, \textit{and } \varrho\colon x_2 = x_3.$$

Where $k^{\tau} = (1-t^{\sigma^{-1}}\Delta_t^{*^{-1}}t)^{-\sigma}\Delta_t^{*^{-1}}t$, $r(k) = (1-t^{\sigma^{-1}}\Delta_t^{*^{-1}}t)^{-1}t^{\sigma^{-1}}\Delta_t^{*^{-1}}$, $s(k) = -(1-t^{\sigma^{-1}}\Delta_t^{*^{-1}}t)^{-\sigma}\Delta_t^{*^{-1}}$. Therefore, we have the following right hyperbolic flock:

$$\Gamma_k^{2,right}\colon (-(1-t^{\sigma^{-1}}\Delta_t^{*^{-1}}t)^{-\sigma}\Delta_t^{*^{-1}})x_1 - (1-t^{\sigma^{-1}}\Delta_t^{*^{-1}}t)^{-\sigma}\Delta_t^{*^{-1}}t)x_2 + \\ +((1-t^{\sigma^{-1}}\Delta_t^{*^{-1}}t)^{-1}t^{\sigma^{-1}}\Delta_t^{*^{-1}})^{\tau}x_3 - x_4 = 0$$

and

$$\varrho\colon x_2 = x_3.$$

The right conical flock is equivalent to

$$x^T = 0^T, y^T = \begin{bmatrix} u & t \\ \theta t^{\sigma^{-1}} & u^{\sigma^{-1}} \end{bmatrix} x^T$$

is

$$\Sigma_t^{3,right}\colon (\theta t^{\sigma^{-1}})x_1 - (0)x_2 + x_3 - t^{\sigma^{-1}}x_4 = 0.$$

The fourth right flock is the transposed version of the above, which is

$$x^T = 0^T, y^T = \begin{bmatrix} u, & \theta t^{\sigma^{-1}} \\ t, & u^{\sigma^{-1}} \end{bmatrix} x^T,$$

and the right conical flock is:

$$\Sigma_t^{4,right}\colon tx_1 - 0x_2 + x_3 - (\theta t^{\sigma^{-1}})^{\sigma^{-1}}x_4 = 0.$$

This flock is nonlinear as σ has order >2 .

The last four right flocks are of types

$$(3)\ N_{hy}^R(2)^{\sigma}, (4) N_{hy}^R(2)^{\sigma,inverse},$$

$$(8)\ N_{co}^R(2)^{\sigma}, (9)\ N_{co}^R(2)^{trans,\sigma}, (11)\ N^R{}_{hy\&co}(2)^{F,\sigma-A}.$$

CHAPTER 44

$1-A$ conical spreads

Now we show that it is possible to construct $1-A$ conical flocks that spreads in the characteristic 2 case.

THEOREM 151. *Let D be any finite quaternion division ring of characteristic 2 over a field F and consider $D_1 = D \otimes F(z)$, which is also a division ring. Consider $F(z)\sqrt{z}$, which is a quadratic (but not Galois) extension of $F(z)$. Now form $D \otimes_{F(z)} F(z)(\sqrt{z}) = D_2$, which is a division ring with center $F(z)$, which is a quadratic extension of D_1. Define $A(u)=du/d\sqrt{z}$, which is a derivative function.*

(1) Then the following matrix set is a $1-A$ left conical semifield spread:

$$x = 0, y = x \begin{bmatrix} u + A(t), & A(u) + t\sqrt{z} \\ t & u \end{bmatrix},$$

for all t, u in D_2.

(2) Since $-A(u^\sigma)^\sigma = A(u)$ and $A^2(u) = 0$, the spread is simultaneous. Furthermore, the left spread is a double left hyperbolic spread, so that there are right and left-acting pseudo-regulus-inducing homology groups.

(3) The transpose of the matrix set is a set of non-singular matrices that provides a right spread.

(4) The transpose of the right spread is a double right hyperbolic spread.

(5) Hence, if D is non-commutative, corresponding to the original spread are eight flocks: four left and four right; two conical flocks and six hyperbolic flocks.

(6) If D is an infinite field, there is a left-acting regulus-inducing homology group, acting on the transpose spread. This then produces four flocks: a conical flock, a hyperbolic flock and two left-inversing hyperbolic flocks.

The next example will deal with the $1-A$-type spreads, both left and right. There can be infinite fields and non-commutative left and right examples. The examples all have characteristic 2,D is a division ring of finite dimension over a field F,$F(z)$ is the rational function field

DOI: 10.1201/9781003767978-44

of F, $D(z)$ is the extension division ring, and $D(z)(\sqrt{z})$ is a central extension of $D(z)$ of dimension 2. When $u = u_0 + u_1\sqrt{z}$, then $A(u) = du/d\sqrt{z}$, and $\sqrt{z}$ is central for u_0, u_1 in $D(z)$.

Then

$$x = 0, y = x\begin{bmatrix} u + A(t) & , A(u) + \sqrt{z}t \\ t & u \end{bmatrix},$$

for all u, t in $D(z)(\sqrt{z})$ is a left simultaneous spread. This spread, when D is a non-commutative skewfield will then generate six flocks as mentioned in part (5) above: three left flocks and three right hyperbolic flocks and one left conical spread and one right conical flock. There will be no finite examples here, but there will be four infinite field examples:

$$(12)\ IN_{hy}(1)^{1-A}, (13)\ IN_{hy}(1)^{1-A,inverse},$$
$$(14)\ IN_{hy\&co}(1)^{1-A,inverse}, (15)\ IN_{co}(1)^{1-A}.$$

Then the left and right flocks:

$$(12)\ N^L_{hy}(1)^{1-A}, (13)\ N^L_{hy}(1)^{1-A,inverse},$$
$$(14)\ N^L_{hy\&co}(1)^{1-A,inverse}, (15)\ N^L_{co}(1)^{1-A}.$$

$$(12)\ N^R_{hy}(1)^{1-A}, (13)\ N^R_{hy}(1)^{1-A,inverse},$$
$$(14)\ N^R_{hy\&co}(1)^{1-A,inverse}, (15)\ N^R_{co}(1)^{1-A}.$$

Super Simultaneous Spreads.

There are various other spreads that arise from the super simultaneous spreads of type 2 and not fake type 2. Since there are conical flock spreads, there are also hyperbolic flocks obtained by changing the expressions. When c is not zero, the only known set of examples is when $\sigma = \sigma^{-1}$, and $c = 1$. Since the transpose spreads admit left-acting pseudo-regulus generating homology groups, there are left-inversing spreads. Since all constants are central, there are corresponding right flock spreads. Since the left spreads are elation double spreads, there are two additional left conical spreads obtained.

CHAPTER 45

Flocks from lifted types

The above matrices for all t, u in $D_1(\theta)$ or $D(\theta)$, use $E(\theta)$ in general.

From the previous material, all left and right flocks of types involving an inner

$$\sigma\text{-derivation function}$$

may be obtained from this type.

When $c^\sigma = -c$, left hyperbolic spreads are generated, which have the form:

$$x = 0, y = 0, y = x \begin{bmatrix} u^\sigma & A_c(u) \\ 0 & u \end{bmatrix}$$

and

$$y = x \begin{bmatrix} t^\sigma & A_c(t) + \theta \\ 1 & t \end{bmatrix} \begin{bmatrix} v^\sigma & A_c(v) \\ 0 & v \end{bmatrix};$$

for all non-zero u and v and for all t in $E(\theta)$. Since all of the conical flock spreads admit left-acting pseudo-regulus-inducing homology groups, the left-inversing spreads of the following form are generated.

$$y = x \begin{bmatrix} -t, & -(t\Delta_t^{-1} A_c(\Delta_t^\sigma) + (1 - t\Delta_t^{-1} t^\sigma)\Delta_t^\sigma) \\ 1, & , \Delta_t^{-1}(A_c(\Delta_t^\sigma) - t^\sigma \Delta_t^\sigma) \end{bmatrix} \begin{bmatrix} u^\sigma & A(u) \\ 0 & u \end{bmatrix}.$$

Since the constants in the matrices are central, the right spreads may be obtained by the codes, with τ as σ^{-1}.

Therefore, the examples obtained are:

Finite:

(1) $F_{hy}(1)^{\sigma-A}$(fake type 1),(2) $F_{hy}(1)^{\sigma-A,inverse}$(fake type 1),

(6) $F_{co}(1)^{\sigma-A}$(fake type1), (10) $F_{hy\&co}(1)^{\sigma-A}$(fake type 1).

Infinite field:

(1) $IN_{hy}(1)^{\sigma-A}$(fake type 1), (2) $IN_{hy}(1)^{\sigma-A,inverse}$(fake type 1),

(6) $IN_{co}(1)^{\sigma-A}$(fake type1), (10) $IN_{hy\&co}(1)^{\sigma-A}$(fake type 1).

Left Non-Commutative:

(1) $N_{hy}^L(1)^{\sigma-A}$(fake type 1), (2) $N_{hy}^L(1)^{\sigma-A,inverse}$(fake type 1),

DOI: 10.1201/9781003767978-45

$$(6)\ N_{co}^{L}(1)^{\sigma-A}(\text{fake type1}), (10)\ N_{hy\&co}^{L}(1)^{\sigma-A}(\text{fake type 1}).$$

Right Non-Commutative:

$$(1)\ N_{hy}^{R}(1)^{\sigma-A}(\text{fake type 1}), (2)\ N_{hy}^{R}(1)^{\sigma-A,inverse}(\text{fake type 1}),$$
$$(6) N_{co}^{R}(1)^{\sigma-A}(\text{fake type1}), (10)\ N_{hy\&co}^{R}(1)^{\sigma-A}(\text{fake type 1}).$$

Examples from the Strings.

First, using the right string of flock spreads will now determine the corresponding right flock spreads as examples of #12, 13, 14, 15. And, from the left flock of spreads, the same types were obtained.

$$12. N_{hy}^{R}(1)^{F,1-A};$$

(A an inner derivation function; "double fake type 1").

$$13. N_{hy}^{R}(1)^{F,1-A,inverse};$$

(A an inner derivation function; a "double fake type 1".)

$$14. N_{hy\&co}^{R}(1)^{F,\ 1-A,inverse};$$

the situation where there is a conical type admitting a left-acting regulus-inducing homology group, thereby obtaining also a hyperbolic type. A an inner derivation function; a "double fake type 1".

$$15. N^{R}(1)_{co}^{F,1-A},$$

(A an inner derivation function; a "double fake type 1".)

And, with the same comments: The same types on the left:

$$12. N_{hy}^{L}(1)^{F,1-A}, 13. N_{hy}^{L}(1)^{F,1-A,inverse},$$
$$14. N_{hy\&co}^{L}(1)^{F,\ 1-A,inverse}, 15. N^{L}(1)_{co}^{F,1-A}.$$

CHAPTER 46

The open types and new directions

We separate here the types of flocks that are open to hopefully stimulate some interest in solving these problems.

Inclusion Problems.

(1) $IN(1)^{trans}_{co}$: $x\begin{bmatrix} u - F(t), & G(t) + A(u) \\ t & u \end{bmatrix}$, the transpose of the #15, if (i) exists (yes) and not isomorphic (Open).

(2) $N^L_{co}(1)^{trans.,\,1-A}$: $x\begin{bmatrix} w - F(t), & G(t) + A(w) - A(F(t)) \\ t & w \end{bmatrix}$, A an outer derivation function. The transpose of #15, if not isomorphic. (i) existence (yes), (ii) not isomorphic (Open).

(3) $N^R_{co}(1)^{1-A^*}$: $\begin{bmatrix} w - R(t) & t \\ S(t) + A^{**}(w) - A^*(R(t)), & w \end{bmatrix} x^T$; A^* is an outer derivation function. The transpose of #15. (i) existence (yes), (ii) not isomorphic (Open).

(4) $N^L(2)^{trans,F,1-A}_{co}$: $x\begin{bmatrix} w - F(t), & G(t) + A(w) - A(F(t)) \\ t & w \end{bmatrix}$, A an inner derivation function; a "fake type 1". Transpose of #20. (Open).

(5) $N^R_{co}(1)^{trans.,F,1-A^*}$: $\begin{bmatrix} w - R(t) & t \\ S(t) + A^{**}(w) - A^*(R(t)), & w \end{bmatrix} x^T$, A^*

A^*

is an inner derivation function; a " fake type 1". Transpose of #20. (Open).

(6) $N^L(3)^{inverse}_{hy}$: $\begin{bmatrix} f(t), & g(t) \\ 1 & t \end{bmatrix}^{-1} \begin{bmatrix} v, & 0 \\ 0 & v \end{bmatrix}$. (Open case).

(7) $N^R_{hy}(3)^{inverse}$: $\begin{bmatrix} v & 0 \\ 0 & v \end{bmatrix} \begin{bmatrix} f(t), & g(t) \\ 1 & t \end{bmatrix}^{-1} x^T$. (Open case).

This type of question comes up during the research simply because this sort of question is natural in other types, but not because there is progress toward finding types like these.

DOI: 10.1201/9781003767978-46

Missing $\sigma - A$, *BothOuter.*

(1) $N^{L}(2)_{co}^{trans,F,1-A}$: $x\begin{bmatrix} w-F(t), & G(t)+A(w)-A(F(t)) \\ t & w \end{bmatrix}$, A an inner derivation function; a "fake type 1". Transpose of #20. (Open).

(2) $N_{hy}^{L,\sigma-A,inverse}(1)$: $x\begin{bmatrix} f(t), & g(t) \\ 1 & t \end{bmatrix}^{-1}\begin{bmatrix} v^{\sigma}, & A(v) \\ 0 & v \end{bmatrix}$; A is a non-trivial σ-outer derivation function, and there is a left-acting pseudo-regulus-inducing homology group. (Open).

(3) $N_{hy}^{L}(1)^{\sigma-A}$: $x\begin{bmatrix} f(t) & ,g(t) \\ 1 & t \end{bmatrix}\begin{bmatrix} v^{\sigma}, & A(v) \\ 0 & v \end{bmatrix}$; $\sigma-A$-outer derivation function, σ non-trivial outer automorphism. (Open case.)

(4) $N_{co}^{L}(1)^{\sigma-A}$: $x\begin{bmatrix} u^{\sigma}+F(t), & G(t)+A(u) \\ t & u \end{bmatrix}$; A is a non-trivial outer $\sigma-$derivation function for σ a non-trivial outer automorphism. (Open case.)

(5) $N_{co}^{L}(1)^{trans,\sigma-A}$: $\begin{bmatrix} w^{\sigma^{-1}}-F(t)^{\sigma^{-1}}, & L \\ t & w \end{bmatrix}$ is a non-trivial outer $\sigma-$derivation function for σ, a non-trivial outer automorphism. Where, $A^{*}(u)=A(u^{\sigma^{-1}})$ and

$$L=G(t)+A^{*}(w)-A^{*}(F(t)).$$

The transpose of #24 (Open case).

(6) $N_{co}^{R}(1)^{trans.,F,1-A^{*}}$: $\begin{bmatrix} w-R(t) & t \\ S(t)+A^{**}(w)-A^{*}(R(t)), & w \end{bmatrix}x^{T}$,

A^{*}

is an inner derivation function; a " fake type 1". (Open).

(7) $N_{hy}^{R}(1)^{\tau-A^{*},inverse}$: $\begin{bmatrix} v, & 0 \\ A^{*}(v) & v^{\tau} \end{bmatrix}\begin{bmatrix} f(t), & 1 \\ g(t) & t^{\tau} \end{bmatrix}^{-1}x^{T}$;

A^{*} is a non-trival outer τ-derivation function,

τ non-trivial outer automorphism. (Open case.)

(8) $N_{hy}^{R}(1)^{\tau-A^{*}}$: $\begin{bmatrix} v & 0 \\ A^{*}(v), & v^{\tau} \end{bmatrix}\begin{bmatrix} r(t), & 1 \\ s(t) & t \end{bmatrix}x^{T}$, $\tau-A^{*}$ outer derivation function. (Open case).

(9) $N_{co}^{R}(1)^{\tau-A^{*}}$: $\begin{bmatrix} u+F(t), & t \\ G(t)+A^{*}(u), & u^{\tau} \end{bmatrix}x^{T}$; A^{*} is a non-trivial outer τ-derivation function for τ, a non-trivial outer automorphism. (Open case.)

(10) $N_{co}^{R}(1)^{\tau^{-1}-A^{**}}$: $\left[\begin{array}{cc} w - R(t)^{\tau} & t \\ S(t) + A^{**}(w) - A^{*}(R(t)), & w^{\tau^{-1}} \end{array}\right] x^{T}$; A^{*} is a non-trivial outer τ-derivation function for τ, a non-trivial outer automorphism. The transpose of #24 (Open case.)

In looking at the literature, there seem to be many types of "derivative functions", most were of the common generalized derivative, and $\sigma - A$ *σ-derivative* functions are mentioned in Cohn [**6**]. However, what is generally found is that there are situations for which such degenerate to inner $\sigma - A$ functions. One of the reasons for this might be that the finite dimensionality of the division ring tends to do that. So perhaps infinite-dimensional division rings might be studied. But, there is now a problem. The spreads in these texts are supposed to consist of derivable nets. There are studies that show that in the infinite-dimensional spreads, it is possible to find derivable nets that are derivable as nets but are not derivable with respect to the entire spread.

This is interesting; maybe someone can find a true outer-outer σ-derivative function.

On the other hand, using the kaleidoscope of derivable nets, it is possible to find inner-outer σ-derivative functions in finite dimensions, where the automorphism is inner but the $A-$ is outer.

In terms of open problems, there are so many ideas that have not been fully developed. We end with just a few of these:

11. The only covering idea that we have for infinite spreads is that of unwrapping, and that only works for additive quasi-fibrations. Are there flocks over non-commutative skewfields that are not somehow connected with semifield spreads?
12. Is there a theory for non-generic spreads? Well, yes. The general theorem works using generic spreads by assuming there is a generic pseudo-regulus net, then using K-mappings and the $*$-notation to create the transforms and return to the original affine points to create a new flock. It is curious in this procedure that there is no mention of what form the generalized quadratics actually take. The non-commutative form must actually be $(xK, yK) = (x^{*}, y^{*})$,
$$x_1^{*}A(x_1^{*\sigma^{-1}} x_3^{\sigma^{-1}}) + x_2^{*}x_1^{*\sigma^{-1}} x_3^{*\sigma^{-1}} = x_4^{*}$$
for hyperbolic and $x_1^{*} = x_4^{*}$. However, this is not necessary for the actual flock theory. Once one knows the generic flock theory, the rest can be obtained using the $*$-procedure. All of this, however, depends on the theory of "four types". We are asking, then, is there some other direction?

13. The multiple replacement theory does not seem to be anything close to complete. In particular, using inner automorphisms in replacement theory does create spreads obtained by a sort of Ostrom-like replacement theory, which feels something like multiple derivation. Take any division ring D over a field F. There is then an André's theory over D. This gets wild very quickly. Is a nearfield construction possible? Take the unusual non-cyclic division rings; the ideas also work here.
14. Suppose there is a type 1 left generic $\sigma - A$ flocktype spread. Apply the K-mappings to create the kaleidoscope of derivable nets. This creates a set of isomorphic left spreads. Are they all of type 1? There are nine types of general $\sigma - A$ types spreads.

    Is there a theory that shows how the K-mappings permute the nine types?
15. Are there proper non-commutative quasifibrations? This is open, in spite of some discussion on why this should not be.
16. What are the applications that use division rings? Take any and see if spreads over division rings work as well. For example, for coding theory, is the division ring or the non-commutative of spreads over division rings that is important?
17. When a conical flock spread admits a left-acting pseudo-regulus-inducing homology group, the inversing mapping creates an associated hyperbolic flock. Is there another flock theory over skewfields that uses the left-acting pseudo-regulus homology group directly? And can it be shown that these and the right theory and the left theory are connected by the inversing mapping?
18. The idea of flocks over infinite-dimensional division rings D seems like a natural one to pursue. It has been mentioned that there could be spreads composed of sets of derivable nets such that "derivation" of the spread does not construct another spread. What would it construct? There are pseudo-regulus D-nets. Are there any spreads that contain them?
19. Are there flock-like spreads over subplane covered nets? We have noted in the first volume that there could be such spreads of type (0) over cyclic division rings D, which would correspond to central extensions of D. There should be central collineation groups that act in the proper way. Is there a theory here?
20. There are a few "deficiency one" results in this text. Develop this in a more concrete manner. This is completely open!

21. For a finite spread of over q^2, there are geometries associated with many groups of orders $q, q-1, q+1$. The q and $q-1$ groups as regulus-inducing central collineation groups connect strongly with the Thas-Walker theory. What about cyclic groups of order $q+1$? This feels like André's theory, and the classical regulus net, as a derivable net may be embedded in $PG(3, GF(q))$ with a special line N indicated. The collineation group of the classical regulus net is $P\Gamma L(4,q)_N$. If a homology group of order $q+1$ is a collineation group of a flock, like spread, there is actually an associated flock of a quadratic cone (see [**34**], p. 126, chapter 7). The idea for this work uses the ideas of the finite André's theory. And then connects with the theory of hyperbolic fibrations. Since in this text, there is an André's theory for any finite-dimensional division ring, there is probably a generalization of hyperbolic fibrations that is connected with a spread admitting an appropriate homology group. The work in both [**31**] and [**34**] concerning hyperbolic fibrations would be background reading for such a generalization.
22. In Jacobson [**18**], (1.8), derivations of purely inseparable extensions E of exponent one over F. For division rings of finite dimension, such fields F, there are derivable nets of the following type: if E has order p over F and

$$D \otimes_F E$$

is a division ring, then there is a corresponding derivable net defined as follows:

$$x = 0, y = x \begin{bmatrix} u, & du/d(\sqrt[p]{z}) \\ 0 & u \end{bmatrix}$$

for all u in $D \otimes_F E$. In this series, there are examples when $p = 2$. Are they others? And do spreads exist containing this derivable net in $PG(3, D \otimes_F E)$?

Bibliography

[1] L. Bader and G. Lunardon, On the flocks of $Q^+(3,q)$, Geom. Ded. **29** (1989), 177-183.

[2] A. Barlotti, On the definition of Baer subplanes of infinite planes, J. Geom. **3** (1979), 87-92.

[3] M. Biliotti, V. Jha and N.L. Johnson, Foundations of Translation Planes. Monographs and Textbooks in Pure and Applied Mathematics, Vol. 243, Marcel Dekker, New York, Basel, 2001, xvi +542 pp.

[4] I. Cardinalli. O. Polverino and R. Trombetti, Semifield planes of order q^4 with kernel F_{q^2} and order F_q, Eur. J. Comb. **27** (2006), no. 6, 940-961.

[5] J. Cofman, Baer subplanes and Baer collineations of derivable projective planes, Abh. Math. Sem. Hamburg **44** (1975), 187-192.

[6] P.M. Cohn, Skew Fields, Theory of General Division Rings. Cambridge University Press, 1995.

[7] P.M. Cohn and W. Dicks, On central extensions of skewfields, J. Algebra, **63** (1980), 143-151.

[8] W. Cherowitzo, N.L. Johnson and O. Vega, α-Flokki and Partial α-Flokki, Innov. Incid. Geom. **15** (2017), 5-29.

[9] J. Dieudonne', Les determinants sur un corp non-communtatif, Bull. Soc. Math. France (1943), 27-45.

[10] P.K. Draxl, Skew Fields, Cambridge University Press, Cambridge, 1983.

[11] D.S. Dummit and R.M. Foote, Abstract Algebra, 3rd Edition 2004, 932 pp.

[12] D.A. Foulser and N.L. Johnson, The translation planes of order q^2 that admit $SL(2,q)$, I. Odd Order, J. Geometry **18** (1982), no. 2 122-139.

[13] D.A. Foulser and N.L. Johnson, The translation planes of order q^2 that admit $SL(2,q)$, II. Even Order, J. Algebra **86** (1984), no. 2 385-406.

[14] D.A. Foulser, N.L. Johnson and T.G. Ostrom, A characterization of the Desarguesian planes of order q^2 by $SL(2,q)$, Int. J. Math. Sci. **6** (1983), 301-317.

[15] H. Gavaert and N.L. Johnson, Flocks of quadratic cones, generalized quadrangles and translation planes, Geom. Dedicata **26** (1988), 301-317.

[16] C. Grover, C. Ling and R. Vehkalahti, Non-Commutative Ring Learning with Errors from Cyclic Algebras, arXiv: 2008.0834.

[17] Y. Hiramine, M. Matsumoto and T. Oyama, On some extension of 1-spread sets, Osaka J. Math **24** (1987), no. 1 123-137.

[18] N. Jacobson, Finite-dimensional Division Algebras over Fields, Springer Books, Heidelberg, 1996.

[19] Vikram Jha and Norman L. Johnson, Infinite flocks of a quadratic cone. J. Geom. **57** (1996), 123-150.

[20] Vikram Jha and Norman L. Johnson, Structure Theory for Point-Baer and Line-Baer Collineation groups in Affine Planes, Mostly finite geometries, Lecture notes in Pure and Applied Mathematics, Vol. 190, Marcel Dekker, 1997, 235-274.

[21] Vikram Jha and Norman L. Johnson, Conical, ruled, and deficiency one translation planes, Bull. Belg. Math. Soc. **6** (1999), 187-218.

[22] Vikram Jha and Norman L. Johnson, Infinite Baer nets, J. Geom. **68** (2000), no. 1-2 114-141.

[23] Vikram Jha and Norman L. Johnson, Lifting quasifibrations II – Non-normalizing Baer involutions, Note Mat. **20** (2001/01), no. 2, 51-68.

[24] Vikram Jha and Norman L. Johnson, Infinite flocks of a quadratic cone – II generalized Fisher flocks, J. Korean Math. Soc. **39** (2002), no. 4, 653-664.

[25] Vikram Jha and Norman L. Johnson, Nuclear fusion in finite semifield planes, Adv. Geom. **4** (2004), no. 4 413-432.

[26] N.L. Johnson, Classification of semi-translation planes, Canad. J. Math **21** (1969), 1332-1387.

[27] N.L. Johnson, Lezioni sui Piani di Traslazione, Quaderni dei Dipartimento di Matematica, dell'Universita di Lecce, Q. **3** 1986 (Italian).

[28] N.L. Johnson, Semifield flocks of quadratic cones, Simon Stevin **61** (1987), 313-326.

[29] N.L. Johnson, Derivable nets can be embedded in nonderivable panes, Trends Math. (1998), Birkhauser Verlag, Basel Switzerland, 123-144.

[30] N.L. Johnson, Subplane covered nets, Pure and Applied Mathematics, Vol. 222, Marcel Dekker, 2000.

[31] N.L. Johnson, Combinatorics of Spreads and Parallelisms, Pure and Applied Mathematics , Vol. 295, Marcel Dekker, 2010.

[32] N.L. Johnson, Twisted hyperbolic flocks, Innov. Incid. Geom. **19** (2021), no. 1 1-23.

[33] N.L. Johnson, Twisted quadrics and σ-flocks, Algebr. Comb., **5** (2022), no. 5 803-826.

[34] N.L. Johnson, Geometry of Derivation with Applications, Taylor/Frances, 2023, 355 pp.

[35] N. L. Johnson, Theory of Skewfield Flocks, Geometry of Derivation, Vol. II.

[36] N.L. Johnson and M. Cordero, Transitive partial hyperbolic flocks of deficiency-one, Note di Mat. **29** (2009) no. 1 89-98.

[37] N.L. Johnson, V. Jha and M. Biliotti, The Handbook of Finite Translation Planes, Pure and Applied Mathematics, Vol. 289, Marcel Dekker, 2007.

[38] N.L. Johnson and V. Jha, Lifting Skewfields, J. Geom. **113** (2022), no. 5.

[39] N.L. Johnson and X. Liu, Flocks of quadratic and semi-elliptic Cones, Mostly finite geometries, Lecture notes in Pure and Applied Mathematics, Vol. 190, Marcel Dekker, 1997, 275-304 pp.

[40] N.L Johnson and R. Pomareda, A maximal partial flock of deficiency – one of the hyperbolic quadric in $PG(3,9)$, Simon Stevin, **64** (1990), no. 2 169-177.

[41] Norman L. Johnson and T.G. Ostrom, Direct products of affine partial linear spaces, J. Combin. Theory Ser. A, **75** (1996) no. 1, 99-140.

[42] N.L. Johnson, R. Pomareda and F. Wilke, j-planes, J. Combin. Theory Ser. A, **56** (1991), 271-284.

[43] W.M. Kantor and T. Penttila, Flokki planes and cubic polynomials, Note di Matematica, Vol. 29, suppl. numero **1** (2009), 219-230.

[44] J. Lahtonen, N. Markin and G. McGuire, Construction of multiblock space – time codes from division algebras with roots of unity as nonnorm elements, IEEE Trans. Inf. Theory, **54** (2008), no. 11 5231-5235.

[45] F. Oggier, J-C. Belfiore and E. Viterbo, Cyclic Division Algebras: A Tool for Space-Time Coding, Found. Trends Commun. Inf. Theory, **4** (2007), no. 1 1-95.

[46] Gordon F. Royle, An orderly algorithm and some applications to finite geometry, Discrete Math. **185** (1998), 105-115.

[47] H.J. Schaeffer, Translationsebenen, auf denen die Grupper $SL(2, p^r)$ Operiert, Diplomarbeit, Univ. Tubingen, 1975.

[48] J.A. Thas, Flocks, maximal exterior sets and inversive planes, Finite Geom. and Combinatorial Designs, Amer. Math. Soc. Providence, RI (1990), 187-218.

[49] M.Walker, A class of translation planes, Geom. Dedicata **5** (1976), 135-166.

[50] M. Walker, On translation planes and their collineation groups, Ph.D thesis, Univ. London, 1973.

Index

For Product Safety Concerns and Information please contact our EU representative GPSR@taylorandfrancis.com
Taylor & Francis Verlag GmbH, Kaufingerstraße 24, 80331 München, Germany

www.ingramcontent.com/pod-product-compliance
Ingram Content Group UK Ltd.
Pitfield, Milton Keynes, MK11 3LW, UK
UKHW021513250926
475702UK00043B/364
* 9 7 8 1 0 4 1 2 9 0 8 8 9 *